# Linear Inverse Problems
# and Tikhonov Regularization

© *2016 by the Mathematical Association of America, Inc.*

*Library of Congress Catalog Card Number 2016935233*

Print ISBN 978-0-88385-141-8
Electronic ISBN 978-1-61444-029-1

*Printed in the United States of America*

Current Printing (last digit):
10 9 8 7 6 5 4 3 2 1

The Carus Mathematical Monographs

Number Thirty-Two

# Linear Inverse Problems and Tikhonov Regularization

Mark S. Gockenbach
*Michigan Technological University*

*Published and Distributed by*
THE MATHEMATICAL ASSOCIATION OF AMERICA

The Carus Mathematical Monographs are an expression of the desire of Mrs. Mary Hegeler Carus and of her son, Dr. Edward H. Carus, to contribute to the dissemination of mathematical knowledge by making accessible a series of expository presentations of the best thoughts and keenest research in pure and applied mathematics. The publication of the first four of these monographs was made possible by Mrs. Carus as sole trustee of the Edward C. Hegeler Trust Fund. The sales from these have resulted in the Carus Monograph Fund, and the Mathematical Association of America has used this as a revolving book fund to publish the succeeding monographs.

The expositions of mathematical subjects that the monographs contain are set forth in a manner comprehensible not only to teachers and students specializing in mathematics, but also to scientific workers in other fields. More generally, the monographs are intended for the wide circle of thoughtful people familiar with basic graduate or advanced undergraduate mathematics encountered in the study of mathematics itself or in the context of related disciplines who wish to extend their knowledge without prolonged and critical study of the mathematical journals and treatises.

The following Monographs have been published:

MAA Service Center
P.O. Box 91112
Washington, DC 20090-1112
1-800-331-1MAA      FAX: 1-301-206-9789

# Contents

# Preface

Inverse problems are of ever-increasing importance in applied mathematics and science. Loosely speaking, an *inverse problem* asks us to infer some cause or causes from effects that we can measure; often the desired causes are inaccessible to direct measurement. To be more precise, we need to understand the concept of a *direct* or *forward* problem. The classical problems of mathematical physics are forward problems; the goal is to create a mathematical model that predicts the outcome of an experiment, assuming that all the experimental conditions are known.

To make this discussion easier to follow, let us consider an image-processing problem, and to keep this discussion as simple as possible, let us assume that the problem is one-dimensional. The desired image is represented by a real-valued function defined on the interval $[0, 1]$. We have a detector for recording values of $x(t)$, $t \in [0, 1]$; however, the detector produces a smeared-out image $y$ defined by

$$y(s) = \frac{1}{\sqrt{2\pi}\sigma} \int_0^1 e^{-(s-t)^2/(2\sigma^2)} x(t)\, dt, \ 0 \le s \le 1. \qquad (1)$$

Here $\sigma$ is a small positive number, which is assumed to be known for this discussion; it should be noted that each value of $y(s)$ is a weighted average of the values of $x(t)$, with the most weight given to $t$ near $s$.

The forward problem is to model the behavior of the detector and thus produce (1). It tells us the outcome of the experiment, namely the image $y$ that will be recorded, given that we know the true image $x$. The inverse problem is to estimate the true image $x$ from the measured data $y$.

A critical feature of forward problems is *stability*: the solution of the problem depends continuously on the parameters defining it. In our example, this means that the measured image $y$ depends continuously on the true image $x$, which implies that small errors in determining $x$ will not greatly affect the calculation of $y$. On the other hand, the defining feature of an inverse problem is *instability*: the quantity to be estimated ($x$) does not depend continuously on the data that is measured ($y$). This instability is often so severe that when an inverse problem is solved by a naive method, the noise in the data results in a computed solution that is completely worthless. However, it is frequently the case that even noisy data contains useful information about the unknown solution, and that it is possible to extract this information by suppressing the effect of the noise in the data. This process is called *regularization*.

The purpose of this book is to examine one of the most important classes of inverse problems—linear inverse problems defined on Hilbert spaces—and the most important general-purpose regularization technique, the method of Tikhonov regularization. A linear inverse problem can be represented as $Tx = y$, where $T : X \rightarrow Y$ is a linear operator, $y$ is a given (data) vector, and $x$ is to be computed. As indicated above, the fundamental problem is that $x$ does not depend continuously on $y$, which means that if $T^{-1}$ exists it is a discontinuous operator. In addition, $T^{-1}$ may fail to exist because $T$ is not one-to-one (injective) or it is not onto (surjective). The goal of regularization is to construct an operator $R : Y \rightarrow X$ such that $x = Ry$ is an acceptable (approximate) solution of $Tx = y$. When the problem is unstable, which is the only case of interest to us, it is not possible to find a single operator $R$ that will produce an acceptable answer for all $y \in Y$; instead, we construct a family $\{R_\lambda : \lambda > 0\}$ such that $x = R_\lambda y$ is an acceptable solution of $Tx = y$ for some $\lambda > 0$. An appropriate value of the *regularization parameter* $\lambda$ depends on the particular data vector $y$. Tikhonov regularization is one popular method for constructing $R_\lambda$.

This is primarily a mathematics book. Although specific applications are mentioned, they are not studied in any detail. Instead, we consider the class of problems that can be expressed as $Tx = y$, where $T : X \rightarrow Y$ is a continuous linear operator and $X, Y$ are Hilbert spaces. Chapter 2 distinguishes between equations $Tx = y$ that are *well-posed* and those that are *ill-posed*, and defines an inverse problem to be a particular type of ill-posed problem. Chapter 3 introduces Tikhonov regularization and presents the basic analysis of the method.

If $T$ happens to be a *compact* operator, then the *singular value expansion* (SVE) allows a more refined analysis. The SVE and its application to inverse problems are presented in Chapter 4. As we will see, the SVE makes it clear why Tikhonov regularization works as well as it does.

A variant of Tikhonov regularization, sometimes called *seminorm regularization* or *Tikhonov regularization with seminorms*, is often used in practice. This is the subject of Chapter 5.

The Epilogue presents a brief discussion of some of the important aspects of inverse problems that are not covered by this book. The reader will find some suggestions for further reading there.

**Acknowledgements**  Most of what I know about Tikhonov regularization was learned from the monograph *The theory of Tikhonov regularization for Fredholm equations of the first kind* [8] by Charles W. Groetsch and the book *Regularization of Inverse Problems* [3] by Engl, Hanke, and Neubauer. I heartily recommend both for readers who want to know more. In particular, [3] considers many methods in addition to Tikhonov regularization, including regularization by projection, iterative regularization methods, and regularization for nonlinear inverse problems. It also includes a chapter on methods for discretizing inverse problems.

I used the *Mathematica* software [16] to perform the computations and prepare the graphics for the numerical examples.

I will maintain a web page for this book, including a list of errata, at

www.math.mtu.edu/~msgocken/tikhbook.

Readers are invited to send me suspected errors by email.

*Mark S. Gockenbach*
msgocken@mtu.edu

# Chapter 1

# Introduction to inverse problems

We are going to study certain linear equations of the form $Tx = y$, where $y$ is given and $x$ is to be computed, that are particularly difficult to solve. When we say "difficult to solve," we mean difficult in the sense that a small error in $y$ leads to a large error in $x$, to the extent that the computed value of $x$ may be completely worthless. In a typical application, the quantity $y$ should be regarded as data that must be measured or collected, which implies that there is inevitably error in $y$. Therefore, the amplification of error in passing from the data to the solution is a critical feature of the problem and, unless there is some way to avoid it, such a problem is impossible to solve in a meaningful way.

A problem with the character just described will be called an inverse problem; we define the concept precisely in Chapter 2. We are going to assume that the operator $T$, which is given, is linear; thus we study only linear inverse problems, although nonlinear inverse problems are certainly of interest. The linearity of the operator implies that $x$ and $y$ belong to appropriate vector spaces.

In a genuine inverse problem, it is impossible to avoid some amplification of error in computing a solution from noisy data. However, as we will see, there are mathematical techniques for reducing this amplification to a manageable level, so that meaningful solutions can be computed. The specific subject of this book is one such method, namely the method

1

of Tikhonov regularization. We will touch on related methods that lend
themselves to similar analysis.

In this introductory chapter, we present two pairs of problems. Each
consists of problems that are related and for which it may not be easy,
at first glance, to determine which is more difficult. The calculations
that we present, however, will show that one of the problems is straightforward (in the sense of no significant amplification of error) and the
other is difficult (exhibiting a catastrophic amplification of the error in
the data). The remainder of the book will describe mathematically the
source of the difficulties in inverse problems and show how they can be
addressed.

## 1.1   Two integral equations

For our first example, we suppose that two real-valued functions $x$ and $y$
are defined on the interval $[0, 1]$ and are related by

$$\int_0^1 k(s, t)x(t)\, dt = y(s), \ 0 \le s \le 1. \tag{1.1}$$

We would like to know $x$, but we cannot measure it directly; instead,
we can only measure the result of the integral operator acting on $x$. This
is a common mathematical model that arises in many applications. For
example, we may have a device (such as a camera) that can, in principle, record a mathematical representation $x$ of an image. Such a device
is likely to have the property that, when trying to record $x(s)$ for a specific value of $s$, it actually averages values of $x(\tilde{s})$ for $\tilde{s}$ near $s$. The true
image is then blurred, and the inverse problem is to deblur the recorded
image. The imaging application is more meaningful for images recorded
on two-dimensional regions but, for convenience, we consider the one-
dimensional version of the problem.

To explore the possibility of estimating $x$ from a (noisy) measurement of $y$, we discretize (1.1), approximating the integral by the simple midpoint quadrature rule on a uniform grid. Define $h = 1/n$, where

$n$ is a positive integer, and consider the grid defined by the subintervals $[0, h], [h, 2h], \ldots, [(n-1)h, 1]$. The midpoints are $s_i = (i - 1/2)h$, $i = 1, 2, \ldots, n$. We also write $t_i = s_i$ for each $i$.

The midpoint rule yields

$$\int_0^1 k(s, t)x(t)\, dt \approx \sum_{j=1}^n k(s, t_j)x(t_j)h, \ \ s \in [0, 1]$$

and therefore

$$\int_0^1 k(s_i, t)x(t)\, dt \approx \sum_{j=1}^n k(s_i, t_j)x(t_j)h, \ \ i = 1, 2, \ldots, n.$$

We define the vector $\mathbf{y} \in \mathbb{R}^n$ by $y_i = y(s_i)$ and write $\mathbf{x} \in \mathbb{R}^n$ for the computed values of $x(t)$: $x_j \approx x(t_j)$. We then replace the equation (1.1) by the system of equations

$$\sum_{j=1}^n k(s_i, t_j)x_j h = y_i, \ \ i = 1, 2, \ldots, n. \tag{1.2}$$

If we define the matrix $A \in \mathbb{R}^{n \times n}$ by $A_{ij} = k(s_i, t_j)h$ for $i, j = 1, 2, \ldots, n$, then (1.2) is equivalent to the matrix-vector equation $A\mathbf{x} = \mathbf{y}$. We can now estimate a solution of (1.1) by solving $A\mathbf{x} = \mathbf{y}$ for $\mathbf{x}$.

Before we do a specific example, we present another example of an integral equation, one that is closely related to the first:

$$x(s) + \int_0^1 k(s, t)x(t)\, dt = y(s), \ \ 0 \le s \le 1. \tag{1.3}$$

Here we can imagine that our measurement of $x$ is contaminated by some additive noise that itself is a blurred image of $x$. We can discretize this by the midpoint rule, just as before, and it should be clear that the corresponding matrix-vector equation is $(I + A)\mathbf{x} = \mathbf{y}$, where $A$ is exactly the same matrix as before (assuming that $k$ is the same function in (1.1) and (1.3)).

To create specific instances of these problems, we will choose $k$ based on a Gaussian function:

$$k(s,t) = \frac{1}{\sqrt{2\pi}\sigma} e^{-(s-t)^2/(2\sigma^2)}, \quad \sigma = 0.05.$$

Let us assume that the exact image we wish to compute is $x(t) = \sin(2\pi t)$, and that $y$ is defined by (1.1) or (1.3) for this $k$ and $x$. We will choose $n = 30$ to define the grid used for the computations.

Let us first consider the second example, based on (1.3). Figure 1.1 shows the exact data and noisy data produced by adding uniformly distributed random numbers to the components of the exact vector $\mathbf{y}$ to produce a noisy vector $\hat{\mathbf{y}}$; the error in $\hat{\mathbf{y}}$ is scaled to 5% in the Euclidean norm. Although both $\mathbf{y}$ and $\hat{\mathbf{y}}$ are Euclidean vectors, we graph $\mathbf{y}$ as a curve (by joining the points representing the components of $\mathbf{y}$) for visual convenience. The same figure shows the exact solution $\mathbf{x}$ and the estimated solution $\hat{\mathbf{x}}$ obtained by solving $A\hat{\mathbf{x}} = \hat{\mathbf{y}}$.

There are no surprises in the results shown in Figure 1.1. It should be noted that the errors in the computed solution $\hat{\mathbf{x}}$ are comparable to the errors in the noisy data $\hat{\mathbf{y}}$.

We find a completely different situation when we consider (1.1). Performing the analogous numerical experiment (with $\hat{\mathbf{y}}$ constructed in the same fashion and $\hat{\mathbf{x}}$ computed in the same way), we obtain the results shown in Figure 1.2. Now we see that the computed solution is completely wrong; it is of an entirely different scale than the true solution, and the signs of the computed components alternate in a way that is unrelated to the behavior of the true solution.

The two examples presented in this section share a superficial similarity, but one of them has the characteristics of an inverse problem, namely that small changes in the data lead to abrupt and large changes in the computed solution. We will see that this phenomenon is not due to a poor computational scheme (that is, we would not have seen better results from a better quadrature scheme or a better way of solving the discretized equations). In a sense that will be thoroughly explored in the next chapter, an inverse problem has the property that the computed solution does not depend continuously on the data. This is a property that cannot be

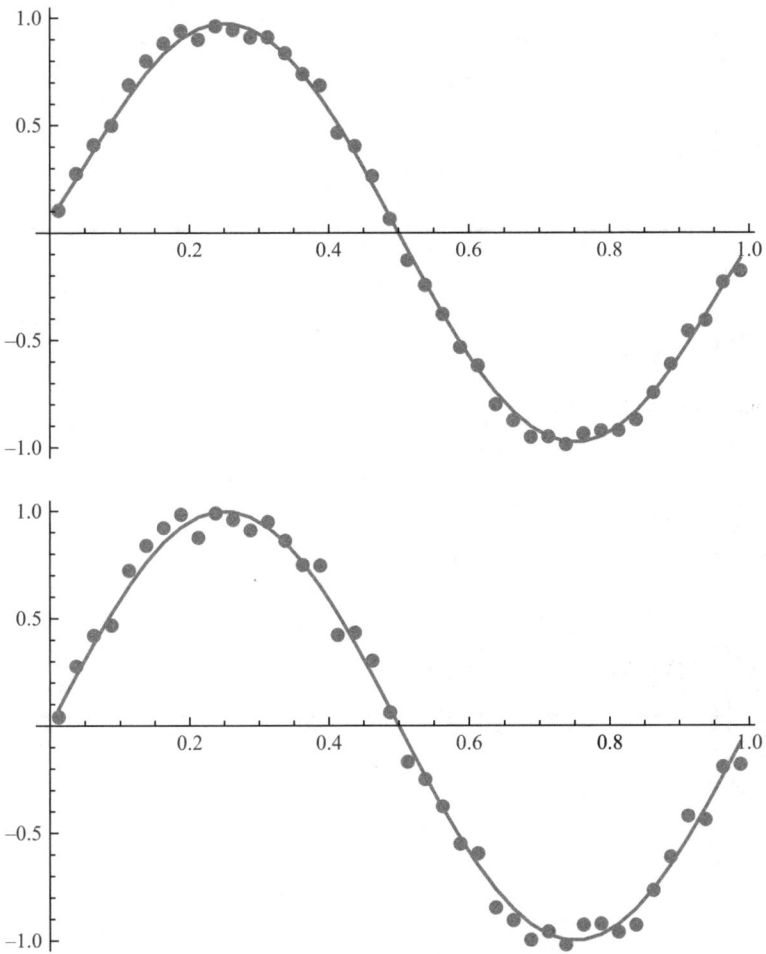

**Figure 1.1.** Top: the exact and noisy data for the example based on (1.3). Bottom: the exact and computed solution. (The exact quantities are displayed as curves.)

eliminated; in a true inverse problem, there is an inevitable loss of information (amplification of error) in passing from the data to the solution. However, *regularization methods*, including Tikhonov regularization, the topic of this book, are able to reduce the amplification of error and

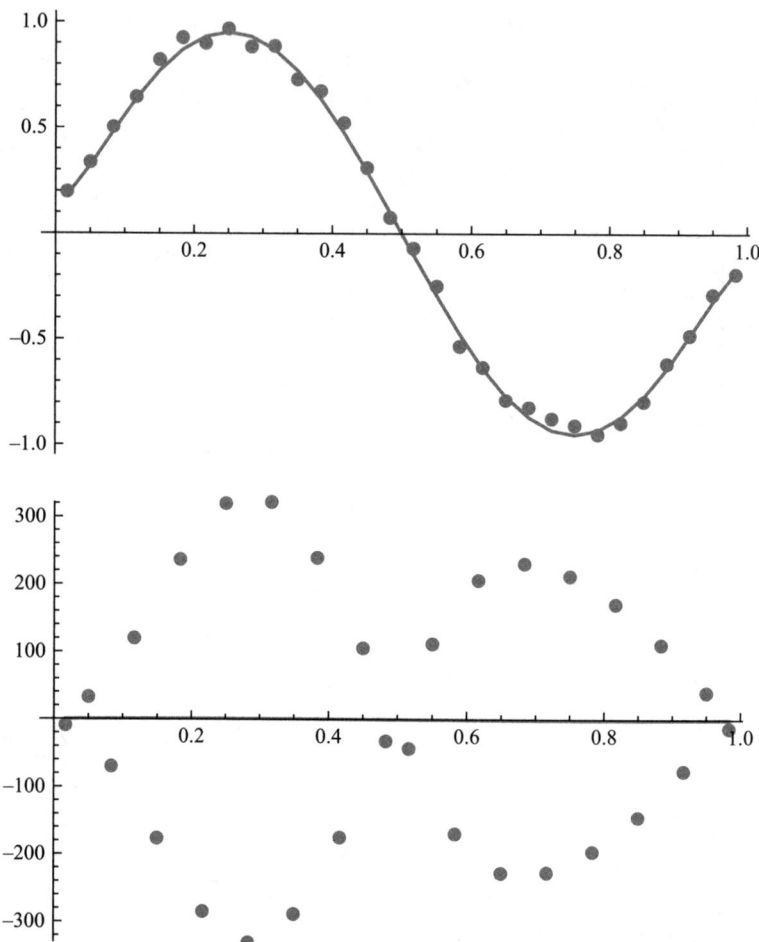

**Figure 1.2.** Top: the exact and noisy data for the example based on (1.1).
Bottom: the exact and computed solution. (The exact quantities are displayed
as curves. The exact solution is not visible in the bottom plot because the scale
of the computed solution is completely wrong.)

preserve the information in the data that is not overwhelmed by error. In Chapter 3 we show that this is possible and in Chapter 4 we use the *singular value expansion* to show how and why Tikhonov regularization works.

## 1.2 A forward problem/inverse problem pair

In the previous section, we used numerical computations to illustrate the difficulties that arise in solving certain linear equations. We now turn to two problems that can be studied analytically. These problems are related in a more fundamental way than was the case with the problems in the previous section; in fact, one is a forward problem (also called a direct problem) and the other is the corresponding inverse problem. In a superficial sense, a forward problem is the problem of computing $Tx$ given $x$, while the corresponding inverse problem is to solve $Tx = y$ for $x$ given $y$ (in both cases, of course, the operator $T$ is given). However, a little thought shows that this cannot be the basis for calling one problem the forward problem and the other the inverse problem. If we define $S = T^{-1}$, then computing $Tx$ from $x$ becomes the problem of solving $Sy = x$ for $y$ given $x$, while solving $Tx = y$ for $x$ given $y$ becomes the problem of computing $Sy$ given $y$.

Instead of the superficial distinction described in the previous paragraph, we distinguish the two problems based on which has the property of an inverse problem, that the solution fails to depend continuously on the data. The examples we study in this section will allow us to explain why it is common, in a pair of problems that are inverse to one another, that one problem exhibits continuous dependence of the solution on the data and the other one does not.

Let us consider an elastic string that is stretched horizontally between two points. If we apply a force, directed vertically downward, along the string, the string will be displaced. We now present a mathematical model for the relationship between the force and the displacement of the string.

We define a coordinate system so that the string always lies in the $xy$-plane, with the two fixed points lying at $(0, 0)$ and $(1, 0)$ (thus the string,

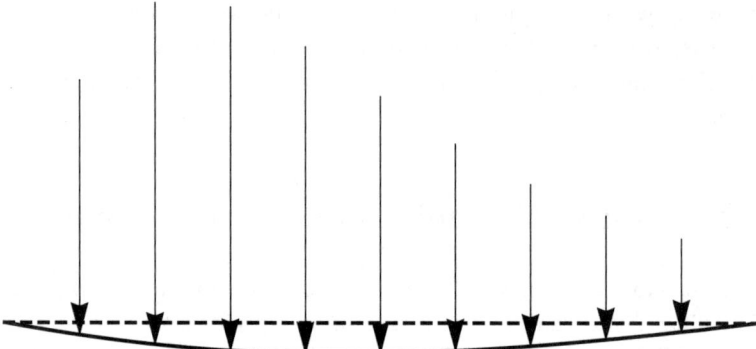

**Figure 1.3.** The string in its initial (dashed line) and displaced (solid curve) positions.

in its initial position, lies on the $x$-axis). If we assume that the applied force is small compared to the tension in the string, it is a reasonable approximation that the point on the string initially found at $(x, 0)$ moves vertically to lie at $(x, u(x))$ (this is a simplifying assumption, since the points on the string do not move solely in the vertical direction).

It can be shown that, to a good approximation, the displacement $u$ is the solution of the following two-point boundary value problem:

$$-u''(x) = f(x), \ x \in (0, 1),$$
$$u(0) = u(1) = 0. \tag{1.4}$$

(Here we have assumed that the tension in the string is 1; otherwise, a constant representing the tension appears in the differential equation as the coefficient of $u''$.) We can now pose two problems based on the BVP (1.4). First, we might be given the displacement $u$ and asked to compute the force $f$ that leads to it. Conversely, we might be given $f$ and asked to predict the resulting displacement $u$. The first problem seems to be the simpler; it is just a matter of computing the (negative) second derivative of the given $u$. On the other hand, the second problem requires solving the differential equation subject to the boundary conditions. Which is the forward problem and which is the inverse problem?

Suppose $u$ and $f$ represent the exact values; that is, $u$ and $f$ together satisfy (1.4). Consider first the problem of computing $f$ from $u$. Measured data always contains errors, and they are often oscillatory (because, when we measure point values, we are as likely to underestimate the correct value as to overestimate it). Therefore, let us assume that $\hat{u}$, the measured value of $u$, differs from the true value by a high-frequency sinusoid: $\hat{u}(x) = u(x) + \epsilon \sin(n\pi x)$, where $\epsilon > 0$ is a small number. Then the computed solution $\hat{f}$ is

$$\hat{f}(x) = -\hat{u}''(x) = -u''(x) + \epsilon n^2\pi^2 \sin(n\pi x)$$
$$= f(x) + \epsilon n^2\pi^2 \sin(n\pi x).$$

The maximum value of the error $|\hat{u}(x) - u(x)|$ in the data is only $\epsilon$, but the maximum error $|\hat{f}(x) - f(x)|$ in the computed solution is $\epsilon n^2\pi^2$. This can be arbitrarily large, depending on the value of $n$.

Although the actual error in measured data would not be as simple as a single sinusoid, Fourier analysis allows us to represent the error as a combination of such sinusoids:

$$\hat{u}(x) = u(x) + \sum_{n=1}^{\infty} \epsilon_n \sin(n\pi x). \tag{1.5}$$

The corresponding value of $f$ is

$$\hat{f}(x) = f(x) + \sum_{n=1}^{\infty} \epsilon_n n^2\pi^2 \sin(n\pi x). \tag{1.6}$$

In a practical problem, we cannot sample infinitely often, but the finer the grid on which we sample the displacement, the higher the frequencies that are introduced into the error. The above formula suggests that this would make the problem of computing $f$ more difficult.

Our analysis shows that the problem of computing $f$ from $u$ has the key property of an inverse problem, namely, an inevitable amplification of errors when computing the solution from the data.

On the other hand, if we wish to compute $u$ from $f$, we can imagine that the error $\hat{f} - f$ in measuring $f$ is of the form $\epsilon \sin(n\pi x)$. Then the solution of the BVP (1.4) is

$$\hat{u}(x) = u(x) + \frac{\epsilon}{n^2 \pi^2} \sin{(n\pi x)}$$

and the error in the solution is less than the error in data—substantially less if $n$ is large. Once again, in a realistic problem, one would not have a measurement error with such a simple form. However, Fourier analysis allows us to represent the measurement error as a superposition of sinusoids:

$$\hat{f}(x) = f(x) + \sum_{n=1}^{\infty} \epsilon_n \sin{(n\pi x)}.$$

The corresponding solution of (1.4) is

$$\hat{u}(x) = u(x) + \sum_{n=1}^{\infty} \frac{\epsilon_n}{n^2 \pi^2} \sin{(n\pi x)}.$$

This shows that there is no difficulty in computing $u$ from $f$, in the sense that errors (in the data) of all frequencies are damped when the solution is computed.

## 1.2.1   Forward versus inverse problems

Using the above example as a guide, we offer two comments about the distinction between forward problems and the corresponding inverse problems. Forward problems are the type that have traditionally been studied in mathematical physics. The general goal is to produce a mathematical model that describes the outcome of some experiment. In deriving the model, it is assumed that the various experimental conditions are known. The resulting model always includes certain parameters—constants or functions—that describe them. In the above, the forward problem is to determine the displacement $u$, and $f$ is one of the parameters, describing the applied force. (The role of the other parameters—the tension and the length of the string—was de-emphasized by assuming that both were 1. This focused the attention on the parameter of interest for our purposes.)

An inverse problem, in contrast to the corresponding forward problem, asks for the determination of some aspect of the experimental conditions from an observation of the outcome of the experiment. In contrast to determining the effect from the causes, as in a forward problem, an inverse problem seeks the determination of the causes (or some aspect of the causes) from the effect.

A second comment about forward problems is in order. From a mathematical point of view, a typical forward problem tends to be stable precisely because high frequency changes in the data—the parameters—are smoothed out in passing from the data to the solution. In other words, a high-frequency change in the data results in little change in the solution. If this is true, it inevitably implies that when the inverse problem is considered instead, high frequencies in the data (which was the solution for the forward problem) must be amplified in passing to the solution (the data of the forward problem). Formulas (1.5) and (1.6) show this explicitly for the model inverse problem examined in this section.

This situation (smoothing of high frequencies in the forward problem) has another specific implication, which we explain in the context of (1.5) and (1.6). Since $u$ is the exact solution of the forward problem, it contains little contribution from high-frequency components. In other words, if we write

$$u(x) = \sum_{n=1}^{\infty} a_n \sin(n\pi x),$$

then $a_n$ tends to decrease rapidly with $n$ and to be quite small for large $n$. If we now imagine that the error in $\hat{u}$ is mostly white noise (equally distributed among the various frequencies), then we have

$$\hat{u}(x) = \sum_{n=1}^{\infty} a_n \sin(n\pi x) + \sum_{n=1}^{\infty} \epsilon_n \sin(n\pi x) = \sum_{n=1}^{\infty} (a_n + \epsilon_n) \sin(n\pi x),$$

where $\epsilon_1, \epsilon_2, \epsilon_3, \ldots$ are all of roughly the same magnitude, at least up to $\epsilon_N$, where $N$ is relatively large. It follows that $\epsilon_n$ is likely to be small compared to $a_n$ when $n$ is small, but not small compared to $a_n$ when $n$ is large.

This suggests that it should still be possible to compute a reasonably accurate solution to the inverse problem if we can find a way to eliminate or
reduce the influence of certain components of the data (namely the high-
frequency components) without greatly affecting the other components
of the data, which contain useful information. Tikhonov regularization is
able to achieve this, as we show explicitly in Chapter 4.

Looking back at the numerical results for the inverse problem from
the previous section, it will be seen that they are consistent with our discussion. With small but high-frequency noise in the data, the computed
solution shows large high-frequency errors. The small and oscillatory
noise was amplified by a large factor in passing from the data to the solution.

We close with two comments. First, the reader may recall our remarks
from the beginning of this section, namely that one cannot identify one
problem as the forward problem and the other as the inverse problem
based on superficial appearances. For the examples discussed in this section, the problem of computing $f$ from $u$ seems more direct (one only has
to compute the second derivative of $u$), while computing $u$ from $f$ seems
like an inverse problem (it is necessary to solve an equation to find $u$).
This superficial analysis misidentifies the problems.

Second, an integral equation of the form

$$\int_0^1 k(s, t)x(t)\, dt = y(s),\ 0 \le s \le 1 \tag{1.7}$$

is typically an inverse problem, while an integral equation of the form

$$x(s) + \int_0^1 k(s, t)x(t)\, dt = y(s),\ 0 \le s \le 1 \tag{1.8}$$

is not an inverse problem. Equation (1.7) represents an inverse problem
because integration (which defines the forward operator) is inherently
a smoothing operation. Moreover, the smoother the kernel $k$, the more
the forward problem smooths high frequencies and the more difficult the

inverse problem is.[1] On the other hand, in the second equation, the inclusion of $x$ itself on the left side preserves high frequencies in the solution, so the left side, regarded as an operator acting on $x$, is not a smoothing operator. As a result, (1.8) is not an inverse problem. We will study equations of the form (1.7) further in Section 4.4.1, and equations of the form (1.8) in Section 4.4.2.

---

[1] If the reader is familiar with the theory of Green's functions, he or she will recall that the solution $u$ of (1.4) can be expressed as

$$u(x) = \int_0^1 G(x, y)f(y)\,dy, \ 0 \le x \le 1,$$

where $G$ is the Green's function of the BVP. This is consistent with the fact that computing $f$ from $u$ is a true inverse problem. However, in this case, $G$ is not particularly smooth (it is continuous, but its first partial derivatives are discontinuous along the line $y = x$) and hence the inverse problem is not a particularly difficult one. Passing from $f$ to $u$ does dampen the high frequencies, but not dramatically because $G$ is not very smooth.

# Chapter 2

# Well-posed, ill-posed, and inverse problems

The purpose of this chapter is to explain the properties that a problem must have to be considered an inverse problem, and to study them in some detail. We are going to restrict ourselves to linear inverse problems defined on Hilbert spaces. Throughout this book, $X$ and $Y$ will denote Hilbert spaces and $T : X \to Y$ will denote a continuous linear operator. We wish to study the equation

$$Tx = y, \tag{2.1}$$

where $y \in Y$ is given and $x \in X$ is to be determined. It may be straightforward or difficult to solve accurately, depending on the properties of $T$. In this chapter, we describe the conditions that make (2.1) *well-posed*, *ill-posed*, or an *inverse problem*. An inverse problem is a special kind of ill-posed problem that is particularly difficult to solve, and such problems are the subject of this book.

Before we begin, we remind the reader that a linear operator $T : X \to Y$ is *bounded* if and only if there exists $M > 0$ such that

$$\|Tx\|_Y \leq M\|x\|_X \text{ for all } x \in X.$$

This concept is important because a linear operator is continuous if and only if it is bounded, and boundedness is usually easier to use than continuity. Basic facts about analysis in Hilbert spaces are reviewed in

15

Appendix A; specifically, Section A.1 reviews facts about linear operators defined on Hilbert spaces.

## 2.1 Well-posed problems

We introduce the concept of a well-posed problem in the context of (2.1), although the concept can be applied to a broader class of problems.

**Definition 2.1.** The problem of finding $x \in X$ to solve $Tx = y$, where $y \in Y$ is given, is called *well-posed*[2] if three conditions are satisfied:

1. *existence:* there exists at least one solution $x$;

2. *uniqueness:* there exists at most one solution $x$;

3. *stability:* the solution $x$ depends continuously on $y$.

The property of stability (also called *continuous dependence*) requires some elaboration. For the question of continuous dependence to be meaningful, it must be the case that for all $z$ in a neighborhood of $y$, $Tu = z$ has a unique solution $u = u(z)$. Then the stability property holds if $u \to x$ as $z \to y$, that is, if

$$\|z - y\|_Y \to 0 \;\Rightarrow\; \|u - x\|_X \to 0.$$

Because the problem under consideration is linear, these properties are actually global, not local. Moreover, it turns out that local existence and uniqueness together imply not only global existence and uniqueness, but global stability as well. In the following proof, we refer to the null space of $T$, which is the solution set of $Tx = 0$:

$$\mathcal{N}(T) = \{x \in X \; : \; Tx = 0\}.$$

---

[2]The phrase "well-posed in the sense of Hadamard" is often used; however, as explained in Maz'ya and Shaposhnikova's biography [20] and as is discussed further in Section 2.3.3, the term "well-posed in the sense of Courant" might be more appropriate. The concept received its first general definition in Courant and Hilbert's *Methods of Mathematical Physics* [2].

Because $T$ is linear, we have the following fundamental fact: $Tx_1 = Tx_2$ if and only if $x_1 - x_2 \in \mathcal{N}(T)$:

$$Tx_1 = Tx_2 \Leftrightarrow Tx_1 - Tx_2 = 0 \Leftrightarrow T(x_1 - x_2) = 0.$$

**Theorem 2.2.** *Suppose there exist $y \in Y$ and $\epsilon > 0$ such that for all $z$ belonging to*

$$B_\epsilon(y) = \{w \in Y \, : \, \|w - y\|_Y < \epsilon\}$$

*there exists a unique $u \in X$ such that $Tu = z$. Then, in fact, for all $w \in Y$, there exists a unique $v \in X$ such that $Tv = w$ and hence $T^{-1} : Y \to X$ is well-defined. Moreover, in this case, $T^{-1}$ is bounded (continuous).*

*Proof.* Assume that $y \in Y$, $\epsilon > 0$, and for all $z \in B_\epsilon(y)$, there exists a unique $u \in X$ such that $Tu = z$. Let $w \in Y$ and define $z = y + \delta(w - y) = (1 - \delta)y + \delta w$, where $\delta < \epsilon / \|w - y\|_Y$. Then

$$\|z - y\|_Y = \delta\|w - y\|_Y < \epsilon \Rightarrow z \in B_\epsilon(y),$$

and hence there exists a unique $u \in X$ such that $Tu = z$. Now define

$$v = \delta^{-1}(u - (1 - \delta)x),$$

where $Tx = y$. Then

$$Tv = \delta^{-1}(Tu - (1 - \delta)Tx) = \delta^{-1}((1 - \delta)y + \delta w - (1 - \delta)y) = w.$$

Thus $Tv = w$ has a solution. If $v_1, v_2 \in X$ satisfy $Tv_1 = Tv_2 = w$, then $v_1 - v_2$ belongs to $\mathcal{N}(T)$, and hence $T(x + v_1 - v_2) = y$, contradicting the fact that the equation $Tx = y$ has a unique solution.

Thus we have shown that for all $w \in Y$, there exists a unique $v \in X$ such that $Tv = w$. It follows that $T^{-1}$ exists and hence, since $X$, $Y$ are Hilbert spaces, the bounded inverse theorem implies that $T^{-1}$ is bounded. $\qquad\square$

In proving Theorem 2.2, we used the powerful *bounded inverse theorem* (see Corollary A.30), which states that if a bounded linear operator $T : X \to Y$ is bijective (both one-to-one and onto), then the inverse $T^{-1}$ has to be a continuous operator. This is a fundamental fact about linear operators on Hilbert spaces.

Theorem 2.2 shows that, for a given $y \in Y$, the problem of finding $x \in X$ to solve $Tx = y$ is well-posed if and only if $T$ is invertible and has a bounded inverse. It also shows that if $Tx = y$ is well-posed for a given $y \in Y$, then it is well-posed for every $y \in Y$. These facts make the concept of well-posedness for $Tx = y$ rather simple.

In the next section, we discuss methods for replacing an ill-posed problem with a related well-posed problem. If the only issue is a lack of uniqueness, there is an effective way to do this. If existence fails, there is another technique for making the problem well-posed, but this is only effective in certain cases. In the difficult case (which, as we shall see, occurs when the range $\mathcal{R}(T)$ of $T$ fails to be closed), the lack of existence is intrinsically connected with the lack of stability. It is problems that fall into this category that we designate as inverse problems.

## 2.2    Ill-posed problems, part 1

The equation $Tx = y$ is well-posed if and only if it satisfies three conditions: existence, uniqueness, and stability (continuous dependence). We now discuss problems in which one or more of these conditions fails to hold and how we might analyze and solve them. Since an ill-posed problem cannot be solved in the usual sense, we have to redefine it so that it has a meaningful solution.

### 2.2.1    The uniqueness condition and the minimum-norm solution

Because $T$ is linear, the uniqueness condition is the simplest of the three properties of a well-posed problem. If $Tx = y$ has a solution, it is unique if and only if $\mathcal{N}(T) = \{0\}$. In fact, we have the following result.

**Theorem 2.3.** *If $y \in Y$ is given and $\hat{x} \in X$ is a solution of $Tx = y$, then the set of all solutions of $Tx = y$ is*

$$\hat{x} + \mathcal{N}(T) = \{\hat{x} + u : u \in \mathcal{N}(T)\}.$$

*Proof.* Assume that $y \in Y$ is given and that $\hat{x} \in X$ satisfies $T\hat{x} = y$. Suppose first that $\bar{x} \in \hat{x} + \mathcal{N}(T)$, say $\bar{x} = \hat{x} + u$ for some $u \in \mathcal{N}(T)$. Then

$$T\bar{x} = T(\hat{x} + u) = T\hat{x} + Tu = y + 0 = y,$$

where we have used the fact that $Tu = 0$ because $u \in \mathcal{N}(T)$. Thus $\bar{x}$ is a solution of $Tx = y$.

Conversely, suppose $\bar{x}$ is any solution of $Tx = y$ and define $u = \bar{x} - \hat{x}$. Then $\bar{x} = \hat{x} + u$ and

$$Tu = T(\bar{x} - \hat{x}) = T\bar{x} - T\hat{x} = y - y = 0,$$

since both $\bar{x}$ and $\hat{x}$ are solutions of $Tx = y$. This shows that $u \in \mathcal{N}(T)$ and hence that $\bar{x} \in \hat{x} + \mathcal{N}(T)$.

We have now shown that $\bar{x} \in \hat{x} + \mathcal{N}(T)$ if and only if $\bar{x}$ is a solution of $Tx = y$. $\qquad\square$

It is important to note that the choice of the representative vector $\hat{x}$ is arbitrary (among all solutions of $Tx = y$). In other words, if $\hat{x}$ and $\bar{x}$ are two solutions of $Tx = y$, then we can write the solution set as either $\hat{x} + \mathcal{N}(T)$ or $\bar{x} + \mathcal{N}(T)$; the two sets are equal.

If $Tx = y$ has a solution $\hat{x}$ and $\mathcal{N}(T)$ is nontrivial, then we can impose uniqueness by choosing some criterion that selects a unique best solution from the solution set $\hat{x} + \mathcal{N}(T)$. A popular criterion is based on defining a linear operator $L : X \to Z$, where $Z$ is another Hilbert space, and declaring that element of $\hat{x} + \mathcal{N}(T)$ to be best that minimizes $\|Lx\|_Z$. The idea is to choose $L$ so that undesirable solutions $x$ correspond to large values of $\|Lx\|_Z$.

The simplest choice is $Z = X$ and $L = I$ (the identity operator), in which case the best solution is defined as the *minimum-norm* solution. It is not difficult to show that there exists a unique solution of

$$\begin{aligned} \min \ & \|x\|_X \\ \text{s.t. } & x \in \hat{x} + \mathcal{N}(T). \end{aligned} \tag{2.2}$$

(Here and below, "s.t." stands for "subject to.") The proof, which is given below, uses the fact that $\mathcal{N}(T)$ is a closed subspace of $X$, a critical property that we use constantly.

When $X$ is a space of functions (the most common situation), a popular choice for $L$ is a derivative operator, in which case forcing $\|Lx\|_Z$ to be as small as possible forces the solution to have small derivatives and thus to be smooth (that is, non-oscillatory). This is often natural because random noise in the data vector $y$ tends to be oscillatory, which causes a naively computed solution of $Tx = y$ to be highly oscillatory as well, and hence not very smooth.

Analyzing the case of a general operator $L$ is challenging. First, we must ensure that $L$ has the property that

$$\min \|Lx\|_Z$$
$$\text{s.t. } x \in \hat{x} + \mathcal{N}(T) \tag{2.3}$$

has a unique solution. Moreover, many of the most useful choices of $L$ (such as derivative operators) are unbounded and only densely defined on $X$ (that is, defined only on a dense subspace of $X$, not on all $X$), which complicates the analysis considerably. For this reason, we study the case $L = I$ first and in detail before turning to the more general case in Chapter 5.

We remark that (2.3) is not the most general way to define a best solution. In some problems, it may be desirable to choose a more general nonlinear functional $R : X \to \mathbb{R}$ and solve

$$\min R(x)$$
$$\text{s.t. } x \in \hat{x} + \mathcal{N}(T).$$

This formulation is beyond the scope of this book, though the most important example, BV-regularization, is briefly described in the Epilogue (see pages 272ff).

The following theorem shows that if $Tx = y$ has a solution, then it has a unique minimum-norm least-squares solution. Two important constructions are used in the theorem and its proof. First, the *orthogonal complement* of a set $S$ is the set of all vectors orthogonal to every vector in $S$:

$$S^\perp = \{u \in X \ : \ \langle u, x \rangle_X = 0 \text{ for all } x \in S\}.$$

Second, given a closed subspace $S$ of a Hilbert space $X$ and a vector $x \in X$, the *projection* of $x$ onto $S$ is defined to be the vector in $S$ closest to $x$, that is, the vector $u \in S$ satisfying $\|u - x\|_X \leq \|w - x\|_X$ for all $w \in S$. The projection of $x$ onto $S$, which is denoted by $\text{proj}_S x$, is guaranteed to exist and to be unique, and it is characterized by the condition that $\langle u - x, w \rangle_X = 0$ for all $w \in S$, that is, by the condition that $u - x \in S^\perp$.

The basic properties of orthogonal complements and projections are derived in Appendix A.2.

**Theorem 2.4.** *Suppose $y \in Y$ and $Tx = y$ has a solution. Then there is a unique minimum-norm solution of $Tx = y$, which is characterized as the unique solution that belongs to $\mathcal{N}(T)^\perp$.*

*Proof.* Let $\hat{x}$ be a solution of $Tx = y$. We know that the set of all solutions is $\hat{x} + \mathcal{N}(T)$. We first show that there is a unique $\bar{x} \in \mathcal{N}(T)^\perp$ that belongs to $\hat{x} + \mathcal{N}(T)$. Define $\bar{x} = \mathrm{proj}_{\mathcal{N}(T)^\perp} \hat{x}$, the projection of $\hat{x}$ onto $\mathcal{N}(T)^\perp$. Then, by the projection theorem (Theorem A.16), $\bar{x} - \hat{x} \in \mathcal{N}(T)^{\perp\perp} = \mathcal{N}(T)$ and hence $\bar{x} \in \hat{x} + \mathcal{N}(T)$. If also $\bar{x}_1 \in \mathcal{N}(T)^\perp \cap (\hat{x} + \mathcal{N}(T))$, then $\bar{x} - \bar{x}_1$ belongs to $\mathcal{N}(T)$ (because $T\bar{x} = T\bar{x}_1$). But $\bar{x} - \bar{x}_1$ also belongs to $\mathcal{N}(T)^\perp$ because $\mathcal{N}(T)^\perp$ is a subspace. Since $\mathcal{N}(T) \cap \mathcal{N}(T)^\perp = \{0\}$, it follows that $\bar{x} - \bar{x}_1 = 0$, that is, $\bar{x} = \bar{x}_1$. Thus we have shown that there is a unique solution of $Tx = y$ that belongs to $\mathcal{N}(T)^\perp$.

We now show that $\bar{x}$ is the minimum-norm solution of $Tx = y$. Since $\bar{x}$ is a solution of $Tx = y$, the set of all solutions can be written as $\bar{x} + \mathcal{N}(T)$ and hence any other solution is of the form $x = \bar{x} + u$, where $u \in \mathcal{N}(T), u \neq 0$. Since $\bar{x} \in \mathcal{N}(T)^\perp$ and $u \in \mathcal{N}(T)$, the Pythagorean theorem implies that

$$\|x\|_X^2 = \|\bar{x}\|_X^2 + \|u\|_X^2 > \|\bar{x}\|_X^2.$$

This shows that $\bar{x}$ is the unique minimum-norm solution of $Tx = y$.    □

## 2.2.2  The existence condition and least-squares solutions

The equation $Tx = y$ has a solution if and only if $y$ belongs to the range of $T$, which is defined by

$$\mathcal{R}(T) = \{Tx \,:\, x \in X\}.$$

The range of $T$ is a subspace of $Y$; however, unlike the null space of $T$, $\mathcal{R}(T)$ may or may not be closed.

If $\mathcal{R}(T)$ is a proper subspace of $Y$, then there exist vectors $y \in Y$ for which $Tx = y$ has no solution. However, it may still make sense to seek a solution. For example, there may exist an exact data vector $y^* \in \mathcal{R}(T)$ for which measurement error produces $y \in Y \setminus \mathcal{R}(T)$. In this case, it is reasonable to seek the vector $x \in X$ such that $Tx$ is as close as possible to $y$, that is, to solve

$$\min_{x \in X} \|Tx - y\|_Y.$$

If such a vector $x$ exists, it is called a *least-squares solution* of $Tx = y$.[3] Although the least-squares approach is natural, the following result shows that $Tx = y$ need not have a least-squares solution.

**Theorem 2.5.**

1. *A least-squares solution of $Tx = y$ exists if and only if $y \in \mathcal{R}(T) \oplus \mathcal{R}(T)^{\perp}$.*

2. *If $\mathcal{R}(T)$ is closed, then $\mathcal{R}(T) \oplus \mathcal{R}(T)^{\perp} = Y$ and $Tx = y$ has a least-squares solution for every $y \in Y$. Otherwise, $\mathcal{R}(T) \oplus \mathcal{R}(T)^{\perp}$ is a proper dense subspace of $Y$.*

*Proof.*

1. First suppose that $y \in \mathcal{R}(T) \oplus \mathcal{R}(T)^{\perp}$ and let $y = \bar{y} + z$, where $\bar{y} \in \mathcal{R}(T)$ and $z \in \mathcal{R}(T)^{\perp}$. Then, by definition of $\mathcal{R}(T)$, there exists $\bar{x} \in X$ such that $T\bar{x} = \bar{y}$. We claim that $\bar{x}$ is a least-squares solution of $Tx = y$, that is, that

$$\|T\bar{x} - y\|_Y \le \|Tx - y\|_Y \text{ for all } x \in X.$$

---

[3]The term "least-squares solution" arises from the finite-dimensional version of the problem. Given an operator $F : \mathbb{R}^n \to \mathbb{R}^m$ (which could be linear or nonlinear) and an equation $F(x) = y$ that has no exact solution, we can define an approximate solution by minimizing $\|F(x) - y\|$, where the norm is the Euclidean norm. Minimizing $\|F(x) - y\|$ is equivalent to minimizing $\|F(x) - y\|^2$ so, writing everything in terms of components, we wish to choose $x_1, \ldots, x_n$ to minimize

$$\sum_{j=1}^{m} (F_j(x_1, \ldots, x_n) - y_j)^2,$$

which explains how the phrase "least-squares" arose.

This is equivalent to proving that $\|\bar{y} - y\|_Y \leq \|w - y\|_Y$ for all $w \in \mathcal{R}(T)$. But, for all $w \in \mathcal{R}(T)$, $\bar{y} - w \in \mathcal{R}(T)$, which implies that $\bar{y} - w$ is orthogonal to $z = y - \bar{y} \in \mathcal{R}(T)^\perp$. Hence, by the Pythagorean theorem,

$$\|w - y\|_Y^2 = \|w - \bar{y} + \bar{y} - y\|_Y^2$$
$$= \|w - \bar{y}\|_Y^2 + \|\bar{y} - y\|_Y^2 \geq \|\bar{y} - y\|_Y^2.$$

This shows that $\bar{x}$ is a least-squares solution of $Tx = y$.

Conversely, suppose $\bar{x}$ is a least-squares solution of $Tx = y$, define $\bar{y} = T\bar{x}$, and define $z = y - \bar{y}$. We will show that $z \in \mathcal{R}(T)^\perp$, which will imply that $y = \bar{y} + z \in \mathcal{R}(T) \oplus \mathcal{R}(T)^\perp$, as desired. Clearly $w \in \mathcal{R}(T)$ if and only if $w = \bar{y} + t\bar{w}$, where $\bar{w} \in \mathcal{R}(T)$ and $t \in \mathbb{R}$. Define $\phi : \mathbb{R} \to \mathbb{R}$ by $\phi(t) = \|y - (\bar{y} + t\bar{w})\|_Y^2$. Then

$$\phi(t) = \|y - \bar{y} - t\bar{w}\|_Y^2 = \|y - \bar{y}\|_Y^2 - 2t\langle y - \bar{y}, \bar{w}\rangle_Y + t^2\|\bar{w}\|_Y^2$$
$$\Rightarrow \phi'(t) = 2t\|\bar{w}\|_Y^2 - 2\langle y - \bar{y}, \bar{w}\rangle_Y$$
$$\Rightarrow \phi'(0) = -2\langle y - \bar{y}, \bar{w}\rangle_Y.$$

By assumption, $\phi(0) \leq \phi(t)$ for all $t \in \mathbb{R}$, which implies that $\phi'(0) = 0$ because $t = 0$ is a minimizer of $\phi$. But this implies that $\langle y - \bar{y}, \bar{w}\rangle_Y = 0$. Moreover, this holds for every $\bar{w} \in \mathcal{R}(T)$, which shows that $y - \bar{y} \in \mathcal{R}(T)^\perp$, as desired.

2. If $\mathcal{R}(T)$ is closed, then $\mathcal{R}(T) = \mathcal{R}(T)^{\perp\perp}$ and it is well-known that

$$Y = \mathcal{R}(T)^{\perp\perp} \oplus \mathcal{R}(T)^\perp = \mathcal{R}(T) \oplus \mathcal{R}(T)^\perp.$$

Otherwise, we have $\mathcal{R}(T)^{\perp\perp} = \overline{\mathcal{R}(T)}$ and hence $Y = \overline{\mathcal{R}(T)} \oplus \mathcal{R}(T)^\perp$. Since $\mathcal{R}(T)$ is a proper dense subspace of $\overline{\mathcal{R}(T)}$, it is straightforward to show that $\mathcal{R}(T) \oplus \mathcal{R}(T)^\perp$ is a proper dense subspace of $Y$. $\qquad\square$

From the previous result, we see that if $\mathcal{R}(T)$ is closed, then $Tx = y$ has a least-squares solution for each $y \in Y$; otherwise, this is true only for $y$ belonging to the dense subspace $\mathcal{R}(T) \oplus \mathcal{R}(T)^\perp$. The issue is that the closed subspace $\overline{\mathcal{R}(T)}$ contains a unique vector closest to each $y \in Y$, namely, $\bar{y} = \text{proj}_{\overline{\mathcal{R}(T)}}y$. There is a least-squares solution of $Tx = y$ if and only if $\bar{y}$ belongs to $\mathcal{R}(T)$.

We will need a characterization of the least-squares solutions of $Tx = y$. The following theorem refers to $T^*$, the *adjoint* of the linear operator $T$. The linear operator $T^*$ is defined by the condition that

$$\langle Tx, y \rangle_Y = \langle x, T^*y \rangle_X \text{ for all } x \in X, y \in Y.$$

The basic facts about $T^*$, including the fundamental relations $\mathcal{R}(T)^\perp = \mathcal{N}(T^*)$ and $\mathcal{N}(T^*)^\perp = \overline{\mathcal{R}(T^*)}$, are reviewed in Section A.4.

**Theorem 2.6.** *If $y \in \mathcal{R}(T) \oplus \mathcal{R}(T)^\perp$, then $x \in X$ is a least-squares solution of $Tx = y$ if and only if $x$ satisfies $T^*Tx = T^*y$.*

*Proof.* If $x$ is a least-squares solution of $Tx = y$, then $Tx$ is the closest vector in $\mathcal{R}(T)$, and hence in $\overline{\mathcal{R}(T)}$, to $y$. Hence $Tx = \text{proj}_{\overline{\mathcal{R}(T)}}y$, which implies that $y - Tx \in \mathcal{R}(T)^\perp = \mathcal{N}(T^*)$. Therefore

$$T^*(y - Tx) = 0 \implies T^*Tx = T^*y.$$

Conversely, if $T^*Tx = T^*y$, then $y - Tx \in \mathcal{N}(T^*) = \overline{\mathcal{R}(T)}^\perp$ and $Tx$ belongs to $\mathcal{R}(T) \subset \overline{\mathcal{R}(T)}$. This implies that $Tx = \text{proj}_{\overline{\mathcal{R}(T)}}y$ and hence that $x$ is a least-squares solution of $Tx = y$. □

In a finite-dimensional problem, $T$ can be represented by a matrix $A$ and $T^*$ by the transpose matrix $A^T$. In this case, the equation $T^*Tx = T^*y$ becomes $A^TAx = A^Ty$. This matrix-vector equation represents a system of linear algebraic equations; these equations are called the *normal equations*. The normal equations express the fact that $y - Ax$ is normal to (that is, orthogonal to) the column space of $A$. We will sometimes refer to $T^*Tx = T^*y$ as the normal equation for (the least-squares problem defined by) $Tx = y$.

Here is a simple result about operator $T^*T$ that we will need.

**Lemma 2.7.** *The operators $T$ and $T^*T$ have the same null space.*

*Proof.* Obviously $Tx = 0$ implies that $T^*Tx = 0$ and hence $\mathcal{N}(T) \subset \mathcal{N}(T^*T)$. Conversely, if $T^*Tx = 0$, then

$$0 = \langle x, T^*Tx \rangle_X = \langle Tx, Tx \rangle_Y = \|Tx\|_Y^2.$$

Thus $Tx = 0$, and we have shown that $\mathcal{N}(T^*T) \subset \mathcal{N}(T)$. □

The following corollary collects the facts derived above about least-squares solutions of $Tx = y$.

**Corollary 2.8.** *There exists a least-squares solution of $Tx = y$ if and only if $y \in \mathcal{R}(T) \oplus \mathcal{R}(T)^{\perp}$. If this holds, then the following conditions on $\bar{x} \in X$ are equivalent and each implies that $\bar{x}$ is a least-squares solution of $Tx = y$:*

1. *$\|T\bar{x} - y\|_Y = \inf\{\|Tu - y\|_Y : u \in X\}$ (this is the definition of least-squares solution);*

2. *$T\bar{x} = \bar{y} = \operatorname{proj}_{\overline{\mathcal{R}(T)}} y$;*

3. *$T^*T\bar{x} = T^*y$.*

*If $\bar{x}$ is a least-squares solution, then the set of all least-squares solutions is $\bar{x} + \mathcal{N}(T)$ and the solution $\bar{x}$ is unique if and only if $\mathcal{N}(T) = \{0\}$.*

*Proof.* All the conclusions have already been noted except the representation of the set of all least-squares solutions. However, since the set of all least-squares solutions is the solution set of $T^*Tx = T^*y$, we see that it is $\bar{x} + \mathcal{N}(T^*T)$, where $\bar{x}$ is any one least-squares solution. By the previous lemma, $\bar{x} + \mathcal{N}(T^*T) = \bar{x} + \mathcal{N}(T)$, and the proof is complete. $\qquad\square$

If $\mathcal{N}(T)$ is nontrivial and $\mathcal{R}(T)$ is a proper subspace of $Y$, then both existence and uniqueness fail. However, for $y \in \mathcal{R}(T) \oplus \mathcal{R}(T)^{\perp}$, a related problem has a unique solution.

**Theorem 2.9.** *If $y \in \mathcal{R}(T) \oplus \mathcal{R}(T)^{\perp}$, then $Tx = y$ has a unique minimum-norm least-squares solution; that is, there is a unique solution of*

$$\min \|x\|_X$$
$$s.t. \ \|Tx - y\|_Y = \inf\{\|Tu - y\|_Y : u \in X\}.$$

*Moreover, the minimum-norm least squares solution is characterized as the unique vector $x$ satisfying $T^*Tx = T^*y$ and $x \in \mathcal{N}(T)^{\perp}$.*

*Proof.* We already know that $y \in \mathcal{R}(T) \oplus \mathcal{R}(T)^{\perp}$ implies that $Tx = y$ has a least-squares solution $\hat{x}$. Then the set of all least-squares solutions is $\hat{x} + \mathcal{N}(T)$. As in the proof of Theorem 2.4, this set contains a unique element $\bar{x}$ of $\mathcal{N}(T)^{\perp}$, and $\bar{x}$ is the minimum-norm element of $\hat{x} + \mathcal{N}(T)$.  $\square$

We have seen that $Tx = y$ is well-posed if and only if $T^{-1}$ exists and is bounded, in which case the solution is $x = T^{-1}y$. When existence or uniqueness fails, we replace the problem of finding an exact solution of $Tx = y$ by that of determining the minimum-norm least-squares solution. The fact that this solution is unique (when it exists) implies there is an operator that determines it.

**Theorem 2.10.** *Let $D(T^{\dagger}) = \mathcal{R}(T) \oplus \mathcal{R}(T)^{\perp}$. Then there exists a linear operator $T^{\dagger} : D(T^{\dagger}) \to \mathcal{N}(T)^{\perp}$ such that for each $y \in D(T^{\dagger})$, $x = T^{\dagger}y$ is the unique minimum-norm least-squares solution of $Tx = y$.*

*Proof.* By the previous theorem, the operator $T^{\dagger}$ is well-defined. It remains only to show that it is a linear operator. Suppose $y_1, y_2 \in D(T^{\dagger})$ and $\alpha_1, \alpha_2 \in \mathbb{R}$. We must show that $T^{\dagger}(\alpha_1 y_1 + \alpha_2 y_2) = \alpha_1 T^{\dagger}y_1 + \alpha_2 T^{\dagger}y_2$. Let us write $x_1 = T^{\dagger}y_1$, $x_2 = T^{\dagger}y_2$. Then we know that $x_1, x_2 \in \mathcal{N}(T)^{\perp}$ and

$$T^*Tx_1 = T^*y_1, \;\; T^*Tx_2 = T^*y_2.$$

Since $\mathcal{N}(T)^{\perp}$ is a subspace, it follows that $\alpha_1 x_1 + \alpha_2 x_2$ belongs to $\mathcal{N}(T)^{\perp}$. Also,

$$T^*T(\alpha_1 x_1 + \alpha_2 x_2) = \alpha_1 T^*Tx_1 + \alpha_2 T^*Tx_2 = \alpha_1 T^*y_1 + \alpha_2 T^*y_2$$
$$= T^*(\alpha_1 y_1 + \alpha_2 y_2).$$

By the previous theorem, it follows that

$$\alpha_1 x_1 + \alpha_2 x_2 = T^{\dagger}(\alpha_1 y_1 + \alpha_2 y_2),$$

as desired.  $\square$

The operator $T^{\dagger}$ is called the *generalized inverse* of $T$, and $D(T^{\dagger})$ denotes its domain. When $Tx = y$ is well-posed, solving the equation means computing $x = T^{-1}y$. If $Tx = y$ either has no solution or has more than one solution, then we define the (unique) desired solution to be $x =$

$T^\dagger y$. As we have seen, this addresses the uniqueness question completely; however, $T^\dagger y$ exists only if $y$ belongs to the subspace $D(T^\dagger)$ of $Y$. Since $D(T^\dagger)$ can be a proper subspace of $Y$, this implies that existence can still fail. In the next section, we address the question of stability.

## 2.3   Ill-posed problems, part 2

In the previous section, we saw that the existence condition is related to the closure of $\mathcal{R}(T)$. If $\mathcal{R}(T)$ is closed, then $Y = D(T^\dagger) = \mathcal{R}(T) \oplus \mathcal{R}(T)^\perp$, and there exists a unique minimum-norm least-squares solution to $Tx = y$ for each $y \in Y$. Although $T^\dagger y$ may not solve the original problem, it is the best approximate solution in a certain well-defined sense. However, if $\mathcal{R}(T)$ is not closed, then $D(T^\dagger)$ is a proper dense subspace of $Y$ and there exist vectors $y \in Y$ such that $Tx = y$ does not have a solution, even in the least-squares sense.

### 2.3.1   Stability and the generalized inverse

We will now see that the closure of $\mathcal{R}(T)$ is also intrinsically connected to the question of the stability of the problem $Tx = y$, that is, to the question of whether the solution $x = T^\dagger y$ depends continuously on the data $y$. Since $T^\dagger$ is a linear operator, this reduces to the question of whether $T^\dagger$ is bounded or not. To explore this question, we need some preliminary results.

**Lemma 2.11.**

1. *For all $x \in X$, $T^\dagger T x = \text{proj}_{\mathcal{N}(T)^\perp} x$.*

2. *For all $y \in D(T^\dagger)$, $T T^\dagger y = \text{proj}_{\overline{\mathcal{R}(T)}} y$.*

*Proof.*

1. Let $x \in X$ be given and define $\bar{x} = \text{proj}_{\mathcal{N}(T)^\perp} x$. We wish to show that $T^\dagger T x = \bar{x}$, that is, that $\bar{x}$ lies in $\mathcal{N}(T)^\perp$ and satisfies the normal equation $T^* T \bar{x} = T^*(Tx)$ (that is, $T^* T \bar{x} = T^* y$ for $y = Tx$). The first condition holds by construction of $\bar{x}$. Also, since $\bar{x} = \text{proj}_{\mathcal{N}(T)^\perp} x$, the projection theorem implies that $\bar{x} - x \in \mathcal{N}(T)$ and hence that

$T\bar{x} = Tx$. It follows that

$$T^*T\bar{x} = T^*(Tx),$$

as desired.

2. Let $y \in D(T^\dagger)$ and suppose $y = \bar{y} + \hat{y}$, where $\bar{y} \in \mathcal{R}(T)$, $\hat{y} \in \mathcal{R}(T)^\perp$. Since $\bar{y} - y \in \mathcal{R}(T)^\perp$, $\bar{y} = \mathrm{proj}_{\overline{\mathcal{R}(T)}} y$; that is, $\bar{y} \in \mathcal{R}(T)$ is the vector in $\overline{\mathcal{R}(T)}$ closest to $y$ (hence $\bar{y}$ is obviously the vector in $\mathcal{R}(T)$ closest to $y$). But, by definition, $TT^\dagger y$ is the vector in $\mathcal{R}(T)$ closest to $y$, that is, $TT^\dagger y = \bar{y}$.                                         $\square$

**Corollary 2.12.**

*1.* $TT^\dagger T = T$ *(that is,* $TT^\dagger Tx = Tx$ *for all* $x \in X$ *).*

*2.* $T^\dagger TT^\dagger = T^\dagger$ *(that is,* $T^\dagger TT^\dagger y = T^\dagger y$ *for all* $y \in D(T^\dagger)$ *).*

*Proof.*

1. Let $x \in X$ and define $\bar{x} = \mathrm{proj}_{\mathcal{N}(T)^\perp} x$. Since $\bar{x} - x \in \mathcal{N}(T)$, $T\bar{x} = Tx$ and hence, by the previous lemma,

$$TT^\dagger Tx = T\bar{x} = Tx.$$

2. Let $y \in D(T^\dagger)$ and suppose $y = \bar{y} + \hat{y}$, where $y \in \mathcal{R}(T)$ and $\hat{y} \in \mathcal{R}(T)^\perp$. Since $\hat{y}$ belongs to $\mathcal{R}(T)^\perp = \mathcal{N}(T^*)$, $T^*\hat{y} = 0$. Therefore, $x = 0$ satisfies $T^*Tx = T^*\hat{y}$, $x \in \mathcal{N}(T)^\perp$. It follows that $T^\dagger \hat{y} = 0$ and hence, by the previous lemma,

$$T^\dagger TT^\dagger y = T^\dagger \bar{y} = T^\dagger(\bar{y} + \hat{y}) = T^\dagger y,$$

as desired.                                         $\square$

We now want to determine the condition(s) under which $T^\dagger$ is bounded. As we have seen, $T^\dagger$ is defined on $\mathcal{R}(T) \oplus \mathcal{R}(T)^\perp$, which is always a dense subspace of $Y$ and may be a proper dense subspace. Therefore, in general, $T^\dagger$ is a *densely defined* operator. We are going to see that although $T^\dagger$ may fail to be bounded, it always has the related property of closedness, a property that is useful for analyzing densely defined operators.

**Definition 2.13.** Let $U, V$ be Hilbert spaces, let $D(L)$ be a subspace of $U$, and let $L : D(L) \to V$ be a linear operator (not necessarily bounded). We say that $L$ is *closed* if the graph of $L$ is a closed subset of $U \times V$, where the graph of $L$ is

$$\text{gr}(L) = \{(u, Lu) \in U \times V : u \in D(L)\}.$$

Since $L$ is linear, $\text{gr}(L)$ is a subspace of $U \times V$, as is easily verified. Also, $\text{gr}(L)$ is closed if and only if the conditions $\{u_n\} \subset D(L)$, $u_n \to u \in U$, $Lu_n \to v \in V$ imply that $(u, v) \in \text{gr}(L)$, that is, $u \in D(L)$ and $v = Lu$. Finally, in a Hilbert space, if a subspace is closed in the strong (norm) topology, then it is closed with respect to *weak sequential convergence* (see Theorem A.35). The sequence $\{u_n\}$ converges weakly to $u \in U$ if and only if for all $x \in U$,

$$\langle u_n, x \rangle_U \to \langle u, x \rangle_U \text{ as } n \to \infty.$$

The weak topology on an infinite-dimensional Hilbert space is useful primarily because every bounded sequence in such a space has a weakly convergent subsequence. The weak topology and weak convergence are reviewed in Section A.6 of Appendix A. Convergence in norm is often called *strong convergence*, to distinguish it from weak convergence.

A fundamental result of functional analysis is the closed graph theorem.

**Theorem 2.14.** *Let $U$ and $V$ be Hilbert spaces and let $L : U \to V$ be a linear operator. Then $L$ is bounded if and only if $L$ is closed.*

*Proof.* Suppose first that $L$ is bounded, and let $\{u_n\} \subset U$, $u \in U$, and $v \in V$ satisfy $u_n \to u$, $Lu_n \to v$. Because $L$ is bounded, it is continuous, and hence $u_n \to u$ implies that $Lu_n \to Lu$. Since $\{Lu_n\}$ has a unique limit, it follows that $v = Lu$ ($u \in D(L)$ is immediate, since $D(L) = U$). This shows that $L$ is closed.

The proof of the converse is based on the bounded inverse theorem (Theorem A.30). Suppose $L$ is closed. Define $P_1 : U \times V \to U$ and $P_2 : U \times V \to V$ by $P_1(u, v) = u$ and $P_2(u, v) = v$ for all $(u, v) \in U \times V$. It should be obvious that $P_1$ and $P_2$ are bounded linear operators, as are their restrictions to $\text{gr}(L)$. Now define $M : U \to \text{gr}(L)$ by $Mu = (u, Lu)$. Now,

$P_1|_{\text{gr}(L)}$ is surjective since $L$ is defined everywhere on $U$. Also, $P_1|_{\text{gr}(L)}$ is injective because, for each $u \in U$, $\text{gr}(L)$ contains a single element of the form $(u, v)$, namely $(u, Lu)$. Thus $P_1|_{\text{gr}(L)}$ is invertible. Since $\text{gr}(L)$ is a closed subspace of the Hilbert space $U \times V$, it is a Hilbert space in its own right. It follows from the bounded inverse theorem that the inverse of $P_1|_{\text{gr}(L)}$ is a bounded linear operator. But clearly $M$ is the inverse of $P_1|_{\text{gr}(L)}$; hence $M$ is bounded. It then follows that $L$, which is the composition of the bounded operators $P_2$ and $M$, is bounded.          □

It is critical to notice that a closed operator is bounded provided its domain is a Hilbert space. There are important examples of closed operators that are only densely defined and are not bounded. In fact, the generalized inverse $T^\dagger$ is such an operator if $\mathcal{R}(T)$ fails to be closed, as the next two results show. (We will encounter other examples of densely defined operators that are closed but unbounded in Chapter 5.)

**Theorem 2.15.** *Let $U$, $V$ be Hilbert spaces, let $D(L)$ be a subspace of $U$, and let $L : D(L) \to V$ be a closed linear operator. Then $L$ is bounded if and only if $D(L)$ is closed.*

*Proof.* If $D(L)$ is closed, then it is a Hilbert space and hence $L$ is bounded by the previous theorem. Conversely, suppose $L$ is bounded and let $u \in \overline{D(L)}$. Then there exists $\{u_n\} \subset D(L)$ such that $u_n \to u$ as $n \to \infty$. It follows that $\{u_n\}$ is Cauchy and hence, since $L$ is bounded, it is easy to see that $\{Lu_n\}$ is also Cauchy. Therefore, there exists $v \in V$ such that $Lu_n \to v$. Since $L$ is closed, it follows that $u \in D(L)$ (and $v = Lu$). This shows that $D(L)$ is closed          □

We have seen above that $T^\dagger : D(T^\dagger) \to X$ is a densely defined linear operator. We now show that it is closed.

**Theorem 2.16.** *$T^\dagger$ is closed.*

*Proof.* Suppose $\{y_n\} \subset D(T^\dagger)$ and $y_n \to y \in Y$, $T^\dagger y_n \to x \in X$. For each $n \in \mathbb{Z}^+$, define $x_n = T^\dagger y_n$; then $x_n \to x$. We must show that $y \in D(T^\dagger)$

and $x = T^{\dagger} y$. Since $T$ is bounded by assumption, $T x_n \to T x$. Also, applying Lemma 2.11, we have

$$T x_n = T T^{\dagger} y_n = \text{proj}_{\overline{\mathcal{R}(T)}} y_n \to \text{proj}_{\overline{\mathcal{R}(T)}} y.$$

This shows that $T x = \text{proj}_{\overline{\mathcal{R}(T)}} y$, from which it follows that $\text{proj}_{\overline{\mathcal{R}(T)}} y \in \mathcal{R}(T)$, $y \in D(T^{\dagger})$, and $x$ is a least-squares solution of $T x = y$. Finally, since each $x_n$ belongs to the closed subspace $\mathcal{N}(T)^{\perp}$, so does the limit $x$, which implies that $x$ is the minimum-norm least-squares solution of $T x = y$, and hence $x = T^{\dagger} y$.                                                            □

Combining the last two theorems, we obtain the following fundamental result.

**Corollary 2.17.** *The generalized inverse $T^{\dagger}$ of $T$ is bounded if and only if $\mathcal{R}(T)$ is closed.*

Since $T^{\dagger}$ is continuous if and only if it is bounded, this result allows us to characterize precisely the problems $T x = y$ that are unstable in the sense that the solution $x$ fails to depend continuously on the data $y$. This is discussed further below.

The following theorem gives further insight into the meaning of $T^{\dagger}$.

**Theorem 2.18.** *Let $S = T|_{\mathcal{N}(T)^{\perp}}$. Then $S$ is a bijection from $\mathcal{N}(T)^{\perp}$ to $\mathcal{R}(T)$ and $S^{-1}$ is a closed densely defined linear operator from $\overline{\mathcal{R}(T)}$ onto $\mathcal{N}(T)^{\perp}$, with $D(S^{-1}) = \mathcal{R}(T)$. Moreover, $S^{-1} = T^{\dagger}|_{\mathcal{R}(T)}$.*

*Proof.* Since $\mathcal{N}(S) = \mathcal{N}(T) \cap \mathcal{N}(T)^{\perp} = \{0\}$, we see that $S$ is injective. Each $x \in X$ can be written as $x = \bar{x} + u$, where $\bar{x} \in \mathcal{N}(T)^{\perp}$ and $u \in \mathcal{N}(T)$. Therefore, $T x = T \bar{x} = S \bar{x}$, which shows that $\mathcal{R}(S) = \mathcal{R}(T)$. Thus, $S$ is a bijection from $\mathcal{N}(T)^{\perp}$ onto $\mathcal{R}(T)$ and $S^{-1}$ is well-defined. Suppose $\{y_n\} \subset \mathcal{R}(T)$ and $y_n \to y \in Y$, $S^{-1} y_n \to x \in X$. With $x_n = S^{-1} y_n$ for all $n$, we have $x_n \to x$, $x \in \mathcal{N}(T)^{\perp}$ because $\mathcal{N}(T)^{\perp}$ is closed, and hence $S x_n \to S x$ because $S$ is bounded. But $S x_n = y_n \to y$. Thus $y = S x = T x$, which implies that $y \in \mathcal{R}(T) = D(S^{-1})$ and $x = S^{-1} y$. This shows that $S^{-1}$ is closed. Finally, Lemma 2.11 shows that $T^{\dagger} T x = x$ for all $x \in \mathcal{N}(T)^{\perp}$, which shows that $T^{\dagger}$ and $S^{-1}$ agree on $\mathcal{R}(T)$.                    □

**Corollary 2.19.** *Let $S = T|_{\mathcal{N}(T)^\perp}$. Then $S^{-1}$ is bounded if and only if $\mathcal{R}(T)$ is closed.*

*Proof.* This follows from the previous theorem and Theorem 2.15 (a closed operator is bounded if and only if its domain is closed).        □

### 2.3.2    Inverse problems

By definition, $Tx = y$ is ill-posed if any of the three conditions of existence, uniqueness, and continuous dependence fails to hold. If $\mathcal{N}(T)$ is nontrivial, then uniqueness fails (for any choice of $y \in Y$), whereas existence fails for certain $y \in Y$ if $\mathcal{R}(T)$ is a proper subspace of $Y$.

However, a lack of uniqueness can be addressed by seeking the minimum-norm solution of $Tx = y$, while a lack of existence can (sometimes) be remedied by choosing a least-squares solution. The challenging—and interesting—case arises when stability fails. Assuming we interpret solving $Tx = y$ as finding the minimum-norm least-squares solution $x = T^\dagger y$, we now know that stability fails precisely when $\mathcal{R}(T)$ fails to be closed. For such a problem, $T^\dagger y$ exists only for $y$ in a proper dense subspace of $Y$ and $T^\dagger y$ does not depend continuously on $y$.

We will call $Tx = y$ an *inverse problem* if $\mathcal{R}(T)$ fails to be closed. In this case, we will be interested in replacing $Tx = y$ with a regularized problem that is somehow similar to the original problem but for which the solution depends continuously on the data vector $y$. We can express this differently as follows: we wish to replace $T^\dagger$ with an operator $R$ that approximates $T^\dagger$ in some sense but with the property that $R$ is bounded (continuous). As we will see (and as might already be evident to the reader), a single bounded operator $R$ cannot approximate the unbounded operator $T^\dagger$ in a satisfactory manner. Instead, we will construct a family $\{R_\lambda : \lambda > 0\}$ of bounded operators such that $R_\lambda$ converges pointwise to $T^\dagger$ (that is, for each $y \in D(T^\dagger)$, $R_\lambda y \to T^\dagger y$) as $\lambda \to 0^+$. Then $\lambda$, which is called the *regularization parameter*, must be chosen in accordance with the properties of the particular data vector $y$ (the primary property being the amount of noise in $y$).

### 2.3.3   History of the concept of inverse problems

As mentioned earlier, the definition of a well-posed problem is often attributed to Hadamard. The original citation is usually [9], in which Hadamard discussed certain partial differential equations and the auxiliary conditions that make them "possible and determined" (that is, that lead to the existence and uniqueness of solutions). However, he did not mention the issue of continuous dependence in [9]. Later, he showed that the Cauchy problem for Laplace's differential equation does not exhibit the continuous dependence property [10, e.g., pages 33–34] but, as he himself stated later [11, page 20], he did not regard this as part of the definition of a well-posed problem. He cited Courant and Hilbert's influential *Methods of Mathematical Physics* [4] [2, Section 3.6], as presenting the full definition, which he then adopted.

Interestingly, Courant and Hilbert suggested rather explicitly that ill-posed problems, particularly what we have called inverse problems, are not of interest:

> A mathematical problem which is to correspond to physical reality should satisfy the following basic requirements:
>
> (1) The solution must exist.
>
> (2) The solution should be uniquely determined.
>
> (3) The solution should depend continuously on the data (requirement of stability).
>
> The first requirement expresses the logical condition that not too much, i.e., no mutually contradictory properties, is demanded of the solution.
>
> The second requirement stipulates completeness of the problem—leeway or ambiguity should be excluded unless

---

[4] *Methods of Mathematical Physics* was actually written by Richard Courant, who named his PhD advisor David Hilbert as honorary co-author because the work was based on Hilbert's lectures. Hilbert was one of the most famous and influential mathematicians of his time.

inherent in the physical situation.[5] The third requirement, particularly incisive, is necessary if the mathematical formulation is to describe observable natural phenomena. Data in nature cannot possibly be conceived as rigidly fixed; the mere process of measuring them involves small errors. For example, prescribed values for space or time coordinates are always given within certain margins of precision. Therefore, a mathematical problem cannot be considered as realistically corresponding to physical phenomena unless a variation of the given data in a sufficiently small range leads to an arbitrary small change in the solution. This requirement of "stability" is not only essential for meaningful problems in mathematical physics, but also for approximation methods.

Any problem which satisfies our three requirements will be called a properly posed problem. [2, page 227]

In contrast to the viewpoint expressed by Courant and Hilbert, ill-posed problems, including inverse problems in which the property of stability fails to hold, certainly do occur as "mathematical formulation[s] [that] describe observable natural phenomena." We gave two simple examples in Chapter 1, namely an image processing problem in which a blurred image is recorded and the problem of finding the force that results in a certain displacement of an elastic string. Although this book is mostly concerned with theory, not with applications, we will describe another example, a very practical application that demonstrates that Courant and Hilbert's viewpoint cannot be held to be correct.

### 2.3.4   An example: medical imaging and computed tomography

Let us consider the problem of imaging a two-dimensional cross-section of the human body. One method is *X-ray transmission tomography*, in

---

[5]Courant and Hilbert's footnote: "Cases do occur in which uniqueness is not a proper requirement. For example, in the case of multiple eigenvalues whole families of solutions of the eigenvalue problem exist."

which a beam of X-rays is aimed at the cross-section. The intensity of the beam at the source, $I_0$, is known, and the intensity of the beam leaving the body, $I_f$, is measured.

The intensity $I(s)$ satisfies *Beer's Law*, which is the initial value problem (IVP)

$$\frac{dI}{ds} = -\mu I, \ I(0) = I_0.$$

Here $\mu$ is the *attenuation coefficient* and $s$ is the distance (in cm, say; this gives units of cm$^{-1}$ for $\mu$). The human body is heterogeneous and therefore $\mu$ is a function of $(x, y)$: $\mu = \mu(x, y)$. Since the attenuation coefficients of common tissues (fat, muscle, bone, and so forth) are known, the function $\mu$ gives an image of the cross-section of the body.

Solving the IVP yields

$$I_f = I_0 \exp^{-\int_0^L \mu \, ds},$$

where $L$ is the distance from the source to the detector. Rearranging yields

$$\int_0^L \mu ds = -\ln \frac{I_f}{I_0}.$$

We will pretend for simplicity that $-\ln(I_f/I_0)$ is the measured data. Defining $\mu$ to be zero outside of $[0, L]$, we can write

$$\int_{-\infty}^{\infty} \mu ds = \int_0^L \mu ds.$$

A *1D projection of* $\mu$ is obtained by repeating the experiment along a line. Figure 2.1 illustrates an experimental setup resulting in a 1D projection of $\mu$. Sources are located on one side of the body, on a line whose angle with the horizontal is $\phi$, and detectors are located on a parallel line on the other side of the body. The coordinate along the line of sources is $p$, while the orthogonal coordinate is $s$. We have

$$x = p \cos \phi - s \sin \phi,$$
$$y = p \sin \phi + s \cos \phi.$$

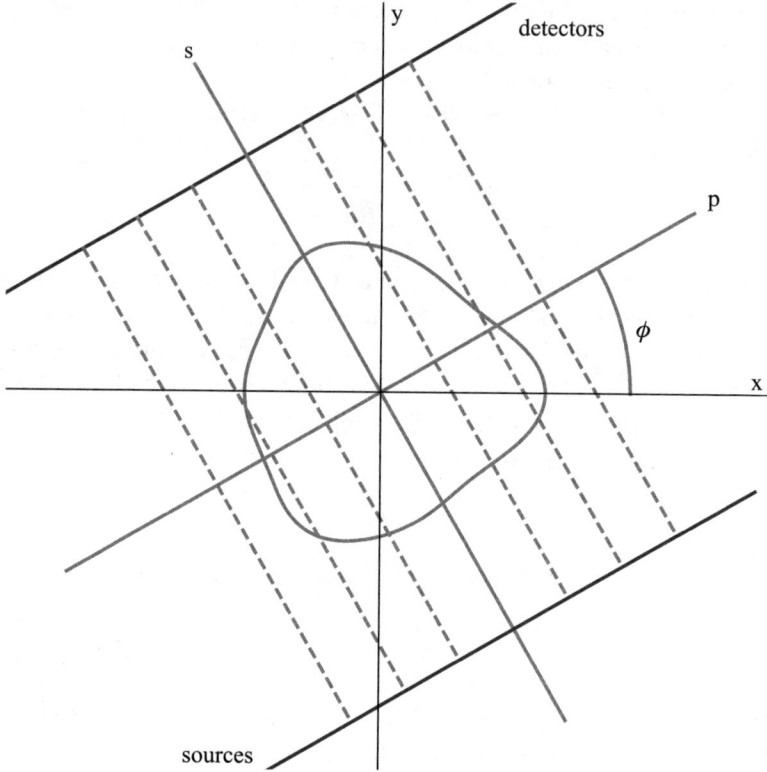

**Figure 2.1.** Using X-rays to measure a 1D projection of the attenuation coefficient of a body.

The 1D projection, as a function of $p$, is given by

$$\lambda_\phi(p) = \int_{-\infty}^{\infty} \mu(p\cos\phi - s\sin\phi, p\sin\phi + s\cos\phi)\,ds, \quad -\infty < p < \infty.$$

The Radon transform $\mathcal{R}\mu$ is the collection of all 1D projections of $\mu$, as a function of $p$ and $\phi$:

$$(\mathcal{R}\mu)(p, \phi) = \lambda_\phi(p), \quad -\infty < p < \infty, \quad -\pi < \phi < \pi.$$

The Radon transformation represents the solution of the forward problem, which is to predict the outcome of the experiment—the intensities recorded at the detectors—given a knowledge of the attenuation coefficient $\mu$. It is not difficult to show that the forward problem is well-posed.

The related inverse problem is to estimate $\mu$ given the X-ray intensities recorded at the detectors. This problem is a genuine inverse problem; although the uniqueness property holds and thus the Radon transform is invertible on its range, the inverse operator is unbounded and hence the attenuation coefficient $\mu$ does not depend continuously on the data. Nevertheless, as shown by the success of computed tomography in modern medicine, mathematical techniques make it possible to compute accurate estimates of $\mu$.

## 2.4  Discretized inverse problems

The primary goal of this book is to analyze the method of Tikhonov regularization for approximately solving inverse problems. We will develop the theory in infinite-dimensional space, that is, for $Tx = y$, where $x$ and $y$ belong to infinite-dimensional Hilbert spaces. However, in any practical application, it is necessary to approximate an infinite-dimensional problem by a finite-dimensional discretization in order to carry out numerical computations. This process of discretization replaces $Tx = y$ by a matrix-vector equation $Ax = y$, where $A$ is a matrix and $\mathbf{x}$, $\mathbf{y}$ are Euclidean vectors.

In this section, we discuss problems of the form $A\mathbf{x} = \mathbf{y}$ and show that, while such a problem cannot be unstable in the sense we have defined, it can be very *ill-conditioned*. An ill-conditioned matrix-vector equation presents many of the same difficulties as a true inverse problem. We will also discuss how the discretization of a true inverse problem produces an ill-conditioned matrix-vector equation and present an example.

### 2.4.1  Ill-conditioned matrix-vector equations

A finite-dimensional Euclidean space is a Hilbert space, and a common linear operator is one defined by matrix multiplication. Specifically, if

$X = \mathbb{R}^n$, $Y = \mathbb{R}^m$, and $A \in \mathbb{R}^{m \times n}$, then $T : X \to Y$ defined by $T\mathbf{x} = A\mathbf{x}$
for all $\mathbf{x} \in \mathbb{R}^n$ is a linear operator. (Conversely, if $T : \mathbb{R}^n \to \mathbb{R}^m$ is a linear
operator, there exists a unique matrix $A \in \mathbb{R}^{m \times n}$ such that $T$ is represented
as multiplication by the matrix $A$.) We can consider the problem of finding
$\mathbf{x} \in \mathbb{R}^n$ to satisfy $T\mathbf{x} = \mathbf{y}$ for a given $\mathbf{y} \in \mathbb{R}^m$. We often omit mention of
$T$ and refer directly to the problem $A\mathbf{x} = \mathbf{y}$.

In this case, the range of $T$, which is also called the column space
of $A$, is necessarily finite-dimensional and hence closed. It follows that
$A\mathbf{x} = \mathbf{y}$ cannot be a true inverse problem in the sense that we have defined.
On the other hand, in Section 1.1, we illustrated the concept of an inverse
problem—specifically, the lack of continuous dependence of the solution
on the data—with a numerical example of the form $A\mathbf{x} = \mathbf{y}$. If such a
problem cannot be a true inverse problem, why did the solution process
appear to be so unstable?

The answer is that while a finite-dimensional problem cannot be unstable, it can be arbitrarily *ill-conditioned* and the solution, while depending continuously on the data, can still be very sensitive to errors in it. We
illustrate this in Figure 2.2, which shows three possible relationships between the data and the solution. In the first, the solution depends continuously on the data and it can be seen that an error in the data will produce
a comparable error in the solution. In the second, the solution depends
discontinuously on the data, and an arbitrarily small error in the data can
produce an abrupt change in the solution. In the third case, the solution
does depend continuously on the data, but nevertheless a small error in the
data can produce a large error in the solution. While it is true that the solution error can be made arbitrarily small by making the error in the data
sufficiently small, this might not be very significant in a real problem.

Let us suppose that $\mathbf{y}^*$ is the exact data, $\mathbf{y}$ is a measurement of $\mathbf{y}^*$, and
$\mathbf{x}^*$, $\mathbf{x}$ satisfy
$$A\mathbf{x}^* = \mathbf{y}^*, \quad A\mathbf{x} = \mathbf{y}.$$

For the sake of this derivation, let us assume that $A$ is square and invertible. Since the problem is linear, we have
$$A(\mathbf{x} - \mathbf{x}^*) = \mathbf{y} - \mathbf{y}^* \Rightarrow \mathbf{x} - \mathbf{x}^* = A^{-1}(\mathbf{y} - \mathbf{y}^*)$$
$$\Rightarrow \|\mathbf{x} - \mathbf{x}^*\|_2 \le \|A^{-1}\| \|\mathbf{y} - \mathbf{y}^*\|_2,$$

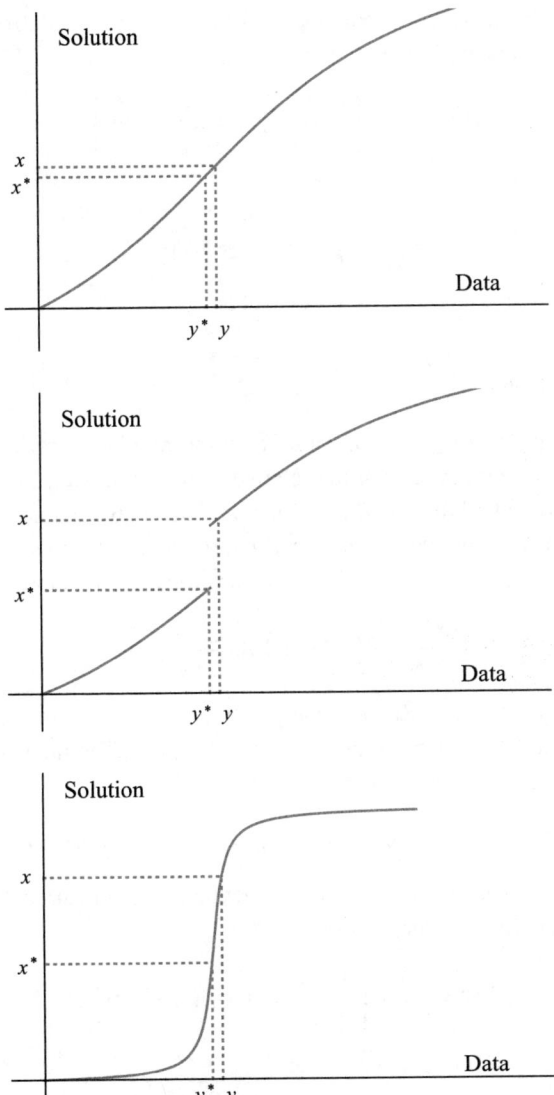

**Figure 2.2.** Illustrations of three possible relationships between the data and solutions: stable and well-conditioned (top); unstable (middle); stable but ill-conditioned (bottom).

where $\| \cdot \|_2$ represents the Euclidean norm for vectors and $\|B\|$, for $B \in \mathbb{R}^{n \times n}$, represents the induced matrix norm:

$$\|B\| = \sup \left\{ \frac{\|Bx\|_2}{\|x\|_2} : x \in \mathbb{R}^n, x \neq 0 \right\}.$$

We also have

$$\mathbf{y}^* = A\mathbf{x}^* \implies \|\mathbf{y}^*\|_2 \leq \|A\| \|\mathbf{x}^*\|_2.$$

Combining the last two inequalities yields

$$\frac{\|\mathbf{x} - \mathbf{x}^*\|_2}{\|A\| \|\mathbf{x}^*\|_2} \leq \frac{\|A^{-1}\| \|\mathbf{y} - \mathbf{y}^*\|_2}{\|\mathbf{y}^*\|_2} \implies \frac{\|\mathbf{x} - \mathbf{x}^*\|_2}{\|\mathbf{x}^*\|_2} \leq \|A\| \|A^{-1}\| \frac{\|\mathbf{y} - \mathbf{y}^*\|_2}{\|\mathbf{y}^*\|_2}.$$

It follows that if $\|A\| \|A^{-1}\|$ is large, then the relative error in the solution can be much larger than the relative error in the data. The quantity $\|A\| \|A^{-1}\|$ is called the *condition number* of the matrix $A$ and is denoted by $\text{cond}(A)$. We say that $A$ is *ill-conditioned* if $\text{cond}(A)$ is large.

The fundamental inequality that bounds the amplification of error is

$$\frac{\|\mathbf{x} - \mathbf{x}^*\|_2}{\|\mathbf{x}^*\|_2} \leq \text{cond}(A) \frac{\|\mathbf{y} - \mathbf{y}^*\|_2}{\|\mathbf{y}^*\|_2}. \tag{2.4}$$

We would like to show that this inequality can be tight (that is, satisfied as an equation) and also discuss when this happens. The inequality (2.4) is derived from the inequalities

$$\|\mathbf{x} - \mathbf{x}^*\|_2 \leq \|A^{-1}\| \|\mathbf{y} - \mathbf{y}^*\|_2, \quad \|\mathbf{y}^*\|_2 \leq \|A\| \|\mathbf{x}^*\|_2,$$

and (2.4) becomes an equation precisely when these inequalities are satisfied as equations. By definition,

$$\|A\| = \max \left\{ \frac{\|A\mathbf{x}\|_2}{\|\mathbf{x}\|_2} : \mathbf{x} \in \mathbb{R}^n, \mathbf{x} \neq 0 \right\}.$$

Therefore, if we choose $\mathbf{x}^*$ such that

$$\|A\| = \frac{\|A\mathbf{x}^*\|_2}{\|\mathbf{x}^*\|_2}$$

and define $\mathbf{y}^* = A\mathbf{x}^*$, then we have $\|\mathbf{y}^*\|_2 = \|A\| \|\mathbf{x}^*\|_2$.

We also choose a vector $\mathbf{z}$ such that

$$\|A^{-1}\| = \frac{\|A^{-1}\mathbf{z}\|_2}{\|\mathbf{z}\|_2}.$$

It follows that $\|A^{-1}\mathbf{z}\|_2 = \|A^{-1}\|\,\|\mathbf{z}\|_2$. Defining $\mathbf{u} = A^{-1}\mathbf{z}$, we obtain $\|\mathbf{u}\|_2 = \|A^{-1}\|\,\|A\mathbf{u}\|_2$ or, equivalently, $\|A\mathbf{u}\|_2 = \|A^{-1}\|^{-1}\|\mathbf{u}\|_2$. If we now define $\mathbf{y} = \mathbf{y}^* + \mathbf{z}$, then $\mathbf{z} = \mathbf{y} - \mathbf{y}^*$ and $\mathbf{u} = \mathbf{x} - \mathbf{x}^*$. For this choice of $\mathbf{y}$, the inequality $\|\mathbf{x} - \mathbf{x}^*\|_2 \leq \|A^{-1}\|\,\|\mathbf{y} - \mathbf{y}^*\|_2$ is satisfied as an equation. Therefore, with $\mathbf{y}^*$ and $\mathbf{y}$ chosen in this fashion, the inequality (2.4) is an equation. This shows that $\mathrm{cond}(A)$ does not overestimate the possible amplification of error in solving $A\mathbf{x} = \mathbf{y}$.

When is $\mathrm{cond}(A)$ large? The key fact that we need is the following: when $A$ is invertible,

$$
\begin{aligned}
\|A^{-1}\| &= \max\left\{ \frac{\|A^{-1}\mathbf{y}\|_2}{\|\mathbf{y}\|_2} \ : \ \mathbf{y} \in \mathbb{R}^n, \mathbf{y} \neq 0 \right\} \\[1mm]
&= \max\left\{ \frac{\|\mathbf{x}\|_2}{\|A\mathbf{x}\|_2} \ : \ \mathbf{x} \in \mathbb{R}^n, \mathbf{x} \neq 0 \right\} \\[1mm]
&= \frac{1}{\min\left\{ \frac{\|A\mathbf{x}\|_2}{\|\mathbf{x}\|_2} \ : \ \mathbf{x} \in \mathbb{R}^n, \mathbf{x} \neq 0 \right\}}.
\end{aligned}
$$

Therefore,

$$
\mathrm{cond}(A) = \|A\|\,\|A^{-1}\| = \frac{\max\left\{ \frac{\|A\mathbf{x}\|_2}{\|\mathbf{x}\|_2} \ : \ \mathbf{x} \in \mathbb{R}^n, \mathbf{x} \neq 0 \right\}}{\min\left\{ \frac{\|A\mathbf{x}\|_2}{\|\mathbf{x}\|_2} \ : \ \mathbf{x} \in \mathbb{R}^n, \mathbf{x} \neq 0 \right\}}.
$$

This might be easier to understand if we rewrite the formulas using unit vectors:

$$
\mathrm{cond}(A) = \frac{\max\left\{ \|A\mathbf{x}\|_2 \ : \ \mathbf{x} \in \mathbb{R}^n, \|\mathbf{x}\|_2 = 1 \right\}}{\min\left\{ \|A\mathbf{x}\|_2 \ : \ \mathbf{x} \in \mathbb{R}^n, \|\mathbf{x}\|_2 = 1 \right\}}.
$$

Thus $\mathrm{cond}(A)$ is large if, as $x$ ranges over all unit vectors, the ratio of the largest value of $\|Ax\|_2$ to its smallest value is large.

## 2.4.2   Discretization of inverse problems

In a true (infinite-dimensional) inverse problem $Tx = y$, the ratio of the largest values of $\|Tx\|_Y$ to the smallest, as $x$ ranges over all unit vectors, is infinite, as the following result shows.

**Theorem 2.20.** *If $X$ and $Y$ are Hilbert spaces and $T : X \to Y$ is a bounded linear operator, then there exists $\gamma > 0$ such that*

$$\|Tx\|_Y^2 \geq \gamma \|x\|_X^2 \text{ for all } x \in \mathcal{N}(T)^\perp \tag{2.5}$$

*if and only if $\mathcal{R}(T)$ is closed. Therefore, if $\mathcal{R}(T)$ fails to be closed, then there exists $\{x_n\} \subset \mathcal{N}(T)^\perp$ such that $\|x_n\|_X = 1$ for all $n$ and $\|Tx_n\|_Y \to 0$ as $n \to \infty$.*

*Proof.* Suppose first that such a $\gamma > 0$ exists, and let $y \in \overline{\mathcal{R}(T)}$. Then there exists $\{x_n\} \subset \mathcal{N}(T)^\perp$ such that $Tx_n \to y$. Since $\|Tx_n\|_Y^2 \geq \gamma \|x_n\|_X$ and the left side of this inequality is bounded (because it is convergent), it follows that $\{\|x_n\|_X\}$ must also be bounded. Without loss of generality, we can assume that there exists $x \in X$ such that $x_n \to x$ weakly. But then $Tx_n \to Tx$ weakly; since we already know that $Tx_n \to y$, it follows that $Tx = y$ and hence that $y \in \mathcal{R}(T)$. This proves that $\mathcal{R}(T)$ is closed.

Conversely, suppose $\mathcal{R}(T)$ is closed. We wish to prove that (2.5) holds for some $\gamma > 0$. We argue by contradiction and suppose that there exists a sequence $\{x_n\} \subset \mathcal{N}(T)^\perp$ such that $\|x_n\|_X = 1$ for all $n \in \mathbb{Z}^+$ and $Tx_n \to 0$ as $n \to \infty$. Since $\mathcal{R}(T)$ is closed, $T^\dagger$ is bounded, and Lemma 2.11 implies that $T^\dagger Tx_n = x_n$ for all $n \in \mathbb{Z}^+$. This yields

$$x_n = T^\dagger Tx_n \to 0 \text{ as } n \to \infty,$$

a contradiction because $\|x_n\|_X = 1$ for all $n$.                    $\square$

The inequality (2.5) can be expressed in terms of the operator $T^*T$:

$$\langle T^*Tx, x \rangle_X \geq \gamma \|x\|_X^2 \text{ for all } x \in \mathcal{N}(T)^\perp, \tag{2.6}$$

This will be useful later.

When $\mathcal{R}(T)$ fails to be closed, Theorem 2.20 implies that

$$\frac{\sup\{\|Tx\|_Y \ : \ x \in \mathcal{N}(T)^\perp, \ \|x\|_X = 1\}}{\inf\{\|Tx\|_Y \ : \ x \in \mathcal{N}(T)^\perp, \ \|x\|_X = 1\}} = \infty.$$

When $Ax = y$ is obtained by discretizing the infinite-dimensional inverse problem $Tx = y$, it is not difficult to believe that it will inherit this property to the extent that

$$\mathrm{cond}(A) = \frac{\max\{\|Ax\|_2 \ : \ x \in \mathbb{R}^n, \|x\|_2 = 1\}}{\min\{\|Ax\|_2 \ : \ x \in \mathbb{R}^n, \|x\|_2 = 1\}}$$

will be very large. Here we assume that when $Tx = y$ is discretized, the matrix $A$ turns out to be invertible. If $A$ turns out not to be invertible, we must consider $Ax = y$ in the least-squares sense. The concept of condition number can be extended to a general matrix (that is, one that is not necessarily square and nonsingular); we present this extension in Section 4.1 (see page 136).

To illustrate these comments, we consider again the inverse problem from Section 1.1. The problem $Tx = y$ is defined by the operator

$$(Tx)(s) = \int_0^1 k(s, t)x(t)\,dt, \ 0 \le s \le 1,$$

where

$$k(s, t) = \frac{1}{\sqrt{2\pi}\sigma}e^{-(s-t)^2/(2\sigma^2)}, \ \sigma = 0.05.$$

(The operator $T$ maps a function $x$ to another function $Tx$; thus the notation $(Tx)(s)$ represents the value of this function $Tx$ at $s \in [0, 1]$. For the purposes of this numerical example, we will not bother to define the function spaces $X$ and $Y$; see Section 4.4.1 for the details.) In the numerical experiment presented in Section 1.1, we discretized the integral operator using the midpoint rule, using $n = 30$ quadrature points, to obtain a matrix-vector equation $A\mathbf{x} = \mathbf{y}$, where $A$ is $30 \times 30$. A direct calculation finds that

$$\text{cond}(A) \approx 24,500,$$

while the given vectors $\mathbf{y}^*, \mathbf{y}, \mathbf{x}^*, \mathbf{x}$, satisfy

$$\frac{\|\mathbf{y} - \mathbf{y}^*\|_2}{\|\mathbf{y}^*\|_2} = 0.05, \quad \text{(by construction)}$$

$$\frac{\|\mathbf{x} - \mathbf{x}^*\|_2}{\|\mathbf{x}^*\|_2} \approx 268.5,$$

which means that the relative error in the data $\mathbf{y}$ is magnified by a factor of approximately 5370 when the solution $\mathbf{x}$ is computed. Though less than the maximum possible amplification factor (which is $\text{cond}(A)$), this still reflects the ill-conditioning of the discretized problem.

What happens if we refine the discretization? We expect the discretized problem on a finer grid to more accurately reflect the underlying infinite-dimensional problem. In this case, since the underlying problem is unstable, we expect the conditioning of the discretized problem to get worse. For instance, if we use $n = 40$ quadrature points instead of $n = 30$, we obtain the results displayed in Figure 2.3. The new matrix $A$ has an approximate condition number of $9.5 \cdot 10^7$, while the amplification of the error is approximately $2.6 \cdot 10^7$.

The reader will notice how, as the grid is refined, the computed solution gets larger (and hence farther from the exact solution). As noted above, with a finer grid, we expect to be able to represent the underlying problem more accurately. In this regard, the following theorem is relevant.

**Theorem 2.21.** *Suppose $\mathcal{R}(T)$ fails to be closed and $y \notin D(T^\dagger)$. If $\{x_n\} \subset X$ is a minimizing sequence for $\|Tx - y\|_Y$, that is, if*

$$\|Tx_n - y\|_Y \to \inf\{\|Tu - y\|_Y : u \in X\} \text{ as } n \to \infty,$$

*then $\|x_n\|_X \to \infty$ as $n \to \infty$.*

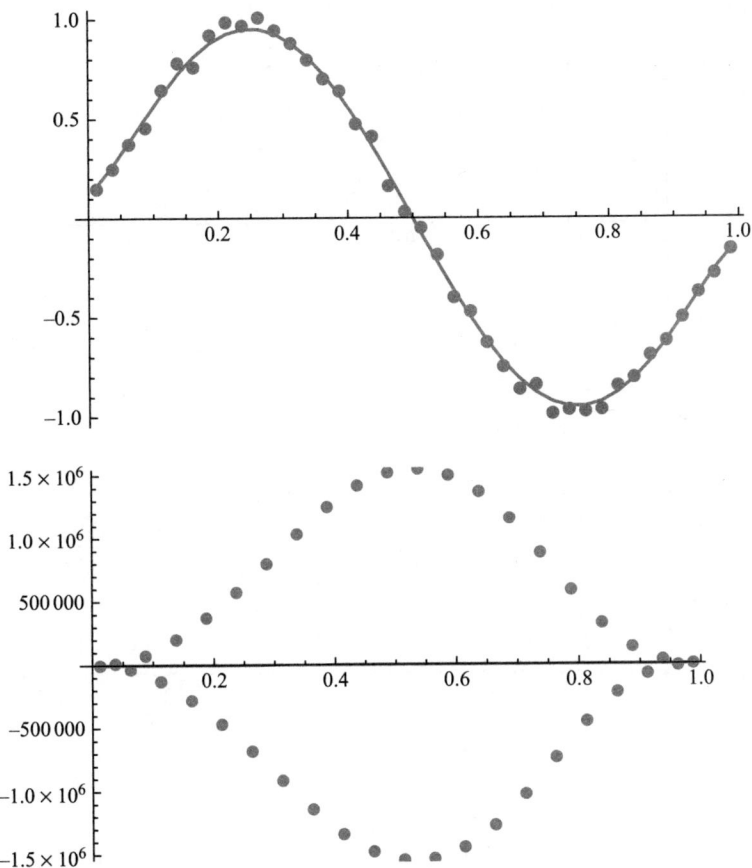

**Figure 2.3.** Top: the exact and noisy data for the inverse problem from Section 1.1, now with $n = 40$ (cf. Figure 1.2). Bottom: the exact and computed solution. (The exact quantities are displayed as curves.)

*Proof.* Obviously

$$\inf\{\|Tu - y\|_Y \ : \ u \in X\} = \inf\{\|v - y\|_Y \ : \ v \in \mathcal{R}(T)\}$$
$$= \inf\{\|v - y\|_Y \ : \ v \in \overline{\mathcal{R}(T)}\}$$
$$= \|\bar{y} - y\|_Y,$$

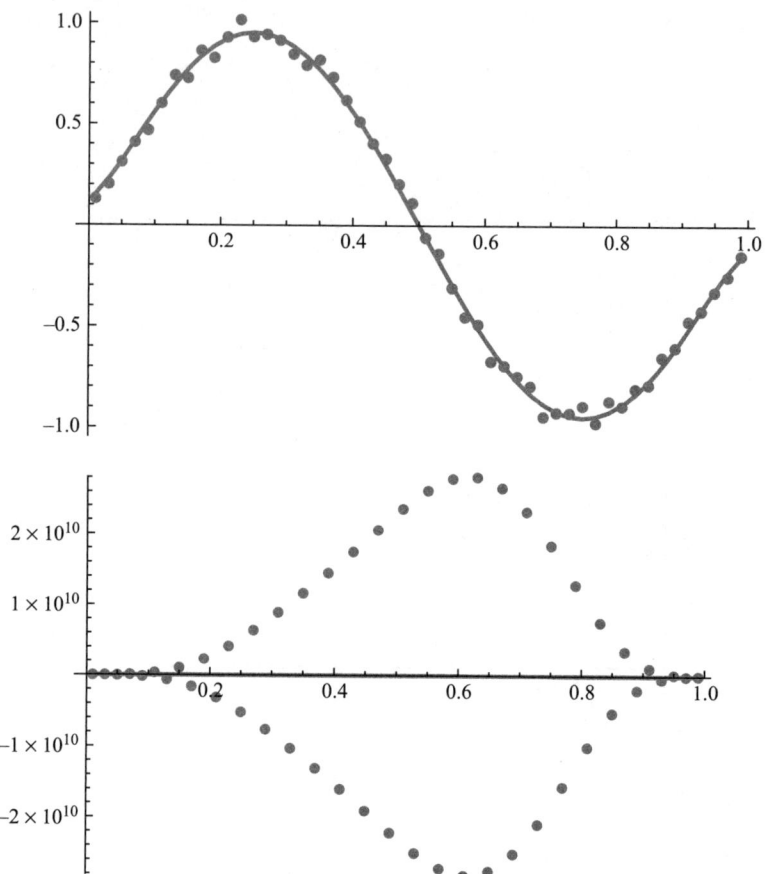

**Figure 2.4.** Top: the exact and noisy data for the inverse problem from Section 1.1 with $n = 50$ (cf. Figures 1.2 and 2.3). Bottom: the exact and computed solutions. (The exact solution is not visible on this scale.)

where $\bar{y} = \text{proj}_{\overline{Ra(T)}} y$. Therefore, since $\{x_n\}$ is a minimizing sequence for $\|Tx - y\|_Y$, it follows that $\|Tx_n - y\|_Y \to \|\bar{y} - y\|_Y$ as $n \to \infty$. Since $Tx_n - \bar{y} \in \overline{\mathcal{R}(T)}$, $\bar{y} - y \in \mathcal{R}(T)^\perp$, the Pythagorean theorem implies that

$$\|Tx_n - y\|_Y^2 = \|Tx_n - \bar{y}\|_Y^2 + \|\bar{y} - y\|_Y^2$$
$$\Rightarrow \|Tx_n - \bar{y}\|_Y^2 = \|Tx_n - y\|_Y^2 - \|\bar{y} - y\|_Y^2.$$

Therefore, $\|Tx_n - \bar{y}\|_Y \to 0$, that is, $Tx_n \to \bar{y}$, as $n \to \infty$.

We now argue by contradiction and assume that $\|x_n\|_X$ does not grow without bound as $n \to \infty$. Then there exists a bounded subsequence $\{x_{n_k}\}$ of $\{x_n\}$ and hence, without loss of generality, there exists $x \in X$ such that $x_{n_k} \to x$ weakly as $k \to \infty$. Since $T$ is bounded, $Tx_{n_k} \to Tx$ weakly in $Y$ as $k \to \infty$. But $Tx_n \to \bar{y}$ strongly, which implies that $Tx = \bar{y}$ and hence that $y \in \mathcal{R}(T) \oplus \mathcal{R}(T)^{\perp} = D(T^{\dagger})$. This contradiction shows that $\|x_n\|_X \to \infty$ as $n \to \infty$ must hold. $\qquad\square$

This is consistent with our numerical results: as we refine the grid and more accurately represent the underlying (unstable) problem, the computed solution grows without bound. To emphasize this, we do one more calculation, this time with $n = 50$. The results are shown in Figure 2.4; the reader should notice the vertical scale on the right-hand graph.

In the next chapter, we introduce the main topic of this book, the method of Tikhonov regularization, which is able to produce good approximate solutions to inverse problems, even in the face of noisy data.

# Chapter 3

# Tikhonov regularization

We continue to study the problem $Tx = y$, where $X$, $Y$ are Hilbert spaces, $T$ is a bounded linear operator, and the interesting case occurs if $\mathcal{R}(T)$ fails to be closed. In this case, $Tx = y$ is a true inverse problem and the desired solution, $x = T^\dagger y$, does not depend continuously on the vector $y$.

Tikhonov regularization replaces the problem of computing the minimum-norm least-squares solution with the following problem:

$$\min_{x \in X} \|Tx - y\|_Y^2 + \lambda \|x\|_X^2. \tag{3.1}$$

The objective function is a weighted combination of the residual $\|Tx - y\|_Y^2$ and the norm of the solution vector $x$. Theorem 2.21 shows that, for $y \notin D(T^\dagger)$, if we try to minimize $\|Tx - y\|_Y^2$, then $\|x\|_X^2$ will necessarily go to $\infty$. Thus the Tikhonov regularization approach is to make $\|Tx - y\|_Y^2$ small (without minimizing it) while not allowing $\|x\|_X^2$ to become too large. The positive real number $\lambda$ is called the *regularization parameter*.

Before we begin the analysis of Tikhonov regularization, we apply it to the inverse problem first presented in Section 1.1 and revisited in Section 2.4. The operator $T$ is defined by

$$(Tx)(s) = \int_0^1 k(s, t)x(t)\, dt, \ 0 \le s \le 1,$$

where the kernel $k$ is

$$k(s,t) = \frac{1}{\sqrt{2\pi}\sigma} e^{-(s-t)^2/(2\sigma^2)}, \quad \sigma = 0.05.$$

We use the same function $x$ (the true solution) as before: $x(t) = \sin(2\pi t)$.

This problem is discretized as in Section 1.1, which results in a matrix-vector equation $A\mathbf{x} = \mathbf{y}$ for $\mathbf{x}$. The Tikhonov approach replaces this equation by

$$\min_{\mathbf{x} \in \mathbb{R}^n} \|A\mathbf{x} - \mathbf{y}\|_2^2 + \lambda \|\mathbf{x}\|_2^2 \tag{3.2}$$

(where $\|\cdot\|_2$ denotes the Euclidean norm). Figure 3.1 shows the result of solving (3.2) for four values of $\lambda$: 2.0, 0.2, 0.05, and 0.001. Of these, $\lambda = 0.2$ produces the best result, with the computed solution lying close to the exact solution over most of the interval. Larger values of $\lambda$, such as $\lambda = 2$, produce a solution that is over-regularized. The effect of the term $\lambda \|\mathbf{x}\|_2^2$, with a large value of $\lambda$, is to force the computed solution to be small. In fact, it can be shown that the computed solution tends to zero as $\lambda \to \infty$. For values of $\lambda$ much smaller than 0.2, the solution is under-regularized. For instance, with $\lambda = 0.05$, the error in the computed solution is still relatively small, but unwanted oscillations are evident. For still smaller values of $\lambda$, such as 0.001, the errors are much larger. It is not difficult to believe that, as $\lambda \to 0^+$, the Tikhonov solution converges to the solution computed in Section 1.1 (see Figure 1.2).

As discussed in Section 1.2.1, the difficulty presented by an inverse problem is that noise in the data results in large and spurious oscillations in the (naively) computed solution. Figure 3.1 shows that Tikhonov regularization is able to suppress this effect, at least for a properly chosen value of the regularization parameter $\lambda$. In this chapter, we will show that Tikhonov regularization works, at least in the sense that the computed solution converges to the exact solution as the error in the data goes to zero, provided the regularization parameters are chosen appropriately. We will also discuss how to choose a good value of the regularization parameter in a practical problem (where the error in the data does not go to zero). In the next chapter, we will use the *singular value expansion* to explain why Tikhonov regularization is so effective.

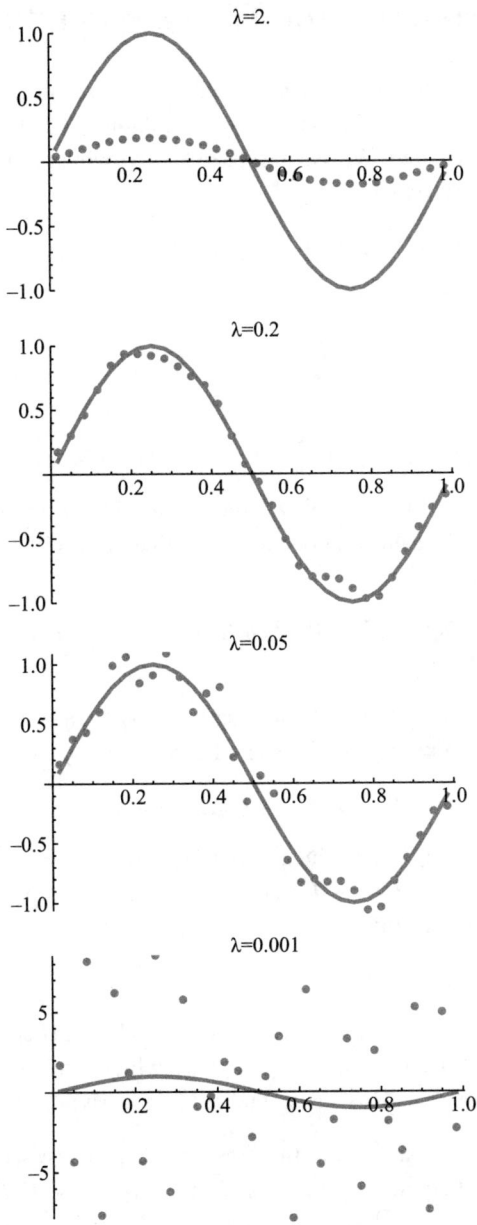

**Figure 3.1.** Solutions of (3.2) for four values of λ. In each graph, the exact solution is plotted as a solid curve, and the computed solution is plotted as a collection of discrete points.

# 3.1    Existence of the Tikhonov solution

We begin our analysis by showing that (3.1) has a unique solution $x_{\lambda,y}$ for each $y \in Y$ and $\lambda > 0$. We can do this by relating (3.1) to a well-posed least-squares problem. We use the space $Y \times X$, which is a Hilbert space under the inner product

$$\langle (y, x), (w, u) \rangle_{Y \times X} = \langle y, w \rangle_Y + \langle x, u \rangle_X.$$

We define $M_\lambda : X \to Y \times X$ by

$$M_\lambda x = (Tx, \sqrt{\lambda}x).$$

Then

$$\|M_\lambda x - (y, 0)\|_{Y \times X}^2 = \|Tx - y\|_Y^2 + \lambda \|x\|_X^2$$

and hence solving (3.1) is equivalent to solving $M_\lambda x = (y, 0)$ in the least-squares sense. The following result implies that this problem has a unique solution for each $y \in Y$.

**Theorem 3.1.** *Suppose $\lambda > 0$. Then $\mathcal{R}(M_\lambda)$ is closed and $\mathcal{N}(M_\lambda)$ is trivial.*

*Proof.* Suppose $\{(y_n, u_n)\} \subset \mathcal{R}(M_\lambda)$ satisfies $(y_n, u_n) \to (y, u)$. Then, by definition, there exists $\{x_n\} \subset X$ such that $M_\lambda x_n = (y_n, u_n)$, that is,

$$(Tx_n, \sqrt{\lambda}x_n) = (y_n, u_n) \text{ for all } n.$$

It follows that $\sqrt{\lambda}x_n \to u$, which implies that $x_n \to x = \lambda^{-1/2}u$. Since $T$ is bounded, we have $Tx_n \to Tx$. But $Tx_n \to y$ by assumption, and hence $y = Tx$. We then see that

$$M_\lambda x = (Tx, \sqrt{\lambda}x) = (y, \sqrt{\lambda}\lambda^{-1/2}u) = (y, u).$$

Thus $(y, u) \in \mathcal{R}(M_\lambda)$ and we have shown that $\mathcal{R}(M_\lambda)$ is closed.

If $x \in \mathcal{N}(M_\lambda)$, then $(Tx, \sqrt{\lambda}x) = (0, 0)$, which implies that $\sqrt{\lambda}x = 0$ and hence that $x = 0$. This shows that $\mathcal{N}(M_\lambda)$ is trivial.    $\square$

Based on our results from the previous chapter, we see that $M_\lambda x = (y, 0)$ has a least-squares solution for all $y \in Y$ (because $\mathcal{R}(M_\lambda)$ is closed and hence $D(M_\lambda^\dagger) = Y \times X$) and the least-squares solution is unique (because $\mathcal{N}(M_\lambda)$ is trivial). Moreover, $M_\lambda^\dagger$ is bounded (because $\mathcal{R}(M_\lambda)$ is closed) and hence $x = M_\lambda^\dagger(y, 0)$ depends continuously on $y$.

Since (3.1) is equivalent to solving $M_\lambda x = (y, 0)$ in the least-squares sense, it follows that (3.1) has a unique solution $x_{\lambda,y}$ for each $y \in Y$ and $\lambda > 0$. We can prove this more directly by examining the normal equation for $M_\lambda x = (y, 0)$. We must first compute the adjoint of $M_\lambda$. For all $x \in X$, $(w, u) \in Y \times X$, we have

$$\begin{aligned}
\langle M_\lambda x, (w, u) \rangle_{Y \times X} &= \langle (Tx, \sqrt{\lambda}x), (w, u) \rangle_{Y \times X} \\
&= \langle Tx, w \rangle_Y + \langle \sqrt{\lambda}x, u \rangle_X \\
&= \langle x, T^*w \rangle_X + \langle x, \sqrt{\lambda}u \rangle_X \\
&= \langle x, T^*w + \sqrt{\lambda}u \rangle_X,
\end{aligned}$$

which shows that

$$M_\lambda^*(w, u) = T^*w + \sqrt{\lambda}u.$$

It follows that

$$M_\lambda^* M_\lambda x = M_\lambda^*(Tx, \sqrt{\lambda}x) = T^*Tx + \lambda x$$

and hence that

$$M_\lambda^* M_\lambda x = M_\lambda^*(y, 0) \iff T^*Tx + \lambda x = T^*y \iff (T^*T + \lambda I)x = T^*y.$$

Since we know that $M_\lambda x = (y, 0)$ has a unique least-squares solution, it follows that the equation $M_\lambda^* M_\lambda x = M_\lambda^*(y, 0)$ has a unique solution. Therefore, the unique solution $x_{\lambda,y}$ of (3.1) is the unique solution of

$$(T^*T + \lambda I)x = T^*y. \tag{3.3}$$

Since we will refer to the operator $T^*T + \lambda I$ frequently, we define the notation $N_\lambda = T^*T + \lambda I$ and note that $N_\lambda$ maps $X$ into $X$.

We will now give another proof that (3.3) has a unique solution $x_{\lambda,y}$ for each $y \in Y$ and that $x_{\lambda,y}$ depends continuously on $y$ by proving that $N_\lambda$ is invertible with a bounded inverse. This alternate, more direct proof is useful because it gives an explicit bound on $\|N_\lambda^{-1}\|$. The proof is based on two facts. First, for all $x \in X$,

$$\langle N_\lambda x, x \rangle_X = \langle (T^*T + \lambda I)x, x \rangle_X = \langle T^*Tx, x \rangle_X + \lambda \langle x, x \rangle_X$$
$$= \langle Tx, Tx \rangle_Y + \lambda \langle x, x \rangle_X$$
$$= \|Tx\|_Y^2 + \lambda \|x\|_X^2$$
$$\geq \lambda \|x\|_X^2.$$

Second, $N_\lambda$ is *self-adjoint*, that is, $N_\lambda^* = N_\lambda$. This follows because the identity operator is obviously self-adjoint, and $\langle T^*Tx, u \rangle_X = \langle Tx, Tu \rangle_Y = \langle x, T^*Tu \rangle_X$ for all $x, u \in X$.

We can now apply the following general theorem to $N_\lambda$.

**Theorem 3.2.** *Let $X$ be a Hilbert space and let $A : X \to X$ be a bounded self-adjoint linear operator. If there exists $\gamma > 0$ such that*

$$\langle Ax, x \rangle_X \geq \gamma \|x\|_X^2 \text{ for all } x \in X, \tag{3.4}$$

*then $A$ is invertible, $A^{-1}$ is bounded, and $\|A^{-1}\| \leq \gamma^{-1}$.*

*Proof.* We begin by noting that (3.4) and the Cauchy-Schwarz inequality imply that

$$\|x\|_X^2 \leq \gamma^{-1} \langle Ax, x \rangle_X \leq \gamma^{-1} \|Ax\|_X \|x\|_X \text{ for all } x \in X$$
$$\Rightarrow \|x\|_X \leq \gamma^{-1} \|Ax\|_X \text{ for all } x \in X.$$

This implies that if $Ax = 0$, then $x = 0$; thus, $\mathcal{N}(A)$ is trivial. Let us suppose that $\{Ax_n\}$ is a sequence in $\mathcal{R}(A)$ that satisfies $Ax_n \to u \in X$. Then we have

$$\|x_n - x_m\|_X \leq \gamma^{-1} \|A(x_n - x_m)\|_X = \gamma^{-1} \|Ax_n - Ax_m\|_X \text{ for all } m, n.$$

Since $\{Ax_n\}$ is convergent and hence Cauchy, this last inequality implies that $\{x_n\}$ is also Cauchy. Therefore, $\{x_n\}$ converges to some $x \in X$, which implies that $Ax_n \to Ax$ because $A$ is bounded. Since $Ax_n \to u$ by assumption, it follows that $u = Ax$; hence $u \in \mathcal{R}(A)$ and we have shown that $\mathcal{R}(A)$ is closed.

Next, we have $\mathcal{R}(A)^\perp = \mathcal{N}(A^*) = \mathcal{N}(A) = \{0\}$. Since $\mathcal{R}(A)$ is closed, this implies that $\mathcal{R}(A) = X$. Thus $A$ defines a bijection from $X$

to itself. Finally, as we have seen before,

$$\|x\|_X \leq \gamma^{-1} \|Ax\|_X \text{ for all } x \in X$$
$$\Rightarrow \|A^{-1}y\|_X \leq \gamma^{-1} \|y\|_X \text{ for all } y \in X.$$

Therefore, $A^{-1}$ is bounded and $\|A^{-1}\| \leq \gamma^{-1}$.                    □

Since $A = N_\lambda$ satisfies the hypotheses of the theorem (with $\gamma = \lambda$), we obtain the following corollary.

**Corollary 3.3.** *If $\lambda > 0$, then $N_\lambda$ is invertible, $N_\lambda^{-1}$ is bounded, and $\|N_\lambda^{-1}\| \leq \lambda^{-1}$.*

We now know that the problem

$$\min_{x \in X} \|Tx - y\|_Y^2 + \lambda \|x\|_X^2$$

has a unique solution $x_{\lambda,y} = N_\lambda^{-1} T^* y$ for each $y \in Y$ and each $\lambda > 0$. Moreover, for a fixed $\lambda > 0$, $x_{\lambda,y}$ depends continuously on $y$ (since $N_\lambda^{-1}$ and $T^*$ are bounded). However, we note that the bound on $\|N_\lambda^{-1}\|$, namely $\|N_\lambda^{-1}\| \leq \lambda^{-1}$, deteriorates as $\lambda \to 0^+$. As we show below, there is a better (smaller) bound on $\|N_\lambda^{-1} T^*\|$, but the bound still blows up as $\lambda \to 0^+$.

In Section 2.3.2, we mentioned that the idea of regularization is to construct a family $\{R_\lambda\}$ of bounded operators that approximates $T^\dagger$ in some sense. In Tikhonov regularization, $R_\lambda$ is defined by $R_\lambda = N_\lambda^{-1} T^*$. As noted above, $\|R_\lambda\|$ is unbounded as $\lambda \to 0^+$; this fact is consistent with the goal that $R_\lambda$ should approximate the unbounded operator $T^\dagger$. In the next sections, we study the convergence of Tikhonov regularization as $\lambda \to 0^+$, showing (among other results) that $R_\lambda$ converges pointwise to $T^\dagger$ on $D(T^\dagger)$. We prepare for that analysis by studying further properties of the operators $N_\lambda$ and $N_\lambda^{-1} T^*$.

**Theorem 3.4.** *If $\lambda > 0$, then $N_\lambda$ defines a bijection from $\mathcal{N}(T)$ onto itself, and also from $\mathcal{N}(T)^\perp$ onto itself. Also, $N_\lambda|_{\mathcal{N}(T)} = \lambda I$.*

*Proof.* We first note that if $x \in \mathcal{N}(T)$, then

$$(T^*T + \lambda I)x = \lambda x \in \mathcal{N}(T)$$

and

$$(T^*T + \lambda I)(\lambda^{-1}x) = x.$$

The first result shows that $N_\lambda$ maps $\mathcal{N}(T)$ into $\mathcal{N}(T)$ and the second shows that $N_\lambda$ maps $\mathcal{N}(T)$ onto $\mathcal{N}(T)$. We already know that $N_\lambda$ is injective, and hence it follows that $N_\lambda$ defines a bijection from $\mathcal{N}(T)$ to itself.

For any $x \in \mathcal{N}(T)^\perp$, $T^*Tx \in \mathcal{R}(T^*) \subset \mathcal{N}(T)^\perp$ and $\lambda x \in \mathcal{N}(T)^\perp$, which imply that $N_\lambda x \in \mathcal{N}(T)^\perp$. Thus $N_\lambda$ maps $\mathcal{N}(T)^\perp$ into itself. Next we show that if $x \in \mathcal{N}(T)^\perp$, then $N_\lambda^{-1}x \in \mathcal{N}(T)^\perp$. Let $x \in \mathcal{N}(T)^\perp$, $u \in \mathcal{N}(T)$. Then

$$\langle u, N_\lambda^{-1}x\rangle_X = \langle N_\lambda^{-1}u, x\rangle_X = \langle \lambda^{-1}u, x\rangle_X = \lambda^{-1}\langle u, x\rangle_X = 0,$$

which shows that $N_\lambda^{-1}x \in \mathcal{N}(T)^\perp$. Therefore, for any $x \in \mathcal{N}(T)^\perp$, $u = N_\lambda^{-1}x$ belongs to $\mathcal{N}(T)^\perp$ and $N_\lambda u = x$. This shows that $N_\lambda$ maps $\mathcal{N}(T)^\perp$ onto $\mathcal{N}(T)^\perp$. Since $N_\lambda$ is injective, this shows that $N_\lambda$ defines a bijection from $\mathcal{N}(T)^\perp$ onto itself. $\qquad\square$

**Corollary 3.5.** *For all $y \in Y$ and all $\lambda > 0$, $x_{\lambda,y} \in \mathcal{N}(T)^\perp$.*

*Proof.* Since $T^*y \in \mathcal{R}(T^*) \subset \mathcal{N}(T)^\perp$ and $x_{\lambda,y} = N_\lambda^{-1}T^*y$, this follows from the previous result. $\qquad\square$

We showed in Chapter 2 that $T^\dagger y \in \mathcal{N}(T)^\perp$ for each $y \in D(T^\dagger)$. We will use the fact that $x_{\lambda,y}$ also lies in $\mathcal{N}(T)^\perp$ in the next section when we prove that $x_{\lambda,y} \to T^\dagger y$ as $\lambda \to 0^+$ (provided $y \in D(T^\dagger)$).

To derive the next result, we need the following simple lemma.

**Lemma 3.6.** *If $y \in Y$, then*

$$\|s\|_Y^2 + \|y - s\|_Y^2 \geq \frac{1}{2}\|y\|_Y^2 \text{ for all } s \in Y.$$

*Proof.* Given a fixed $y \in Y$, define $J : Y \to \mathbb{R}$ by

$$J(s) = \|s\|_Y^2 + \|y - s\|_Y^2.$$

Since $J$ is quadratic, it is easy to prove directly that the global minimizer of $J$ is $s = (1/2)y$ and that $J((1/2)y) = (1/2)\|y\|_Y^2$. $\qquad\square$

The following theorem provides bounds on the norms of various operators related to $N_\lambda^{-1}$, including the regularization operator $N_\lambda^{-1} T^*$.

**Theorem 3.7.** *If $\lambda > 0$, then*

1. $\|N_\lambda^{-1} T^*\| \le \frac{1}{2} \lambda^{-1/2}$,

2. $\|T N_\lambda^{-1}\| \le \frac{1}{2} \lambda^{-1/2}$,

3. $\|T N_\lambda^{-1} T^*\| \le 1$.

*Proof.* Since $x_{\lambda,y}$ is the unique least-squares solution of $M_\lambda x = (y, 0)$, $M_\lambda x_{\lambda,y}$ is the projection of $(y, 0)$ onto $\mathcal{R}(M_\lambda)$ and therefore, by the Pythagorean theorem,

$$\|M_\lambda x_{\lambda,y}\|_{Y \times X}^2 + \|M_\lambda x_{\lambda,y} - (y, 0)\|_{Y \times X}^2 = \|(y, 0)\|_{Y \times X}^2$$
$$\Rightarrow \|T x_{\lambda,y}\|_Y^2 + \lambda \|x_{\lambda,y}\|_X^2 + \|T x_{\lambda,y} - y\|_Y^2 + \lambda \|x_{\lambda,y}\|_X^2 = \|y\|_Y^2$$
$$\Rightarrow \|T x_{\lambda,y}\|_Y^2 + \|T x_{\lambda,y} - y\|_Y^2 + 2\lambda \|x_{\lambda,y}\|_X^2 = \|y\|_Y^2.$$

By the preceding lemma (applied to $s = T x_{\lambda,y}$), we obtain

$$\frac{1}{2} \|y\|_Y^2 + 2\lambda \|x_{\lambda,y}\|_X^2 \le \|y\|_Y^2$$
$$\Rightarrow \|x_{\lambda,y}\|_X^2 \le \frac{1}{4} \lambda^{-1} \|y\|_Y^2$$
$$\Rightarrow \|N_\lambda^{-1} T^* y\|_X \le \frac{1}{2} \lambda^{-1/2} \|y\|_Y.$$

This holds for all $y \in Y$, which proves the first result.

The second result follows from the first and the fact that, for any bounded linear operator $A$ defined on Hilbert space, $\|A^*\| = \|A\|$ (see Theorem A.22 in Section A.4).

To prove the third result, note that, for all $y \in Y$,

$$\|T x_{\lambda,y}\|_Y^2 + \|T x_{\lambda,y} - y\|_Y^2 + 2\lambda \|x_{\lambda,y}\|_X^2 = \|y\|_Y^2$$
$$\Rightarrow \|T x_{\lambda,y}\|_Y^2 \le \|y\|_Y^2$$
$$\Rightarrow \|T x_{\lambda,y}\|_Y \le \|y\|_Y$$
$$\Rightarrow \|T N_\lambda^{-1} T^* y\|_Y \le \|y\|_Y. \qquad \square$$

When we study the rate at which $x_{\lambda,y} \to x_{0,y}$, we will need a bound on $\|N_\lambda^{-1} T^* T\|$. To derive it, we need a couple of preliminary results.

**Lemma 3.8.** *Let $U$ be a Hilbert space. Then, for any $u \in U$,*

$$\|u\|_U = \sup\{\langle u, v \rangle_U \,:\, v \in U, \|v\|_U = 1\}.$$

*Proof.* By the Cauchy-Schwarz inequality, we have

$$\langle u, v \rangle_U \leq \|u\|_U \|v\|_U = \|u\|_U \quad \text{for all } v \in V, \ \|v\|_U = 1.$$

Moreover, with $v = u/\|u\|_U$, we obtain $\langle u, v \rangle_U = \|u\|$.    □

**Lemma 3.9.** *Let $U$ be a Hilbert space, and let $L : U \to U$ be a bounded self-adjoint operator. Then*

$$\|L\| = \sup\{|\langle Lu, u \rangle_U| \,:\, u \in U, \|u\|_U = 1\}.$$

*Proof.* By definition,

$$\|L\| = \sup\{\|Lu\|_U \,:\, u \in U, \|u\|_U = 1\}.$$

By the Cauchy-Schwarz inequality, if $\|u\|_U = 1$, then

$$\|Lu\|_U = \|Lu\|_U \|u\|_U \geq |\langle Lu, u \rangle_U|.$$

Therefore,

$$\|L\| \geq \sup\{|\langle Lu, u \rangle_U| \,:\, u \in U, \|u\|_U = 1\}.$$

Write $M = \sup\{|\langle Lu, u \rangle_U| \,:\, u \in U, \|u\|_U = 1\}$. Then $\|L\| \geq M$, and we must show that $\|L\| \leq M$.

Now, we have $|\langle Lu, u \rangle_U| \leq M\|u\|_U^2$ for all $u \in U$. Since $L$ is self-adjoint, we have

$$\langle L(u \pm v), u \pm v \rangle_U = \langle Lu, u \rangle_U \pm 2\langle Lu, v \rangle_U + \langle Lv, v \rangle_U,$$

and therefore

$$
\begin{aligned}
4\,|\langle Lu, v \rangle_U| &= |\langle L(u+v), u+v \rangle_U - \langle L(u-v), u-v \rangle_U| \\
&\leq M(\|u+v\|_U^2 + \|u-v\|_U^2) \\
&= 2M(\|u\|_U^2 + \|v\|_U^2),
\end{aligned}
$$

the last step following from the parallelogram law (Lemma A.15). Thus

$$|\langle Lu, v \rangle_U| \leq \frac{M}{2}(\|u\|_U^2 + \|v\|_U^2) \text{ for all } u, v \in U. \tag{3.5}$$

It follows that

$$\|Lu\|_U = \sup\{|\langle Lu, v \rangle_U| : v \in V, \|v\|_U = 1\} \leq \frac{M}{2}\left(\|u\|_U^2 + 1\right),$$

and hence that

$$\|L\| = \sup\{\|Lu\|_U : u \in U, \|u\|_U = 1\} \leq \frac{M}{2}(1 + 1) = M. \qquad \square$$

We can now give the desired bound on $\|N_\lambda^{-1} T^* T\|$.

**Theorem 3.10.** *For all $\lambda > 0$, $N_\lambda^{-1} T^* T$ is self-adjoint, with*

$$N_\lambda^{-1} T^* T = I - \lambda N_\lambda^{-1}$$

*and $\|N_\lambda^{-1} T^* T\| \leq 1$.*

*Proof.* Since we can write $T^* T = N_\lambda - \lambda I$, we have

$$N_\lambda^{-1} T^* T = N_\lambda^{-1}(N_\lambda - \lambda I) = I - \lambda N_\lambda^{-1},$$

which shows that $N_\lambda^{-1} T^* T$ is self-adjoint and yields the first result. The fact that $\|N_\lambda^{-1}\| \leq \lambda^{-1}$ implies that

$$\langle \lambda N_\lambda^{-1} x, x \rangle_X \leq \|\lambda N_\lambda^{-1} x\|_X \|x\|_X \leq \|x\|_X^2 \text{ for all } x \in X.$$

On the other hand, for $x \in X$, we can define $u = N_\lambda^{-1} x$, in which case $x = N_\lambda u$ and

$$\langle \lambda N_\lambda^{-1} x, x \rangle_X = \lambda \langle u, N_\lambda u \rangle_X = \lambda \langle u, (T^* T + \lambda I) u \rangle_X$$
$$= \lambda \left(\|Tu\|_Y^2 + \lambda \|u\|_X^2\right) > 0.$$

Therefore, we see that $0 < \langle \lambda N_\lambda^{-1} x, x \rangle_X \leq \|x\|_X^2$ for all $x \in X$, and hence

$$\left|\langle N_\lambda^{-1} T^* T x, x \rangle_X\right| = \left|\langle (I - \lambda N_\lambda^{-1}) x, x \rangle_X\right|$$
$$= \left|\|x\|_X^2 - \langle \lambda N_\lambda^{-1} x, x \rangle_X\right| \leq \|x\|_X^2.$$

It now follows from Lemma 3.9 that $\|N_\lambda^{-1} T^* T\| \leq 1$. $\qquad \square$

## 3.2    Convergence of Tikhonov regularization

The purpose of this section is to prove that if $y \in D(T^{\dagger})$ then $x_{\lambda,y} \to T^{\dagger}y$ as $\lambda \to 0^{+}$. In preparation for this, we now study the Tikhonov functional

$$\|Tx - y\|_Y^2 + \lambda \|x\|_X^2$$

and its minimizer $x_{\lambda,y}$. We begin by identifying a certain trivial case for $Tx = y$, namely, the case in which $y \in \mathcal{R}(T)^{\perp}$. In this case, the element of $\mathcal{R}(T)$ closest to $y$ is the zero vector and it is clear that the minimum-norm least-squares solution of $Tx = y$ is $x = 0$. Moreover, since $x_{\lambda,y} = N_{\lambda}^{-1}T^*y$ and $\mathcal{N}(T^*) = \mathcal{R}(T)^{\perp}$, the condition $y \in \mathcal{R}(T)^{\perp}$ implies that $x_{\lambda,y} = 0$ for all $\lambda > 0$. On the other hand, we can draw the following conclusions if $y \notin \mathcal{R}(T)^{\perp}$.

**Theorem 3.11.**

*1. Let $y \in Y$ be given. If $y \notin \mathcal{R}(T)^{\perp}$, then $x_{\lambda_1,y} \neq x_{\lambda_2,y}$ for all $\lambda_1 \neq \lambda_2$.*

*2. Let $y \in D(T^{\dagger})$. If $y \notin \mathcal{R}(T)^{\perp}$, then $x_{\lambda,y} \neq T^{\dagger}y$ for all $\lambda > 0$.*

*Proof.* In both cases, it is convenient to prove the contrapositive.

1. We have $(T^*T + \lambda)x_{\lambda,y} = T^*y$ for all $\lambda > 0$. Therefore, if $x_{\lambda_1,y} = x_{\lambda_2,y}$ then

$$(T^*T + \lambda_1 I)x_{\lambda_1,y} = (T^*T + \lambda_2 I)x_{\lambda_1,y}$$
$$\Rightarrow \lambda_1 x_{\lambda_1,y} = \lambda_2 x_{\lambda_1,y}$$
$$\Rightarrow (\lambda_1 - \lambda_2)x_{\lambda_1,y} = 0.$$

If $\lambda_1 \neq \lambda_2$, then it follows that $x_{\lambda_1,y} = 0$. But then, since $N_{\lambda_1}^{-1}$ is nonsingular, it must be the case that $T^*y = 0$, which implies that $y \in \mathcal{N}(T^*) = \mathcal{R}(T)^{\perp}$.

2. We have $(T^*T + \lambda I)x_{\lambda,y} = T^*y = T^*T(T^{\dagger}y)$. Hence, if $x_{\lambda,y} = T^{\dagger}y$, then $x_{\lambda,y} = 0$ and the result follows as in the first part of the theorem.    $\square$

We can now derive some properties of the terms related to the Tikhonov functional, namely, the *residual* $\|Tx_{\lambda,y} - y\|_Y$ and the solution norm $\|x_{\lambda,y}\|_X$. Since $x_{\lambda,y}$ is the vector that minimizes

$$\|Tx - y\|_Y^2 + \lambda \|x\|_X^2,$$

we can predict the result of increasing the regularization weight $\lambda$: a larger value of $\lambda$ should force $\|x_{\lambda,y}\|_X$ to be smaller (larger values of $\|x\|_X$ are discouraged by the greater weight $\lambda$) and therefore force $\|Tx_{\lambda,y} - y\|_Y$ to be larger ($x_{\lambda,y}$ has to be chosen from a smaller set of vectors, since $\|x\|_X$ cannot be too large). This is the content of the following theorem.

**Theorem 3.12.**

1. *For all $y \in Y$, $\|x_{\lambda,y}\|_X$ is a nonincreasing function of $\lambda$. If $y \notin \mathcal{R}(T)^\perp$, $\|x_{\lambda,y}\|_X$ is a strictly decreasing function of $\lambda$.*

2. *For all $y \in Y$, $\|Tx_{\lambda,y} - y\|_Y$ is a nondecreasing function of $\lambda$. If $y \notin \mathcal{R}(T)^\perp$, $\|Tx_{\lambda,y} - y\|_Y$ is a strictly increasing function of $\lambda$.*

*Proof.* Suppose $0 < \lambda_1 < \lambda_2$. Then, by definition of $x_{\lambda_1,y}$ and $x_{\lambda_2,y}$ as minimizers of the Tikhonov functional, we have

$$\|Tx_{\lambda_1,y} - y\|_Y^2 + \lambda_1 \|x_{\lambda_1,y}\|_X^2 \le \|Tx_{\lambda_2,y} - y\|_Y^2 + \lambda_1 \|x_{\lambda_2,y}\|_X^2,$$
$$\|Tx_{\lambda_2,y} - y\|_Y^2 + \lambda_2 \|x_{\lambda_2,y}\|_X^2 \le \|Tx_{\lambda_1,y} - y\|_Y^2 + \lambda_2 \|x_{\lambda_1,y}\|_X^2.$$

Adding the inequalities and canceling like terms yields

$$\lambda_1 \|x_{\lambda_1,y}\|_X^2 + \lambda_2 \|x_{\lambda_2,y}\|_X^2 \le \lambda_1 \|x_{\lambda_2,y}\|_X^2 + \lambda_2 \|x_{\lambda_1,y}\|_X^2$$
$$\Rightarrow 0 \le (\lambda_2 - \lambda_1)\left(\|x_{\lambda_1,y}\|_X^2 - \|x_{\lambda_2,y}\|_X^2\right)$$
$$\Rightarrow 0 \le \|x_{\lambda_1,y}\|_X^2 - \|x_{\lambda_2,y}\|_X^2,$$

where the last step follows from the assumption that $\lambda_2 > \lambda_1$. This proves that $\|x_{\lambda,y}\|_X$ is a nonincreasing function of $\lambda$. If $y \notin \mathcal{R}(T)^\perp$, then $x_{\lambda_1,y} \ne x_{\lambda_2,y}$ and the original inequalities above, and hence all of the subsequent inequalities, are strict. In this case, we see that $\|x_{\lambda,y}\|_X$ is a strictly decreasing function of $\lambda$.

We then have

$$\|Tx_{\lambda_1,y} - y\|_Y^2 + \lambda_1 \|x_{\lambda_1,y}\|_X^2 \le \|Tx_{\lambda_2,y} - y\|_Y^2 + \lambda_1 \|x_{\lambda_2,y}\|_X^2$$
$$\Rightarrow \lambda_1 \left( \|x_{\lambda_1,y}\|_X^2 - \|x_{\lambda_2,y}\|_X^2 \right) \le \|Tx_{\lambda_2,y} - y\|_Y^2 - \|Tx_{\lambda_1,y} - y\|_Y^2$$
$$\Rightarrow 0 \le \|Tx_{\lambda_2,y} - y\|_Y^2 - \|Tx_{\lambda_1,y} - y\|_Y^2,$$

since $\|x_{\lambda_1,y}\|_X^2 - \|x_{\lambda_2,y}\|_X^2 \ge 0$. Thus $\|Tx_{\lambda,y} - y\|_Y^2$ is a nondecreasing function of $\lambda$. Once again, if $y \notin \mathcal{R}(T)^\perp$, we can conclude that the inequalities are strict and hence that $\|Tx_{\lambda,y} - y\|_Y^2$ is strictly increasing as a function of $\lambda$. $\qquad\square$

Our first result about $x_{\lambda,y}$ as $\lambda \to 0^+$ is valid for all $y$, not just $y$ belonging to $D(T^\dagger)$.

**Theorem 3.13.** *For each $y \in Y$, $Tx_{\lambda,y} \to \bar{y} = \text{proj}_{\overline{\mathcal{R}(T)}} y$ as $\lambda \to 0^+$ and hence $\|Tx_{\lambda,y} - y\|_Y \to \|\bar{y} - y\|_Y$ as $\lambda \to 0^+$.*

*Proof.* We have $\|Tx_{\lambda,y} - y\|_Y \ge \|\bar{y} - y\|_Y$ for all $\lambda > 0$ (because $Tx_{\lambda,y} \in \mathcal{R}(T)$), and we know that $\|Tx_{\lambda,y} - y\|_Y$ is nonincreasing as $\lambda \to 0^+$. Thus

$$L = \lim_{\lambda \to 0^+} \|Tx_{\lambda,y} - y\|_Y^2$$

exists, and $L \ge \|\bar{y} - y\|_Y^2$. Suppose, by way of contradiction, that $L > \|\bar{y} - y\|_Y^2$, say $L - \|\bar{y} - y\|_Y^2 = \epsilon > 0$. Choose $\bar{x} \in X$ such that $\|T\bar{x} - y\|_Y^2 < L - \epsilon/2$. Then, for all $\lambda > 0$ sufficiently small,

$$\|T\bar{x} - y\|_Y^2 + \lambda \|\bar{x}\|_X^2 < L - \frac{\epsilon}{2} + \lambda \|\bar{x}\|_X^2 < L \le \|Tx_{\lambda,y} - y\|_Y^2 + \lambda \|x_{\lambda,y}\|_X^2,$$

a contradiction. This shows that $\lim_{\lambda \to 0^+} \|Tx_{\lambda,y} - y\|_Y^2$ must equal $\|\bar{y} - y\|_Y^2$.

Since $Tx_{\lambda,y} \in \mathcal{R}(T)$ and $\bar{y} - y \in \mathcal{R}(T)^\perp$, the Pythagorean theorem implies that

$$\|Tx_{\lambda,y} - y\|_Y^2 = \|Tx_{\lambda,y} - \bar{y}\|_Y^2 + \|\bar{y} - y\|_Y^2$$

and hence $\|Tx_{\lambda,y} - y\| \to \|\bar{y} - y\|_Y$ implies that $\|Tx_{\lambda,y} - \bar{y}\|_Y \to 0$. This shows that $Tx_{\lambda,y} \to \bar{y}$. $\qquad\square$

We need one more preliminary result.

**Theorem 3.14.** *For each* $y \in D(T^{\dagger})$, $\|x_{\lambda,y}\|_X \leq \|T^{\dagger}y\|_X$ *for all* $\lambda > 0$. *If* $y \notin \mathcal{R}(T)^{\perp}$, *then* $\|x_{\lambda,y}\|_X < \|T^{\dagger}y\|_X$ *for all* $\lambda > 0$.

*Proof.* If $y \in \mathcal{R}(T)^{\perp}$, then $x_{\lambda,y} = 0$ for all $\lambda > 0$ and $T^{\dagger}y = 0$; therefore the result is obviously true in this case. Suppose $y \notin \mathcal{R}(T)^{\perp}$ and $\lambda > 0$. We know that $\|TT^{\dagger}y - y\|_Y < \|Tx_{\lambda,y} - y\|_Y$ (because $T^{\dagger}y \neq x_{\lambda,y}$ by Theorem 3.11). Hence, if $\|T^{\dagger}y\|_X < \|x_{\lambda,y}\|_X$, it follows that

$$\|TT^{\dagger}y - y\|_Y^2 + \lambda\|T^{\dagger}y\|_X^2 < \|Tx_{\lambda,y} - y\|_Y^2 + \lambda\|x_{\lambda,y}\|_X^2,$$

a contradiction. Therefore $\|T^{\dagger}y\|_X \geq \|x_{\lambda,y}\|_X$ must hold. Moreover, since $\|x_{\lambda,y}\|_X$ is strictly increasing as $\lambda$ decreases, the inequality must be strict. $\qquad\square$

We can now prove the main result of this section.

**Theorem 3.15.** *If* $y \in D(T^{\dagger})$, *then* $x_{\lambda,y} \to T^{\dagger}y$ *as* $\lambda \to 0^+$.

*Proof.* The previous theorem implies that $\{x_{\lambda,y} : \lambda > 0\}$ is bounded. Suppose that $\{\lambda_n\}$ is a sequence of positive real numbers such that $\lambda_n \to 0$ and there exists $\bar{x} \in X$ such that $x_{\lambda_n,y} \to \bar{x}$ weakly in $X$. Since $x_{\lambda_n,y} \in \mathcal{N}(T)^{\perp}$ for all $n$ and $\mathcal{N}(T)^{\perp}$ is closed with respect to weak sequential convergence (Theorem A.35), it follows that $\bar{x} \in \mathcal{N}(T)^{\perp}$. Also, we have

$$(T^*T + \lambda_n I)x_{\lambda_n,y} = T^*Tx_{\lambda_n,y} + \lambda_n x_{\lambda_n,y} \to T^*T\bar{x} \text{ weakly},$$

while $(T^*T + \lambda_n I)x_{\lambda_n,y} = T^*y$ for all $n$. This implies that $T^*T\bar{x} = T^*y$. Thus $\bar{x}$ is a least-squares solution of $Tx = y$ that belongs to $\mathcal{N}(T)^{\perp}$; this implies that $\bar{x} = T^{\dagger}y$. Thus we have shown that $x_{\lambda_n,y} \to T^{\dagger}y$ weakly. Using the fact that the norm is weakly lower semicontinuous (see Theorem A.38 in Section A.6) and the previous lemma, we have

$$\|T^{\dagger}y\|_x \leq \liminf_{n\to\infty} \|x_{\lambda_n,y}\|_X \leq \limsup_{n\to\infty} \|x_{\lambda_n,y}\|_X \leq \|T^{\dagger}y\|_X.$$

This shows that $\lim_{n\to\infty} \|x_{\lambda_n,y}\|_X = \|T^{\dagger}y\|_X$, which, together with the fact that $x_{\lambda_n,y} \to T^{\dagger}y$ weakly, implies that $x_{\lambda_n,y} \to T^{\dagger}y$ strongly in $X$.

We can now use the above reasoning to show that every sequence $\{x_{\lambda_n,y}\}$, where $\lambda_n \to 0^+$, has a subsequence converging (strongly) to $T^{\dagger}y$. This proves that $x_{\lambda,y} \to T^{\dagger}y$ strongly as $\lambda \to 0^+$. $\qquad\square$

We will use the notation $x_{0,y} = T^\dagger y$ for $y \in D(T^\dagger)$. The previous result shows that this notation is natural.

In the following example, $L^2(0, 1)$ represents the Hilbert space of all real-valued Lebesgue measurable functions $x$ defined on $(0, 1)$ that are *square-integrable*:

$$\int_0^1 |x(t)|^2 \, dt < \infty.$$

Two square-integrable functions that differ only on a set of Lebuesgue measure zero are regarded as the same element of $L^2(0, 1)$. The inner product on $L^2(0, 1)$ is

$$\langle x, y \rangle_{L^2} = \int_0^1 x(t)y(t) \, dt \text{ for all } x, y \in L^2(0, 1).$$

This inner product is the natural generalization of the Euclidean dot product. To see this, suppose we establish a uniform grid on $[0, 1]$, defining

$$0 = t_0 < t_1 < t_2 < \cdots < t_n = 1,$$

where $t_i = i\Delta t$, $\Delta t = 1/n$, and approximate a real-valued function $x$ defined on $[0, 1]$ by the vector obtained by sampling $x$ on the grid: $\mathbf{x} = (x(t_1), x(t_2), \ldots, x(t_n))$. For two such functions $x$, $y$, the dot product of the corresponding vectors is

$$\mathbf{x} \cdot \mathbf{y} = \sum_{i=1}^n x(t_i)y(t_i).$$

Weighting this by $\Delta t$ (so that we obtain a quantity that converges as $n \to \infty$), we obtain a Riemann sum approximating $\langle x, y \rangle_{L^2}$:

$$(\mathbf{x} \cdot \mathbf{y})\Delta t = \sum_{i=1}^n x(t_i)y(t_i)\Delta t \approx \int_0^1 x(t)y(t) \, dt = \langle x, y \rangle_{L^2}.$$

**Example 3.16.** Let $T : L^2(0, 1) \to L^2(0, 1)$ be defined by

$$(Tx)(s) = \int_0^1 k(s, t)x(t) \, dt, \ 0 < s < 1,$$

where

$$k(s, t) = \begin{cases} s(1 - t), & s < t, \\ t(1 - s), & s \geq t. \end{cases}$$

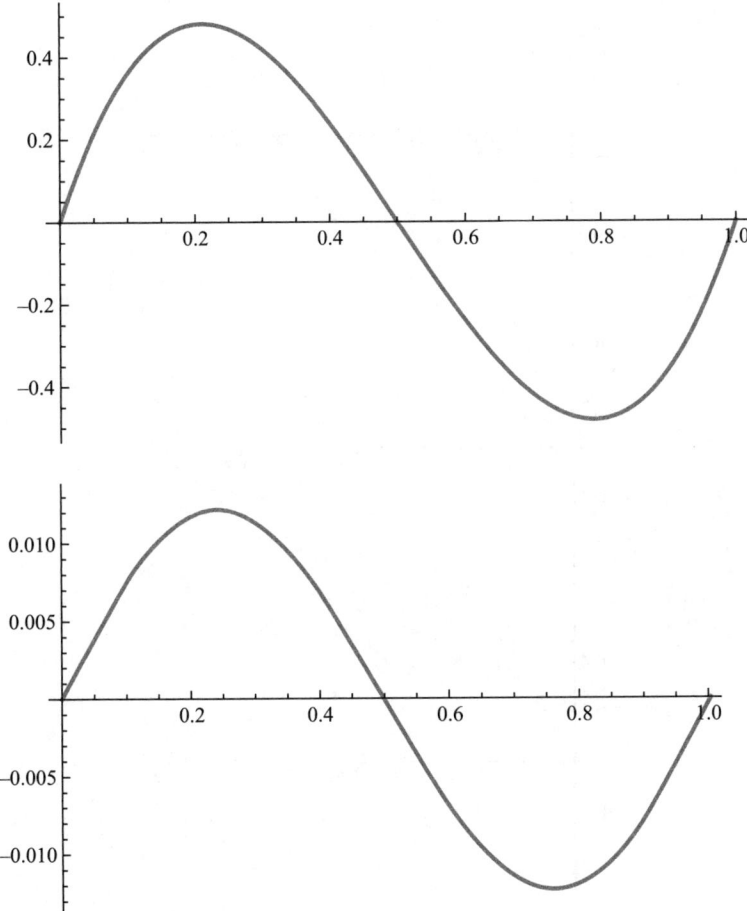

**Figure 3.2.** The exact solution (top) and the exact data (bottom) in Example 3.16.

If we define

$$x^*(t) = 10t \left( \frac{1}{2} - t \right) (1 - t), \; y^*(t) = \frac{1}{12} \left( t - 10t^3 + 15t^4 - 6t^5 \right),$$

then $Tx^* = y^*$ holds; we will think of $x^*$ and $y^*$ as the exact solution and data, respectively, for this example. The functions $x^*$ and $y^*$ are shown in Figure 3.2.

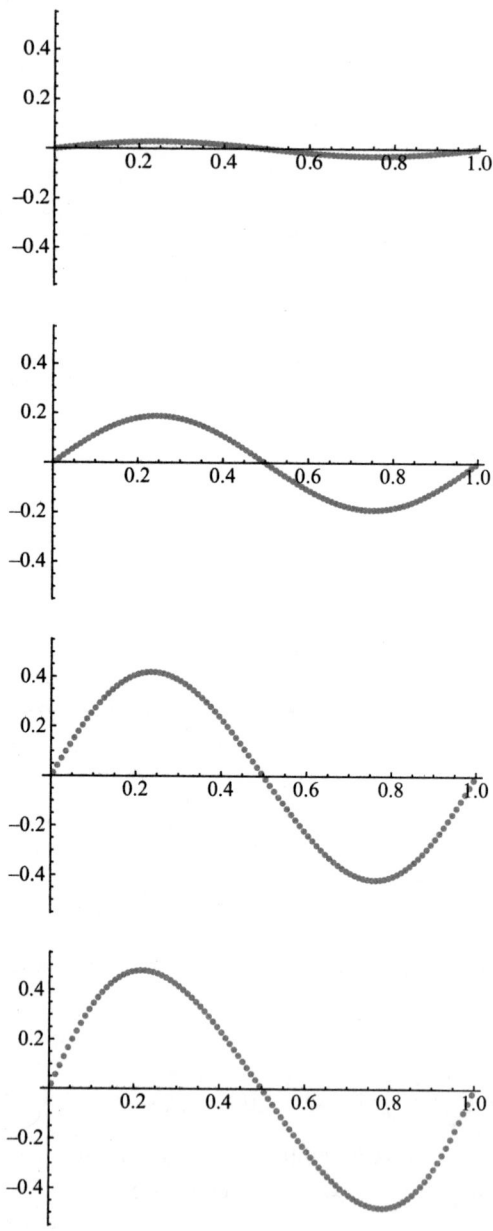

**Figure 3.3.** The Tikhonov solutions from Example 3.16 corresponding to (from top to bottom) $\lambda = 10^{-2}$, $\lambda = 10^{-3}$, $\lambda = 10^{-4}$, and $\lambda = 10^{-5}$.

Let us now consider the Tikhonov solution $x_{\lambda,y^*}$. Theorem 3.15 states that $x_{\lambda,y^*} \to x^*$ (in the $L^2$ norm) as $\lambda \to 0^+$. Figure 3.3 shows $x_{\lambda,y^*}$ for $\lambda = 10^{-2}$, $\lambda = 10^{-3}$, $\lambda = 10^{-4}$, and $\lambda = 10^{-5}$. The relative errors in these Tikhonov solutions are approximately 94%, 62%, 16%, and 3.9%. If we reduce $\lambda$ to $10^{-6}$, then $x_{\lambda,y^*}$ is essentially indistinguishable from $x^*$ on the scale of Figure 3.3 and the relative error in the $L^2$ norm is less than 1%.

The regularization operator underlying Tikhonov regularization is $R_\lambda = N_\lambda^{-1} T^*$. Theorem 3.15 shows that $R_\lambda \to T^\dagger$ pointwise on $D(T^\dagger)$; it is in this sense that $R_\lambda$ approximates $T^\dagger$. The final result of this section shows how $R_\lambda$ behaves when applied to $y \notin D(T^\dagger)$.

**Theorem 3.17.** *If $y \in Y$, $y \notin D(T^\dagger)$, then $\|x_{\lambda,y}\|_X \to \infty$ as $\lambda \to 0^+$.*

*Proof.* If the conclusion fails, then there exists a sequence $\{\lambda_n\}$ of positive real numbers such that $\lambda_n \to 0$ and $\{x_{\lambda_n,y}\}$ is bounded. Then, without loss of generality, there exists $\bar{x} \in X$ such that $x_{\lambda_n,y} \to \bar{x}$ weakly, which implies that $Tx_{\lambda_n,y} \to T\bar{x}$ weakly. But we already know that $Tx_{\lambda_n,y} \to \text{proj}_{\overline{\mathcal{R}(T)}} y$ as $n \to \infty$. Therefore, $\text{proj}_{\overline{\mathcal{R}(T)}} y = T\bar{x} \in \mathcal{R}(T)$, contradicting that $y \notin D(T^\dagger)$. $\square$

**Example 3.18.** This is a continuation of Example 3.16. To illustrate Theorem 3.17, we construct a noisy data vector $\hat{y}$ as $\hat{y} = y^* + z$, where $z$ is the discontinuous function

$$z(t) = \begin{cases} t, & 0 < t < \frac{1}{2}, \\ 2 - 2t, & \frac{1}{2} < t < 1. \end{cases}$$

The exact and noisy data are shown in Figure 3.4. The function $\hat{y}$ does not belong to $D(T^\dagger)$, and hence Theorem 3.17 applies: $\|x_{\lambda,\hat{y}}\|_{L^2}$ should increase without bound as $\lambda \to 0^+$. This is illustrated in Figure 3.5, where the solutions $x_{\lambda,\hat{y}}$ corresponding to $\lambda = 10^{-6}$, $\lambda = 10^{-7}$, $\lambda = 10^{-8}$, and $\lambda = 10^{-9}$ are displayed (the reader should pay attention to the vertical scale in these graphs). The relative errors, in the $L^2$ norm, are approximately 32%, 74%, 173%, and 410%, respectively.

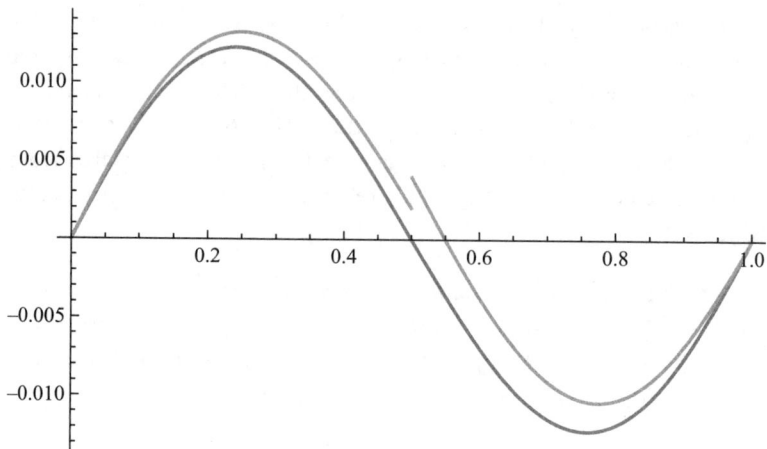

**Figure 3.4.** The exact and noisy data from Example 3.18 (the noisy data is the discontinuous function).

For $y \in D(T^\dagger)$, Tikhonov regularization approximates $T^\dagger y$ by $x_{\lambda,y} = N_\lambda^{-1} T^* y$, where $\lambda$ is a small positive number. We can express this differently by saying that Tikhonov regularization approximates the (unbounded) operator $T^\dagger$ by the bounded operator $R_\lambda = N_\lambda^{-1} T^*$. As noted above, $R_\lambda \to T^\dagger$ pointwise on $D(T^\dagger)$ as $\lambda \to 0^+$. This convergence cannot be uniform if $\mathcal{R}(T)$ fails to be closed. Uniform convergence is equivalent to convergence in the operator norm; since $T^\dagger$ is unbounded, it is not possible for $R_\lambda$ to converge to $T^\dagger$ in the operator norm.

We now know that $x_{\lambda,y}$ converges to $T^\dagger y$ as $\lambda \to 0^+$, provided $y \in D(T^\dagger)$. This leads to two questions:

1. How do we choose a good value of $\lambda$? Since $R_\lambda$ does not converge uniformly to $T^\dagger$, the answer must depend on the particular vector $y$. A single value of $\lambda$ cannot be a good choice for every $y$.

2. How does the method perform if $y \notin D(T^\dagger)$? Since, in the interesting case that $\mathcal{R}(T)$ fails to be closed, $D(T^\dagger)$ is a dense subspace of $Y$, measurement error will almost surely cause $y$ to fail to lie in $D(T^\dagger)$. Theorem 3.17 shows that, in such a case, we must be careful not to choose $\lambda$ to be too small.

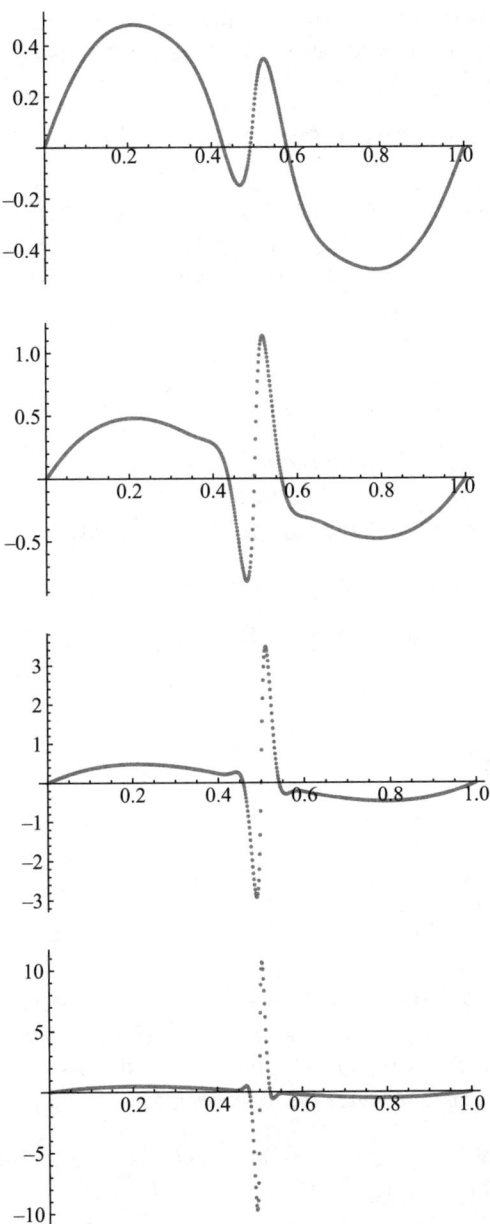

**Figure 3.5.** The Tikhonov solutions from Example 3.18 corresponding to (from top to bottom) $\lambda = 10^{-6}$, $\lambda = 10^{-7}$, $\lambda = 10^{-8}$, and $\lambda = 10^{-9}$.

In the next section, we consider the case in which $y$ is a noisy data vector and begin to provide answers to these questions.

## 3.3 Convergence for noisy data

We now consider an exact data vector $y^* \in D(T^\dagger)$ and corresponding exact solution $x^* = x_{0,y^*}$. If $y$ is an estimate (measurement) of $y^*$, then $x_{\lambda,y} = N_\lambda^{-1} T^* y$ is an estimate of $x^*$. We wish to determine conditions under which $x_{\lambda,y} \to x^*$ as $y \to y^*$ and $\lambda \to 0^+$.

We have

$$x_{\lambda,y} - x^* = x_{\lambda,y} - x_{\lambda,y^*} + x_{\lambda,y^*} - x_{0,y^*}.$$

We call $x_{\lambda,y} - x_{\lambda,y^*}$ the *perturbation error* and $x_{\lambda,y^*} - x_{0,y^*}$ the *regularization error*, while $x_{\lambda,y} - x^*$ is the *total error* in $x_{\lambda,y}$. The perturbation error depends directly on the error in the data. In fact, we have

$$x_{\lambda,y} - x_{\lambda,y^*} = N_\lambda^{-1} T^* y - N_\lambda^{-1} T^* y^* = N_\lambda^{-1} T^* (y - y^*)$$

$$\Rightarrow \|x_{\lambda,y} - x_{\lambda,y^*}\|_X \leq \|N_\lambda^{-1} T^*\| \|y - y^*\|_Y$$

$$\Rightarrow \|x_{\lambda,y} - x_{\lambda,y^*}\|_X \leq \frac{1}{2} \lambda^{-1/2} \|y - y^*\|_Y.$$

Therefore, we have

$$\|x_{\lambda,y} - x_{0,y^*}\|_X \leq \frac{1}{2} \lambda^{-1/2} \|y - y^*\|_Y + \|x_{\lambda,y^*} - x_{0,y^*}\|_X. \qquad (3.6)$$

We already know (by Theorem 3.15) that the regularization error satisfies

$$\|x_{\lambda,y^*} - x_{0,y^*}\|_X \to 0 \text{ as } \lambda \to 0^+.$$

On the other hand, the bound on the perturbation error goes to zero only if the regularization weight $\lambda$ is properly chosen with respect to the error $\|y - y^*\|_Y$. The following theorem provides a sufficient condition for the perturbation error, and hence the total error, to go to zero.

**Theorem 3.19.** *Let $y^* \in D(T^\dagger)$ and assume $\{y_n\} \subset Y$ satisfies $y_n \to y^*$. Let $\{\delta_n\} \subset \mathbb{R}^+$ satisfy $\|y_n - y^*\|_Y \leq \delta_n$ for all $n$ and $\delta_n \to 0$ as $n \to \infty$,* *and suppose $\{\lambda_n\} \subset \mathbb{R}^+$ satisfies*

$$\lambda_n \to 0^+, \ \frac{\delta_n^2}{\lambda_n} \to 0 \ as \ n \to \infty.$$

*Then $x_{\lambda_n, y_n} \to x_{0,y^*}$ as $n \to \infty$.*

*Proof.* By (3.6), we have, for each $n$,

$$\|x_{\lambda_n, y_n} - x_{0,y^*}\|_X \leq \frac{\|y_n - y^*\|_Y}{2\sqrt{\lambda_n}} + \|x_{\lambda_n, y^*} - x_{0,y^*}\|_X$$

$$\leq \frac{\delta_n}{2\sqrt{\lambda_n}} + \|x_{\lambda_n, y^*} - x_{0,y^*}\|_X.$$

Theorem 3.15 implies that the second term on the right goes to zero as $n \to \infty$, while the assumption $\delta_n^2 / \lambda_n \to 0$ implies that the first term also goes to zero. $\qquad\square$

The hypothesis $\delta_n^2 / \lambda_n \to 0$ means that $\lambda_n$ goes to zero more slowly than $\delta_n^2$. Given that $\|x_{\lambda,y}\|_X \to \infty$ as $\lambda \to 0^+$ if $y \notin D(T^\dagger)$ (Theorem 3.17), it is natural that $x_{\lambda_n, y_n} \to x_{0,y^*}$ can only be guaranteed if the regularization weight is not too small (that is, if enough regularization is applied). The preceding theorem quantifies how large $\lambda_n$ must be to obtain convergence.

The convergence in Theorem 3.19 is strong convergence, or convergence in norm ($\|x_{\lambda,y} - x_{0,y^*}\|_X \to 0$). The following result gives conditions under which $x_{\lambda,y} \to x_{0,y^*}$ weakly as $y \to y^*$.

**Theorem 3.20.** *Let $y^* \in D(T^\dagger)$ and assume $\{y_n\} \subset Y$ satisfies $y_n \to y^*$. Let $\{\delta_n\} \subset \mathbb{R}^+$ satisfy $\|y_n - y^*\|_Y \leq \delta_n$ and suppose $\{\lambda_n\} \subset \mathbb{R}^+$ satisfies $\lambda_n \to 0$ and $\lambda_n \geq k\delta_n^2$ for some $k > 0$. Then $x_{\lambda_n, y_n} \to x_{0,y^*}$ weakly as $n \to \infty$.*

*Proof.* The assumption, together with (3.6), implies that $\{x_{\lambda_n, y_n}\}$ is bounded. It suffices, by the usual reasoning, to prove that if $x_{\lambda_n, y_n} \to \bar{x} \in X$ weakly, then $\bar{x} = x_{0,y^*}$. For each $n$, we have

$$(T^*T + \lambda_n I)x_{\lambda_n, y_n} = T^*y_n.$$

Moreover, we have $T^*y_n \to T^*y$ strongly (because $y_n \to y$ strongly and $T^*$ is bounded) and $T^*Tx_{\lambda_n,y_n} \to T^*T\bar{x}$ weakly (because $x_{\lambda_n,y_n} \to \bar{x}$ weakly and $T^*T$ is bounded). Since $\{x_{\lambda_n,y_n}\}$ is bounded, $\lambda_n x_{\lambda_n,y_n} \to 0$ strongly as $n \to \infty$. It follows that $T^*T\bar{x} = T^*y$ and hence that $\bar{x}$ is a least-squares solution of $Tx = y^*$. We have seen that $x_{\lambda_n,y_n} \in \mathcal{N}(T)^\perp$ for all $n$ (Corollary 3.5) and hence, because $\mathcal{N}(T)^\perp$ is closed, $\bar{x} \in \mathcal{N}(T)^\perp$. It follows that $\bar{x}$ is the minimum-norm least-squares solution of $Tx = y^*$, that is, $\bar{x} = x_{0,y^*}$. $\qquad\square$

We now have some answers to the questions we raised at the end of the last section. To obtain convergence of $x_{\lambda,y}$ to $x_{0,y^*}$ as $\lambda \to 0^+$ and $y \to y^*$, $\lambda$ cannot go to zero too quickly compared to the rate at which $\|y - y^*\|_Y$ goes to zero. Specifically, we obtain strong convergence if

$$\lambda \to 0^+ \text{ and } \frac{\delta^2}{\lambda} \to 0,$$

where $\|y - y^*\|_Y \le \delta$. We obtain weak convergence in the borderline case where

$$\lambda \to 0^+ \text{ and } \frac{\delta^2}{\lambda} \text{ is bounded as } \delta \to 0.$$

If $\lambda$ goes to zero too quickly compared to $\delta$, specifically if $\delta^2/\lambda$ is unbounded as $\delta \to 0$, then we cannot guarantee convergence. However, the proof of this requires technical results that are given in Section 3.6, and therefore we save it for that section (see Theorem 3.54 in Section 3.6).

The results presented in this section address the question of convergence, but not the rate of convergence of $x_{\lambda,y}$ to $x_{0,y^*}$. We would like information about the size of $\|x_{\lambda,y} - x_{0,y^*}\|_X$ compared to the size of $\|y - y^*\|_Y$. Our approach has been to analyze the total error in terms of the perturbation error and the regularization error:

$$\|x_{\lambda,y} - x_{0,y^*}\|_X \le \|x_{\lambda,y} - x_{\lambda,y^*}\|_X + \|x_{\lambda,y^*} - x_{0,y^*}\|_X.$$

As we showed, the perturbation error can be bounded in a useful way:

$$\|x_{\lambda,y} - x_{\lambda,y^*}\|_X \le \frac{1}{2}\lambda^{-1/2}\|y - y^*\|_Y.$$

However, it turns out that the regularization error cannot be bounded in general; in fact, depending on the value of $y^*$, $\|x_{\lambda,y^*} - x_{0,y^*}\|_X$ can go to zero arbitrarily slowly with respect to $\lambda$ (see Theorem 3.44 in Section 3.6). In the next section, we show that, by restricting the allowable values of $y^*$ (or, equivalently, of $x_{0,y^*}$), it is possible to derive useful bounds on $\|x_{\lambda,y^*} - x_{0,y^*}\|_X$.

## 3.4   Rates of convergence

We now study the possibility of determining the rates at which $\|x_{\lambda,y^*} - x_{0,y^*}\|_X$ and $\|x_{\lambda,y} - x_{0,y^*}\|_X$ converge to zero. We might expect to bound $\|x_{\lambda,y^*} - x_{0,y^*}\|_X$ in terms of $\lambda$ and possibly $\|y^*\|_Y$. However, we recall that

$$N_\lambda x_{\lambda,y^*} - N_\lambda x_{0,y^*} = T^* y^* - T^* y^* - \lambda x_{0,y^*} = -\lambda x_{0,y^*}$$
$$\Rightarrow x_{\lambda,y^*} - x_{0,y^*} = -\lambda N_\lambda^{-1} x_{0,y^*}.$$

We know that $\|N_\lambda^{-1}\| \leq \lambda^{-1}$ and therefore we obtain only the bound

$$\|x_{\lambda,y^*} - x_{0,y^*}\|_X \leq \|x_{0,y^*}\|_X.$$

Since we already know from Theorem 3.14 that $\|x_{\lambda,y^*} - x_{0,y^*}\|_X \to 0$ as $\lambda \to 0^+$, this bound is not very informative.

However, if we restrict the class of exact solutions $x^*$ under consideration, we can derive some useful results. We consider two possibilities: $x^* \in \mathcal{R}(T^*)$ and $x^* \in \mathcal{R}(T^*T)$. For the first, we need a preliminary result. Theorem 3.7 implies that $\lambda^{1/2}\|N_\lambda^{-1} T^*\| \leq 1/2$ for all $\lambda > 0$; the following theorem gives more information about the operator $\lambda^{1/2} N_\lambda^{-1} T^*$.

**Theorem 3.21.** *For all $y \in Y$, $\lambda^{1/2} N_\lambda^{-1} T^* y \to 0$ as $\lambda \to 0^+$.*

*Proof.* As in the proof of Theorem 3.7, we have

$$\|T x_{\lambda,y}\|_Y^2 + \|T x_{\lambda,y} - y\|_Y^2 + 2\lambda \|x_{\lambda,y}\|_X^2 = \|y\|_Y^2$$
$$\Rightarrow 2\lambda \|x_{\lambda,y}\|_X^2 = \|y\|_Y^2 - \|T x_{\lambda,y}\|_Y^2 - \|T x_{\lambda,y} - y\|_Y^2.$$

But for each $y \in Y$, $Tx_{\lambda,y} \to \bar{y} = \text{proj}_{\overline{\mathcal{R}(T)}} y$ as $\lambda \to 0^+$ and, by the Pythagorean theorem, $\|y\|_Y^2 = \|\bar{y}\|_Y^2 + \|y - \bar{y}\|_Y^2$. Therefore, as $\lambda \to 0^+$,

$$2\lambda\|x_{\lambda,y}\|_X^2 = \|y\|_Y^2 - \|Tx_{\lambda,y}\|_Y^2 - \|Tx_{\lambda,y} - y\|_Y^2 \to \|y\|_Y^2 - \|\bar{y}\|_Y^2 - \|y - \bar{y}\|_Y^2$$
$$= 0. \qquad \square$$

Theorem 3.21 shows that $\lambda^{1/2} N_\lambda^{-1} T^* \to 0$ in the pointwise sense; it does not imply that $\lambda^{1/2}\|N_\lambda^{-1} T^*\| \to 0$.

Now we can derive a convergence rate for $\|x_{\lambda,y} - x_{0,y}\|_X$ in the case that $x_{0,y} \in \mathcal{R}(T^*)$.

**Theorem 3.22.** *If $y \in D(T^\dagger)$ and $x_{0,y} \in \mathcal{R}(T^*)$, then $\|x_{\lambda,y} - x_{0,y}\|_X = o(\lambda^{1/2})$ as $\lambda \to 0^+$.*

*Proof.* Suppose $w_{0,y} \in \mathcal{N}(T^*)^\perp$ and $x_{0,y} = T^* w_{0,y}$. We have

$$N_\lambda(x_{\lambda,y} - x_{0,y}) = -\lambda x_{0,y} = -\lambda T^* w_{0,y}$$
$$\Rightarrow x_{\lambda,y} - x_{0,y} = -\lambda N_\lambda^{-1} T^* w_{0,y} = -\lambda^{1/2}\left(\lambda^{1/2} N_\lambda^{-1} T^* w_{0,y}\right).$$

Theorem 3.21 shows that $\lambda^{1/2} N_\lambda^{-1} T^* w_{0,y} \to 0$, and hence $\|x_{\lambda,y} - x_{0,y}\|_X = o(\lambda^{1/2})$ as $\lambda \to 0^+$. $\qquad \square$

We get a faster convergence rate if $x_{0,y} \in \mathcal{R}(T^*T)$.

**Theorem 3.23.** *If $y \in D(T^\dagger)$ and $x_{0,y} \in \mathcal{R}(T^*T)$, then $\|x_{\lambda,y} - x_{0,y}\|_X = O(\lambda)$ as $\lambda \to 0^+$. More precisely, if $u_{0,y} \in \mathcal{N}(T)^\perp$ and $x_{0,y} = T^*Tu_{0,y}$, then*

$$\|x_{\lambda,y} - x_{0,y}\|_X \leq \lambda\|u_{0,y}\|_X.$$

*Proof.* We have

$$N_\lambda(x_{\lambda,y} - x_{0,y}) = -\lambda x_{0,y} = -\lambda T^*Tu_{0,y}$$
$$\Rightarrow x_{\lambda,y} - x_{0,y} = -\lambda N_\lambda^{-1} T^*Tu_{0,y}.$$

By Theorem 3.10, $\|N_\lambda^{-1} T^*T\| \leq 1$, and the result follows. $\qquad \square$

As we discussed earlier, $T$ and $T^*$ are smoothing operators in a typical inverse problem. Therefore, the conditions $x^* \in \mathcal{R}(T^*)$ and $x^* \in \mathcal{R}(T^*T)$

can be regarded as abstract smoothness conditions and the previous results show that we can bound the regularization error provided the exact solution is sufficiently smooth.

To apply the previous two results to bound the total error, it is convenient to consider first the general case in which the perturbation error satisfies

$$\|x_{\lambda,y^*} - x_{0,y^*}\|_X = O(\lambda^p)$$

for some $p > 0$. Then the total error satisfies

$$\|x_{\lambda,y} - x_{0,y^*}\|_X \leq \frac{1}{2}\|y - y^*\|_Y \lambda^{-1/2} + C\lambda^p$$

for some $C > 0$. We assume that $\|y - y^*\|_Y \leq \delta$ for the error in the data and also that the regularization weight $\lambda$ is chosen as $\lambda = k\delta^q$ for some positive constants $k$ and $q$. We then have

$$\|x_{\lambda,y} - x_{0,y^*}\|_X \leq \frac{k}{2}\delta^{1-q/2} + Ck^p\delta^{pq}$$

and it follows that

$$\|x_{\lambda,y} - x_{0,y^*}\|_X = O\left(\delta^{\min\{1-q/2,pq\}}\right).$$

The total error (or, to be more precise, the bound on the total error) will be as small as possible if we choose $q$ to maximize the exponent $\min\{1 - q/2, pq\}$. A graph shows that this implies choosing $q$ to satisfy $1 - q/2 = pq$ (see Figure 3.6). The result is $q = 2/(2p + 1)$, which implies that the optimal order of convergence is

$$1 - \frac{q}{2} = pq = \frac{2p}{2p + 1}.$$

Thus it is possible (in theory) to choose $\lambda$ to obtain

$$\|x_{\lambda,y} - x_{0,y^*}\|_X = O\left(\delta^{2p/(2p+1)}\right),$$

where $\delta = \|y - y^*\|_Y$. Regardless of the value of $p$, we see that $2p/(2p + 1) < 1$ and hence this bound suggests that there is an inevitable loss of information (that is, amplification of error). This is true, as we will see,

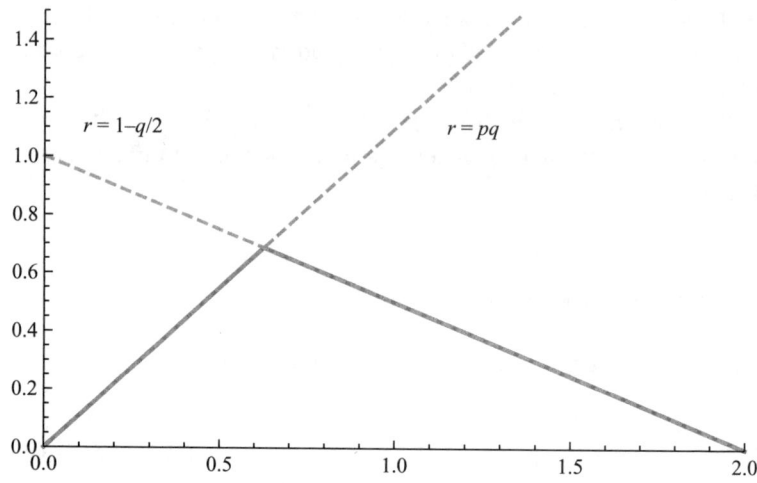

**Figure 3.6.** The graph of $r = \min\{1 - q/2, \, pq\}$.

and it is a consequence of the fact that the problem is an inverse problem. If $T$ is invertible with a bounded inverse $T^{-1}$, then

$$Tx = y, \; Tx^* = y^* \Rightarrow x - x^* = T^{-1}(y - y^*)$$
$$\Rightarrow \|x - x^*\|_X \leq \|T^{-1}\|\|y - y^*\|_Y$$

and we see that $\|x - x^*\|_X$ is (at worst) proportional to $\|y - y^*\|_Y$. (The same is true if $T$ is not invertible but $\mathcal{R}(T)$ is closed, so that $T^\dagger$ is bounded.)

We can now apply the above results to the two specific cases mentioned above, $x_{0,y^*} \in \mathcal{R}(T^*)$ and $x_{0,y^*} \in \mathcal{R}(T^*T)$.

**Theorem 3.24.** *Suppose $y^* \in D(T^\dagger)$.*

*1. If $x_{0,y^*} \in \mathcal{R}(T^*)$ and we choose $\lambda = k\delta$, where $\delta = \|y - y^*\|_Y$, then*

$$\|x_{\lambda,y} - x_{0,y^*}\|_X = O(\delta^{1/2}).$$

2. *If $x_{0,y^*} \in \mathcal{R}(T^*T)$ and we choose $\lambda = k\delta^{2/3}$, where $\delta = \|y - y^*\|_Y$, then*

$$\|x_{\lambda,y} - x_{0,y^*}\|_X = O(\delta^{2/3}).$$

Since $2p/(2p+1)$ increases with $p$, it is natural to ask if we can find a more restricted class of solutions $x^* = x_{0,y^*}$ such that $\|x_{\lambda,y} - x_{0,y^*}\|_X = O(\lambda^p)$ for some $p > 1$. The following result shows that this is not possible (except in the trivial case $y^* \in \mathcal{R}(T)^\perp$).

**Theorem 3.25.** *If $y^* \in D(T^\dagger)$ and $\|x_{\lambda,y^*} - x_{0,y^*}\|_X = o(\lambda)$, then $y^* \in \mathcal{R}(T)^\perp$ and $x_{0,y^*} = 0$.*

*Proof.* If $y^* \in \mathcal{R}(T)^\perp$, then $x_{\lambda,y^*} = x_{0,y^*} = 0$ for all $\lambda > 0$ and hence we have $\|x_{\lambda,y^*} - x_{0,y^*}\|_X = 0 = o(\lambda)$. Now suppose $y^* \notin \mathcal{R}(T)^\perp$ and $\|x_{\lambda,y^*} - x_{0,y^*}\|_X = o(\lambda)$. We show that this leads to a contradiction. We have

$$N_\lambda(x_{\lambda,y^*} - x_{0,y^*}) = -\lambda x_{0,y^*}.$$

Let us define

$$u_\lambda = \frac{x_{\lambda,y^*} - x_{0,y^*}}{\|x_{\lambda,y^*} - x_{0,y^*}\|_X} \quad \text{for all } \lambda > 0$$

(note that $y^* \notin \mathcal{R}(T)^\perp$ implies that $x_{\lambda,y^*} - x_{0,y^*} \neq 0$ for all $\lambda > 0$). We then have

$$\|N_\lambda u_\lambda\|_X \leq \|N_\lambda\| \|u_\lambda\|_X \leq \|T\|^2 + \lambda \rightarrow \|T\|^2 \quad \text{as } \lambda \rightarrow 0^+.$$

However,

$$N_\lambda u_\lambda = -\frac{\lambda}{\|x_{\lambda,y^*} - x_{0,y^*}\|_X} x_{0,y^*}$$

and, since $\|x_{\lambda,y^*} - x_{0,y^*}\|_X = o(\lambda)$ by assumption,

$$\frac{\lambda}{\|x_{\lambda,y^*} - x_{0,y^*}\|_X} \rightarrow \infty \quad \text{as } \lambda \rightarrow 0^+.$$

Since $x_{0,y^*} \neq 0$, this is a contradiction. $\qquad\square$

Thus we see that, for Tikhonov regularization, the optimal bound on the total error is

$$\|x_{\lambda,y^*} - x_{0,y^*}\|_X = O\left(\delta^{2/3}\right),$$

which is attained if $\lambda = k\delta^{2/3}$ and $x_{0,y^*} \in \mathcal{R}(T^*T)$.

Before presenting some examples, we derive the following result, which gives a lower bound on the rate at which the error $\|x_{\lambda,y^*} - x_{0,y^*}\|_X$ converges to zero. This will help us to interpret the examples given below.

**Lemma 3.26.** *Suppose* $y^* \in D(T^\dagger)$, $y^* \notin \mathcal{R}(T)^\perp$, $\{y_n\} \subset Y$, $y_n \to y^*$, *and* $\{\lambda_n\} \subset (0, \infty)$ *is chosen so that*

$$\lambda_n \to 0, \quad \frac{\delta_n}{\lambda_n} \to 0 \text{ as } n \to \infty,$$

*where* $\|y_n - y^*\|_Y \leq \delta_n$ *for all n. Then there exists* $c > 0$ *such that*

$$\|x_{\lambda_n,y_n} - x_{0,y^*}\|_X \geq c\lambda_n \text{ for all n sufficiently large.}$$

*Proof.* We have

$$N_{\lambda_n}(x_{0,y^*} - x_{\lambda_n,y_n}) = N_{\lambda_n}x_{0,y^*} - N_{\lambda_n}x_{\lambda_n,y_n}$$
$$= T^*Tx_{0,y^*} + \lambda_n x_{0,y^*} - T^*y_n$$
$$= T^*(y^* - y_n) + \lambda_n x_{0,y^*}.$$

By the reverse triangle inequality, this implies that

$$\left|\lambda_n\|x_{0,y^*}\|_X - \|T^*(y^* - y_n)\|_X\right| \leq \|N_{\lambda_n}(x_{0,y^*} - x_{\lambda_n,y_n})\|_X$$
$$\leq \|N_{\lambda_n}\|\|x_{0,y^*} - x_{\lambda_n,y_n}\|_X.$$

Since $\|T^*(y^* - y_n)\|_X \leq \|T\|\delta_n = o(\lambda_n)$, we have

$$\|x_{0,y^*} - x_{\lambda_n,y_n}\|_X \geq \frac{1}{2}\|N_{\lambda_n}\|^{-1}\|x_{0,y^*}\|_X\lambda_n \text{ for all } n \text{ sufficiently large.}$$

This yields the desired result because the hypothesis $y^* \notin \mathcal{R}(T)^\perp$ implies that $x_{0,y^*} \neq 0$. $\qquad\qquad\square$

If we choose $\lambda = c\delta^p$, then the hypothesis of Lemma 3.26 is satisfied if $0 < p < 1$.

### 3.4.1 Examples

The following examples involve the operator $T$ of Example 3.16 in Section 3.2: $T : L^2(0, 1) \to L^2(0, 1)$ is defined by

$$(Tx)(s) = \int_0^1 k(s, t)x(t)\, dt, \quad 0 < s < 1,$$

where

$$k(s, t) = \begin{cases} s(1-t), & s < t, \\ t(1-s), & s \geq t. \end{cases}$$

For each example, we construct an exact solution $x^* = x^*(t)$, the corresponding exact data $y^* = y^*(t)$ $(y^* = Tx^*)$, and a sequence $\{y_n\}$ of noisy data vectors converging to $y^*$. Moreover, we will choose $\lambda_n = k\delta_n^{2/3}$, for some constant $k$, in each example. We consider three cases:

1. (Example 3.27) The exact solution $x^*$ does not belong to $\mathcal{R}(T^*)$. In this case, we know that $\|x_{\lambda_n,y_n} - x_{0,y^*}\|_X \to 0$ as $n \to \infty$, but the convergence may be arbitrarily slow. On the other hand, Lemma 3.26 implies that there exists $c > 0$ such that $\|x_{\lambda_n,y_n} - x_{0,y^*}\|_X \geq c\lambda_n = c\delta_n^{2/3}$ for all $n$ sufficiently large.

2. (Example 3.28) The exact solution $x^*$ is chosen to belong to $\mathcal{R}(T^*)$ but not to $\mathcal{R}(T^*T)$. It follows that $\|x_{\lambda,y^*} - x_{0,y^*}\| = O(\lambda^{1/2})$ and therefore, as explained on page 75,

$$\|x_{\lambda,y} - x_{0,y^*}\|_X = O\left(\delta^{\min\{1-q/2,pq\}}\right),$$

where $p = 1/2$ and $q = 2/3$; that is,

$$\|x_{\lambda,y} - x_{0,y^*}\|_X = O\left(\delta^{\min\{2/3,1/3\}}\right) = O(\delta^{1/3}).$$

Combining this result with Lemma 3.26, we see that there exist constants $c > 0$, $C > 0$ such that

$$c\delta_n^{2/3} \leq \|x_{\lambda_n,y_n} - x_{0,y^*}\|_X \leq C\delta_n^{1/3} \text{ for all } n \text{ sufficiently large.}$$

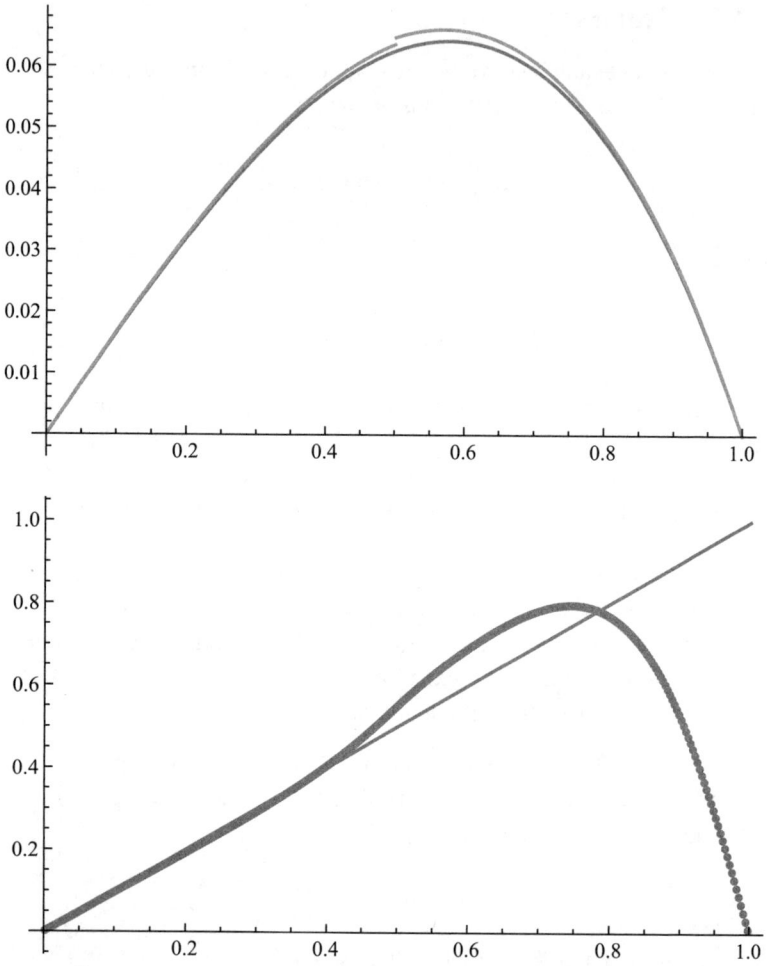

**Figure 3.7.** Example 3.27: (top) the exact and noisy data; (bottom) the exact and computed (Tikhonov) solutions. Both graphs correspond to the noise level $\delta = 10^{-3}$.

3. (Example 3.29) The exact solution $x^*$ is chosen to belong to $\mathcal{R}(T^*T)$. Therefore, by Theorem 3.24 and Lemma 3.26, there exist constants $c > 0, C > 0$ such that

$$c\delta_n^{2/3} \leq \|x_{\lambda_n, y_n} - x_{0, y^*}\|_X \leq C\delta_n^{2/3} \text{ for all } n \text{ sufficiently large.}$$

In this case, we know the exact rate of convergence of the error to zero.

It should be noted that this integral operator is convenient for constructing examples because it is straightforward to generate examples satisfying the conditions described above, and therefore to illustrate the convergence theorems.

In each example, we choose $\{y_n\}$ so that $\delta_n = \|y_n - y^*\|_Y$ and $\delta_{n+1} = \delta_n/10$ for all $n$. Specifically, we define $y_n = y^* + \delta_n z$, where

$$z(t) = \begin{cases} \alpha t, & t < \frac{1}{2}, \\ \alpha(2 - 2t), & t > \frac{1}{2}, \end{cases}$$

and $\alpha$ is chosen so that $\|z\|_{L^2} = 1$. It should be noted that the perturbation $z$ is discontinuous.

If we assume that $\|x_{\lambda_n, y_n} - x_{0, y^*}\|_X \approx a\delta_n^p$ for some $p$, then

$$\frac{\|x_{\lambda_n, y_n} - x_{0, y^*}\|_X}{\|x_{\lambda_{n+1}, y_{n+1}} - x_{0, y^*}\|_X} \approx \frac{a\delta_n^p}{a(\delta_n/10)^p} = 10^p.$$

We can therefore estimate $p$ (assuming that such a $p$ exists) by

$$p \approx \log_{10}\left(\frac{\|x_{\lambda_n, y_n} - x_{0, y^*}\|_X}{\|x_{\lambda_{n+1}, y_{n+1}} - x_{0, y^*}\|_X}\right).$$

In Example 3.27, we expect to find $p \leq 2/3$, in Example 3.28, we should have $1/3 \leq p \leq 2/3$, and in Example 3.29, $p = 2/3$ should hold.

**Example 3.27.** In this example, $x^*(t) = t$ and $y^*(t) = (t - t^3)/6$. As noted above, $x^* \notin \mathcal{R}(T^*)$. Table 3.1 shows the results of applying Tikhonov regularization with $\delta$ ranging from $10^{-1}$ to $10^{-6}$; as stated above, $\lambda$ is chosen as $k\delta^{2/3}$, and we see that $\|x_{\lambda_n, y_n} - x_{0, y^*}\|_X \approx c\delta_n^{1/12}$ as $\delta \to 0$. Figure 3.7 shows the exact and noisy data, and also the exact and computed solution, for $\delta = 10^{-3}$.

**Table 3.1.** The performance of Tikhonov regularization in
Example 3.27.

| $\delta$ | $\lambda$ | $\|x_{\lambda_n, y_n} - x_{0,y^*}\|_X$ | $\log_{10}\left(\frac{\|x_{\lambda_n, y_n} - x_{0,y^*}\|_X}{\|x_{\lambda_{n+1}, y_{n+1}} - x_{0,y^*}\|_X}\right)$ |
|---|---|---|---|
| $10^{-1}$ | $2.15443 \cdot 10^{-3}$ | 0.761624 | 0.4237340 |
| $10^{-2}$ | $4.64159 \cdot 10^{-4}$ | 0.287083 | 0.0924041 |
| $10^{-3}$ | $1.00000 \cdot 10^{-4}$ | 0.232062 | 0.0864615 |
| $10^{-4}$ | $2.15443 \cdot 10^{-5}$ | 0.190170 | 0.0835683 |
| $10^{-5}$ | $4.64159 \cdot 10^{-6}$ | 0.156882 | 0.0833420 |
| $10^{-6}$ | $1.00000 \cdot 10^{-6}$ | 0.129489 | 0.0833579 |

**Example 3.28.** We now take $x^*(t) = t - t^3$ and $y^*(t) = (7t - 10t^3 + 3t^5)/60$; the exact solution $x^*$ lies in $\mathcal{R}(T^*)$ but not in $\mathcal{R}(T^*T)$. We expect that the error will lie between $c\delta^{2/3}$ and $C\delta^{1/3}$. The results of applying Tikhonov regularization, for $\delta$ ranging from $10^{-1}$ to $10^{-6}$, are shown in Table 3.2; they are consistent with our prediction.

Figure 3.8 shows the exact and noisy data, as well as the exact and Tikhonov solution, corresponding to $\delta = 10^{-3}$.

**Example 3.29.** In this example, $x^*(t) = 7t - 10t^3 + 3t^5$ and $y^*(t) = (31t - 49t^3 + 21t^5 - 3t^7)/42$, and the exact solution $x^*$ lies in $\mathcal{R}(T^*T)$.

**Table 3.2.** The performance of Tikhonov regularization in
Example 3.28.

| $\delta$ | $\lambda$ | $\dfrac{\|x_{\lambda_n, y_n} - x_{0,y^*}\|_X}{\|x_{\lambda_{n+1}, y_{n+1}} - x_{0,y^*}\|_X}$ | $\log_{10}\left(\dfrac{\|x_{\lambda_n, y_n} - x_{0,y^*}\|_X}{\|x_{\lambda_{n+1}, y_{n+1}} - x_{0,y^*}\|_X}\right)$ |
|---|---|---|---|
| $10^{-1}$ | $2.15443 \cdot 10^{-3}$ | 0.7509270 | 0.9009170 |
| $10^{-2}$ | $4.64159 \cdot 10^{-4}$ | 0.0943368 | 0.793200 |
| $10^{-3}$ | $1.00000 \cdot 10^{-4}$ | 0.0151873 | 0.588282 |
| $10^{-4}$ | $2.15443 \cdot 10^{-5}$ | 0.00391922 | 0.505028 |
| $10^{-5}$ | $4.64159 \cdot 10^{-6}$ | 0.00122510 | 0.432564 |
| $10^{-6}$ | $1.00000 \cdot 10^{-6}$ | 0.000452488 | 0.420001 |

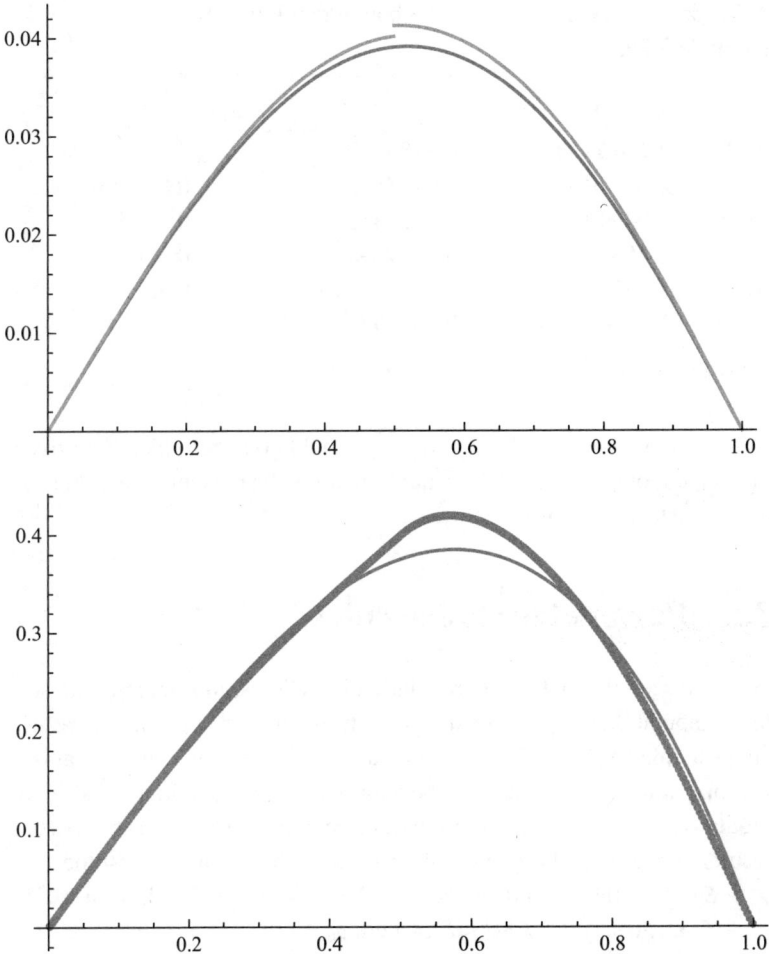

**Figure 3.8.** Example 3.28: (top) the exact and noisy data; (bottom) the exact and computed (Tikhonov) solutions. Both graphs correspond to the noise level $\delta = 10^{-3}$.

**Table 3.3.** The performance of Tikhonov regularization in
Example 3.29.

| $\delta$ | $\lambda$ | $\frac{\|x_{\lambda_n, y_n} - x_{0,y^*}\|_X}{\|x_{\lambda_{n+1}, y_{n+1}} - x_{0,y^*}\|_X}$ | $\log_{10}\left(\frac{\|x_{\lambda_n, y_n} - x_{0,y^*}\|_X}{\|x_{\lambda_{n+1}, y_{n+1}} - x_{0,y^*}\|_X}\right)$ |
|---|---|---|---|
| $10^{-1}$ | $2.15443 \cdot 10^{-3}$ | 0.5184060 | 1.019600 |
| $10^{-2}$ | $4.64159 \cdot 10^{-4}$ | 0.0495532 | 0.582969 |
| $10^{-3}$ | $1.00000 \cdot 10^{-4}$ | 0.0129450 | 0.533578 |
| $10^{-4}$ | $2.15443 \cdot 10^{-5}$ | 0.0037890 | 0.628532 |
| $10^{-5}$ | $4.64159 \cdot 10^{-6}$ | 0.000891236 | 0.655026 |
| $10^{-6}$ | $1.00000 \cdot 10^{-6}$ | 0.000197227 | 0.675105 |

It follows that the error $\|x_{\lambda,y} - x_{0,y^*}\|_X$ should go to zero like $\delta^{2/3}$ (given that we choose $\lambda_n = \delta_n^{2/3}$). The numerical results presented in Table 3.3 are consistent with this conclusion.

## 3.5   Parameter choice rules

We have seen that Tikhonov regularization allows, in principle, the stable computation of approximate solutions to an inverse problem. We say "in principle" because there is a critical issue—the choice of the regularization parameter $\lambda$—that must be addressed in order to make Tikhonov regularization a practical computational method. We have some partial results concerning the choice of $\lambda$; to describe them, we assume that $y^* \in D(T^\dagger)$ is the exact data vector, $x^* = x_{0,y^*} = T^\dagger y^*$, and, for $n \in \mathbb{Z}^+$, $y_n \in Y$ satisfies $\|y_n - y^*\|_Y \le \delta_n$, where $\delta_n \to 0^+$.

- If $\{\lambda_n\} \subset \mathbb{R}^+$ satisfies $\lambda_n \to 0$ and $\delta_n^2/\lambda_n \to 0$ as $n \to \infty$, then $x_{\lambda_n, y_n} \to x_{0,y^*}$ as $n \to \infty$ (Theorem 3.19).

- If $\{\lambda_n\} \subset \mathbb{R}^+$ satisfies $\lambda_n \to 0$ and $\delta_n^2/\lambda_n$ is bounded as $n \to \infty$, then $x_{\lambda_n, y_n} \to x_{0,y^*}$ weakly as $n \to \infty$ (Theorem 3.20).

- If $x_{0,y^*} \in \mathcal{R}(T^*)$ and $\lambda_n = k\delta_n$, then $\|x_{\lambda_n, y_n} - x_{0,y^*}\|_X = o(\delta_n^{1/2})$ (Theorem 3.24). Here $k$ is a positive constant.

- If $x_{0,y^*} \in \mathcal{R}(T^*T)$ and $\lambda_n = k\delta_n^{2/3}$, then $\|x_{\lambda_n,y_n} - x_{0,y^*}\|_X = O(\delta_n^{2/3})$ (Theorem 3.24). Here $k$ is a positive constant.

In a typical application, we do not have a sequence of data vectors converging to the exact data vector; instead, we have one particular noisy data vector $y$, with $\|y - y^*\|_Y$ unknown. In many cases, we might have an upper bound $\delta$ on $\|y - y^*\|_Y$. However, even if $\delta$ is relatively accurate, the results given above do not provide much guidance because they do not completely specify $\lambda$. For example, to apply the last result, we must choose $\lambda = k\delta^{2/3}$ for some constant $k$, but it should be clear that the choice of $k$ is critical, and the theoretical result provides no guidance. In addition, we may not know whether the assumption $x_{0,y^*} \in \mathcal{R}(T^*T)$ is true or not.

For these reasons, the theoretical results that we have derived in this chapter are of little use in choosing a parameter value for Tikhonov regularization in a specific problem. Below we present two parameter choice methods that have been shown to work well in applications. In order to discuss these and other rules, we establish the following terminology.

**Definition 3.30.** A *parameter choice rule* is a function $\lambda : \mathbb{R}^+ \times Y \to [0, \infty)$ that produces a value $\lambda = \lambda(\delta, y)$ for the regularization weight given an estimate $y \in Y$ of the exact data $y^*$ and a bound $\delta$ on the error in $y$. Moreover:

- $\lambda$ is an *a priori* rule if it depends only on $\delta$: $\lambda = \lambda(\delta)$;

- $\lambda$ is an *a posteriori* rule if it depends on $\delta$ and $y$: $\lambda = \lambda(\delta, y)$;

- $\lambda$ is a *purely a posteriori* rule if it depends only on $y$: $\lambda = \lambda(y)$.

If, for a particular $\delta > 0$ and $y \in Y$, $\lambda(\delta, y) = 0$, this is interpreted as

$$x_{\lambda(\delta,y),y} = x_{0,y} = T^\dagger y.$$

Of course, this is a valid regularization weight only if $y \in D(T^\dagger)$.

A good parameter choice rule must be convergent in the following sense.

**Definition 3.31.** Let $\lambda : \mathbb{R}^+ \times Y \to [0, \infty)$ be a parameter choice rule for Tikhonov regularization. We say that $\lambda$ is a *convergent* rule if whenever

$$y^* \in D(T^\dagger), \ \{y_n\} \subset Y, \ \{\delta_n\} \subset \mathbb{R}^+$$

satisfy $\delta_n \to 0$, $\|y_n - y^*\|_Y \le \delta_n$ for all $n \in \mathbb{Z}^+$, then $\lambda_n = \lambda(\delta_n, y_n)$ satisfies $\lambda_n \to 0$ and $x_{\lambda_n, y_n} \to x_{0, y^*}$.

We have already identified convergent *a priori* rules. For instance, given any $k > 0$ and $p \in (0, 2)$, the rule $\lambda(\delta) = k\delta^p$ is convergent. However, as we noted above, the fact that this rule is convergent does not imply that it produces a good value of the regularization parameter for any particular estimate $y$ of $y^*$.

It would seem that a purely *a posteriori* rule would be most useful, since such a rule does not require knowing a bound on the error in the data. However, the following fundamental result shows that no purely *a posteriori* rule can be convergent, except in the trivial case that $\mathcal{R}(T)$ is closed and hence $T^\dagger$ is bounded.

**Theorem 3.32.** *Let $\lambda : Y \to [0, \infty)$ be a purely a posteriori parameter choice rule for Tikhonov regularization. If $\lambda$ is convergent when applied to $Tx = y$, then $\mathcal{R}(T)$ is closed.*

*Proof.* Let $y^* \in D(T^\dagger)$ be given and define $\{y_n\} \subset Y$ by $y_n = y^*$ for all $n \in \mathbb{Z}^+$. Since the parameter choice rule $\lambda$ is convergent, we have $x_{\lambda_n, y_n} \to x_{0, y^*}$ as $n \to \infty$, where $\lambda_n = \lambda(y_n)$. Since $y_n$ is constant with respect to $n$, so is $\lambda_n$ and hence also $x_{\lambda_n, y_n}$. But then $x_{\lambda_n, y_n} \to x_{0, y^*}$ implies that $x_{\lambda_n, y_n} = x_{0, y^*}$, which is only possible if $\lambda_n = 0$. It follows that $\lambda(y^*)$ must be 0 for all $y^* \in D(T^\dagger)$.

Now consider any $y^* \in D(T^\dagger)$ and any sequence $\{y_n\} \subset D(T^\dagger)$ such that $y_n \to y^*$. By assumption, we have $x_{\lambda(y_n), y_n} \to x_{0, y^*}$ as $n \to \infty$ and $\lambda(y_n) = 0$ for all $n \in \mathbb{Z}^+$ (since $y_n \in D(T^\dagger)$ for all $n$). Therefore, $x_{\lambda(y_n), y_n} = T^\dagger y_n$ and we obtain $T^\dagger y_n \to T^\dagger y^*$. Since this holds for all $y^* \in D(T^\dagger)$ and all sequences $\{y_n\}$ in $D(T^\dagger)$ converging to $y^*$, it follows that $T^\dagger$ is continuous and hence, by Corollary 2.17, that $\mathcal{R}(T)$ is closed. $\square$

Based on the previous theorem, a purely *a posteriori* parameter choice rule will be, at best, a heuristic: it may work well for many problems, but it cannot be proven to be effective for all problems. We are going to present a heuristic rule, namely, the *L-curve method*. First, however, we will describe an *a posteriori* rule that can be proved to be convergent.

## 3.5.1 The discrepancy principle

For reasons that will become clear, the discrepancy principle assumes that $y^*$ belongs to $\mathcal{R}(T)$ and hence that $x^* = x_{0,y^*} = T^\dagger y^*$ is the minimum-norm solution (rather than least-squares solution) of $Tx = y^*$. The discrepancy principle is based on the following simple idea: if all we know is that $\|y - y^*\|_Y$ is bounded by $\delta$, then there is no reason to try to reduce $\|Tx - y\|_Y$ to less than $\delta$, especially if this requires a large $x$ (after all, we are trying to compute the minimum-norm solution of $Tx = y$). Therefore, it makes sense to replace the original problem by

$$\min \|x\|_X$$
$$\text{s.t. } \|Tx - y\|_Y \le \delta. \tag{3.7}$$

**Lemma 3.33.** *If $y^* \in \mathcal{R}(T)$ and $\|y - y^*\|_Y \le \delta$, then (3.7) has a unique solution.*

*Proof.* We first show that (3.7) has a solution. The *feasible set*

$$F_\delta = \{x \in X \ : \ \|Tx - y\|_Y \le \delta\}$$

is nonempty because $x^*$ is feasible (that is, satisfies the constraint for the problem (3.7)). Let $\{x_n\} \subset F_\delta$ be a minimizing sequence for (3.7); then it is clear that $\{x_n\}$ is bounded and hence we can assume, without loss of generality, that $x_n \to \bar{x}$ weakly. We then have $\|Tx_n - y\|_Y \le \delta$ for all $n \in \mathbb{Z}^+$ and $Tx_n - y \to T\bar{x} - y$ weakly. It follows from the weak lower semicontinuity of the norm that

$$\|T\bar{x} - y\|_Y \le \liminf_{n \to \infty} \|Tx_n - y\|_Y \le \delta,$$

and hence $\bar{x} \in F_\delta$. Moreover, the weak semicontinuity of the norm also yields

$$\|\bar{x}\|_X \leq \liminf_{n \to \infty} \|x_n\|_X = \inf \{\|x\| : x \in F_\delta\}.$$

This shows that $\bar{x}$ is a solution of (3.7). The uniqueness of the solution follows from the strict convexity of the norm; if there exist two solutions $\bar{x}_1, \bar{x}_2 \in F_\delta$ of (3.7), $\bar{x}_1 \neq \bar{x}_2$, then, for each $\bar{x} = \theta \bar{x}_1 + (1 - \theta)\bar{x}_2$, $\theta \in (0, 1)$,

$$\|\bar{x}\|_X < \theta \|\bar{x}_1\|_X + (1 - \theta)\|\bar{x}_2\|_X = \|\bar{x}_1\|_X,$$

a contradiction. Thus the solution of (3.7) is unique. $\qquad \square$

We now show that (3.7) is equivalent to

$$\begin{aligned} &\min \|x\|_X \\ &\text{s.t. } \|Tx - y\|_Y = \delta. \end{aligned} \tag{3.8}$$

**Lemma 3.34.** *If $y^* \in \mathcal{R}(T)$ and $\|y - y^*\|_Y \leq \delta$, then (3.7) and (3.8) are equivalent.*

*Proof.* It suffices to show that the solution $\bar{x}$ of (3.7) cannot satisfy $\|T\bar{x} - y\|_Y < \delta$. We argue by contradiction and suppose that this strict inequality does hold. Then, by continuity, there exists $\epsilon > 0$ such that $\|T(\tau\bar{x}) - y\|_Y \leq \delta$ for all $\tau \in [1 - \epsilon, 1 + \epsilon]$. But this implies that $(1 - \epsilon)\bar{x}$ belongs to $F_\delta$, a contradiction since

$$\|(1 - \epsilon)\bar{x}\|_X = (1 - \epsilon)\|\bar{x}\|_X < \|\bar{x}\|_X. \qquad \square$$

We are now going to show that (3.8) (and hence (3.7)) amounts to an *a posteriori* parameter choice rule for Tikhonov regularization. We assume that $y^*$ belongs to $\mathcal{R}(T)$, $y \in Y$ satisfies $\|y - y^*\|_Y \leq \delta$, and $\|y\|_Y > \delta$. The last assumption is natural; if it fails, it implies that the error in $y$ could be 100%, in which case one would not expect to compute from $y$ any valid estimate of $T^\dagger y^*$.

**Theorem 3.35.** *Suppose $y^* \in \mathcal{R}(T)$, $y \in Y$, $\|y - y^*\|_Y \leq \delta$, $\|y\|_Y > \delta$, and $y \notin \mathcal{R}(T)^{\perp}$. Then there exists a unique $\lambda \geq 0$ such that $x_{\lambda,y}$ is the unique solution of (3.8), where*

$$x_{\lambda,y} = \begin{cases} N_{\lambda}^{-1} T^* y, & \text{if } \lambda > 0, \\ T^{\dagger} y, & \text{if } \lambda = 0. \end{cases}$$

*Moreover, $\lambda = 0$ holds if and only if $\|y - y^*\|_Y = \delta$ and $y - y^* \in \mathcal{R}(T)^{\perp}$.*

*Proof.* First suppose that $\|y - y^*\|_Y = \delta$ and $y - y^* \in \mathcal{R}(T)^{\perp}$. Then $\text{proj}_{\overline{\mathcal{R}(T)}} y = y^*$, which implies that $T T^{\dagger} y = y^*$, and for any $x \in X$,

$$\|Tx - y\|_Y^2 = \|Tx - y^*\|_Y^2 + \|y^* - y\|_Y^2 \text{ (by the Pythagorean theorem)}$$
$$= \|Tx - y^*\|_Y^2 + \delta^2.$$

This shows that the only solutions of $\|Tx - y\|_Y = \delta$ belong to the set $T^{\dagger} y + \mathcal{N}(T)$, and clearly the solution of (3.8) is $x_{0,y} = T^{\dagger} y$.

We now suppose that $\|y - y^*\|_Y < \delta$ or $y - y^* \notin \mathcal{R}(T)^{\perp}$. In either case, $\|\text{proj}_{\overline{\mathcal{R}(T)}} y - y\|_Y < \delta$ must hold. This will be used below. We use basic Lagrange multiplier theory. The problem (3.8) can be written as

$$\min f(x)$$
$$\text{s.t. } g(x) = 0,$$

where

$$f(x) = \frac{1}{2} \|x\|_X^2 = \frac{1}{2} \langle x, x \rangle_X,$$
$$g(x) = \frac{1}{2} \|Tx - y\|_Y^2 - \frac{1}{2} \delta^2 = \frac{1}{2} \langle Tx - y, Tx - y \rangle_Y - \frac{1}{2} \delta^2.$$

Since $f$ and $g$ are convex functions and (3.8) is equivalent to the convex problem (3.7), it follows that the first-order optimality conditions are both necessary and sufficient. Moreover, since $f$ is strictly convex, any solution is unique. The optimality conditions are

$$\nabla f(x) = \theta \nabla g(x),$$
$$g(x) = 0,$$
$$\theta \leq 0,$$

where $\theta$ is the Lagrange multiplier. We have $\nabla f(x) = x$, $\nabla g(x) = T^*Tx - T^*y$. Therefore, the solution $\bar{x}$ is characterized by

$$\bar{x} = \theta \left( T^*T\bar{x} - T^*y \right),$$

$$\|T\bar{x} - y\|_Y = \delta,$$

$$\theta \le 0.$$

Moreover, $\theta = 0$ implies that $\bar{x} = 0$, which in turn implies that $\|y\|_Y \le \delta$, contradicting the hypothesis that $\|y\|_Y > \delta$. Thus $\theta < 0$ must hold, and defining $\lambda = -\theta^{-1}$, the Lagrange multiplier equation can be rearranged to yield

$$(T^*T + \lambda I)\bar{x} = T^*y.$$

Therefore, there exists $\lambda > 0$ such that the solution $\bar{x}$ equals the Tikhonov solution $x_{\lambda,y}$.                                                                    $\square$

The special case mentioned in Theorem 3.35 ($\|y - y^*\|_Y = \delta$, $y - y^* \in \mathcal{R}(T)^\perp$) can be avoided by assuming that $\|y - y^*\|_Y$ is strictly less than $\delta$ or, equivalently, posing (3.7) as

$$\min \|x\|_X$$

$$\text{s.t. } \|Tx - y\|_Y \le \tau\delta,$$

where $\tau > 1$ is a constant.

As Theorem 3.35 shows, the discrepancy principle can be viewed as an *a posteriori* parameter choice rule for Tikhonov regularization. We now wish to address two questions:

- Is this parameter choice rule convergent?

- If the true solution $x^* = T^\dagger y^*$ has some extra smoothness, that is, if $x^*$ belongs to $\mathcal{R}(T^*)$ or $\mathcal{R}(T^*T)$, can we find the rate at which $\|x_{\lambda,y} - x^*\|_X$ converges to 0?

We begin by proving convergence of the method.

**Theorem 3.36.** *Suppose $y^* \in \mathcal{R}(T)$, $\{\delta_n\} \subset \mathbb{R}^+$, $\{y_n\} \subset Y$, $\|y_n - y^*\|_Y \le \delta_n$ and $\|y_n\|_Y > \delta_n$ for all $n \in \mathbb{Z}^+$, and $\delta_n \to 0$ as $n \to \infty$. Let $\lambda_n > 0$*

*satisfy the discrepancy principle, that is, $\|Tx_{\lambda_n, y_n} - y_n\|_Y = \delta_n$ for all $n \in \mathbb{Z}^+$. Then $x_{\lambda_n, y_n} \to T^\dagger y^*$ as $n \to \infty$.*

*Proof.* We will write $x_n = x_{\lambda_n, y_n}$. We begin by showing that $x_n \to x^* = T^\dagger y^*$ weakly. To do this, it suffices to show that every subsequence of $\{x_n\}$ has a subsequence converging weakly to $x^*$. For each $n$, we have

$$\|Tx_n - y_n\|_Y^2 + \lambda_n \|x_n\|_X^2 \leq \|Tx - y_n\|_Y^2 + \lambda_n \|x\|_X^2 \text{ for all } x \in X$$

and hence

$$\|Tx_n - y_n\|_Y^2 + \lambda_n \|x_n\|_X^2 \leq \|Tx^* - y_n\|_Y^2 + \lambda_n \|x^*\|_X^2$$
$$\leq \delta_n^2 + \lambda_n \|x^*\|_X^2$$

(notice that $\|Tx^* - y_n\|_Y = \|y^* - y_n\|_Y^2 \leq \delta_n$). Since $\|Tx_n - y_n\|_Y^2 = \delta_n^2$, it follows that

$$\|x_n\|_X \leq \|x^*\|_X \text{ for all } n \in \mathbb{Z}^+.$$

This shows that $\{x_n\}$ is bounded and hence every subsequence of $\{x_n\}$ has a subsequence, which we still denote by $\{x_n\}$, such that $\{x_n\}$ converges weakly to a vector $\bar{x} \in X$. Since $T$ is bounded, it follows that $Tx_n \to T\bar{x}$ weakly in $Y$. On the other hand, we have

$$\|Tx_n - y^*\|_Y \leq \|Tx_n - y_n\|_Y + \|y_n - y^*\|_Y \leq 2\delta_n \to 0 \text{ as } n \to \infty.$$

This shows that $Tx_n \to y^*$ strongly and hence that $T\bar{x} = y^*$. Moreover, $x_n = x_{\lambda_n, y_n}$ belongs to $\mathcal{N}(T)^\perp$ for all $n$ and therefore, since $\mathcal{N}(T)^\perp$ is closed and hence closed with respect to weak sequential convergence (Theorem A.35), we have $\bar{x} \in \mathcal{N}(T)^\perp$. It follows that $\bar{x} = T^\dagger y^* = x^*$. We have now shown that every subsequence of $\{x_n\}$ has a subsequence that converges weakly to $x^*$, and hence $\{x_n\}$ itself converges weakly to $x^*$.

It remains to prove that $\{x_n\}$ converges to $x^*$ strongly. We showed above that $\|x_n\|_X \leq \|x^*\|_X$ for all $n \in \mathbb{Z}^+$. Using this fact and the fact that the norm is weakly lower semicontinuous, we have

$$\|x^*\|_X \leq \liminf_{n \to \infty} \|x_n\|_X \leq \limsup_{n \to \infty} \|x_n\|_X \leq \|x^*\|_X.$$

This shows that $\lim_{n \to \infty} \|x_n\|_X = \|x^*\|_X$, which completes the proof that $x_n \to x^*$ strongly. $\qquad \square$

Before we continue with the analysis, we present an example showing that the discrepancy method is effective, given a good estimate of the level of noise in the data.

**Example 3.37.** We define $T : L^2(0, 1) \to L^2(0, 1)$ by

$$(Tx)(s) = \int_0^1 k(s, t)x(t)\,dt, \; 0 < s < 1,$$

where the kernel $k$ is defined by $k(s, t) = \sin(\pi st)$. To construct a test problem, we choose the exact solution to be

$$x^*(t) = \frac{\sin(\pi t) - \pi t \cos(\pi t)}{\pi^2 t^2}$$

and the exact data to be $y^* = Tx^*$. The function $y^*$ can be expressed in terms of the sine integral,

$$\mathrm{Si}(t) = \int_0^t \frac{\sin(\tau)}{\tau}\,d\tau,$$

as

$$y^*(t) = \frac{s(\mathrm{Si}(\pi - \pi s) + \mathrm{Si}(\pi + \pi s))}{2\pi}.$$

We construct a noisy data vector $y$ as

$$y(t) = y^*(t) + \frac{\sin(10\pi t)}{50\sqrt{2}};$$

the exact and noisy data are shown in Figure 3.9. The error in $y$ is $10^{-2}$ in the $L^2$ norm.

We then apply the discrepancy principle to choose the regularization weight. If we use $\delta = 10^{-2}$ (the exact value) in the method, we obtain

$$\lambda = 2.40575 \cdot 10^{-3}$$

and the Tikhonov solution shown on the left in Figure 3.10. If we use the overestimate $\delta = 2 \cdot 10^{-2}$ (a more likely scenario, since we are unlikely to know the exact size of the error in a real application), we obtain

$$\lambda = 3.40652 \cdot 10^{-2}$$

and the Tikhonov solution shown on the right in Figure 3.10.

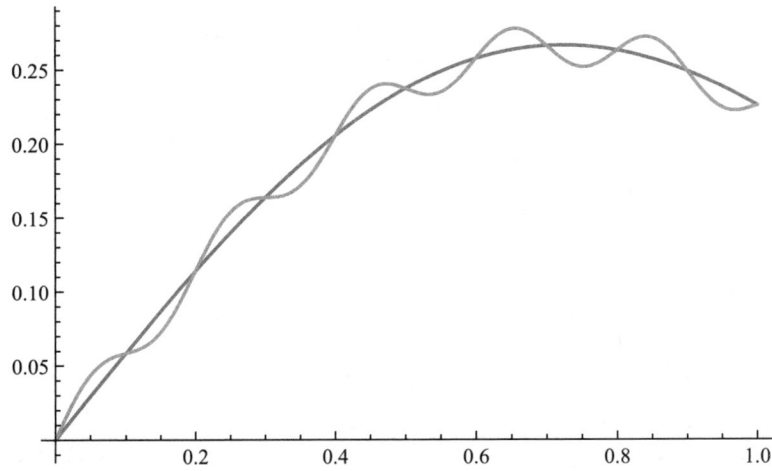

**Figure 3.9.** The exact and noisy data for Example 3.37.

With $\delta$ equal to the exact norm of the error, we obtain a very good approximation to the true solution, while the overestimate leads to a reasonable but much less accurate approximation.

It should be noted that if $\delta$ happens (by mistake) to be an underestimate of the error in the data, it may happen that no value of $\lambda$ satisfies $\|Tx_{\lambda,y} - y\|_Y = \delta$ and hence that the discrepancy principle fails to identify a value for $\lambda$.

Now we wish to derive the rate at which $x_{\lambda(\delta,y),y}$ converges to $x^*$ as $\delta \to 0$, when $\lambda(\delta, y)$ denotes the value of $\lambda$ chosen by the discrepancy principle. We will see that, for $x^* \in \mathcal{R}(T^*)$, the optimal rate of $O(\delta^{1/2})$ is attained. In Section 3.6, we show that this rate cannot be improved, even if $x^* \in \mathcal{R}(T^*T)$.

**Theorem 3.38.** *Suppose* $y^* \in \mathcal{R}(T)$ *and* $x^* = T^\dagger y^* \in \mathcal{R}(T^*)$. *If* $\{y_n\} \subset Y$ *and* $\{\delta_n\} \subset \mathbb{R}^+$ *satisfy* $\|y_n\|_Y > \delta_n$, $\|y_n - y^*\|_Y \leq \delta_n$, $\delta_n \to 0$ *and if, for each* $n \in \mathbb{Z}^+$, $\lambda_n$ *is chosen by the discrepancy principle* ($\|Tx_{\lambda_n,y_n} - y_n\|_Y = \delta_n$), *then*

$$\|x_{\lambda_n,y_n} - x^*\|_X = O(\delta_n^{1/2}).$$

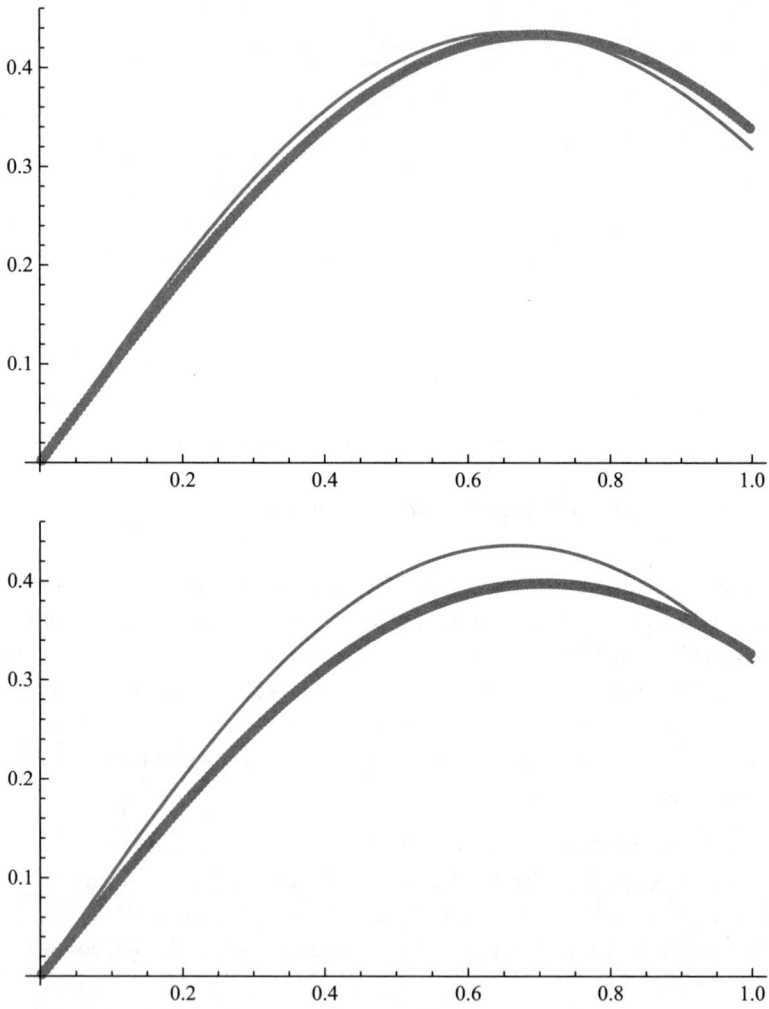

**Figure 3.10.** Example 3.37: (top) the Tikhonov solution computed by the discrepancy principle with $\delta$ equal to the exact norm of the error; (bottom) the Tikhonov solution computed by the discrepancy principle with $\delta$ equal to twice the exact norm of the error.

*Proof.* Let us write $x_n = x_{\lambda_n, y_n}$ for each $n \in \mathbb{Z}^+$. We saw in the proof of Theorem 3.36 that $\|x_n\|_X \leq \|x^*\|_X$ for all $n$. Therefore,

$$
\begin{aligned}
\|x_n - x^*\|_X^2 &= \langle x_n - x^*, x_n - x^* \rangle_X \\
&= \|x_n\|_X^2 - 2\langle x_n, x^* \rangle_X + \|x^*\|_X^2 \\
&\leq 2 \left( \|x^*\|_X^2 - \langle x_n, x^* \rangle_X \right) \\
&= 2\langle x^* - x_n, x^* \rangle_X.
\end{aligned}
$$

Thus, if $x^* = T^* v$, $v \in Y$, we obtain

$$
\begin{aligned}
\|x_n - x^*\|_X^2 &\leq 2\langle x^* - x_n, T^* v \rangle_X \\
&= 2\langle Tx^* - Tx_n, v \rangle_Y \\
&= 2 \left( \langle Tx^* - y_n, v \rangle_Y + \langle y_n - Tx_n, v \rangle_Y \right) \\
&\leq 2 \left( \|Tx^* - y_n\|_Y \|v\|_Y + \|y_n - Tx_n\|_Y \|v\|_Y \right) \\
&\leq 2\delta_n \|v\|_Y
\end{aligned}
$$

(using $\|Tx^* - y_n\|_Y = \|y^* - y_n\|_Y \leq \delta_n$ and $\|Tx_n - y_n\|_Y = \delta_n$). It follows that

$$
\|x_n - x^*\|_X \leq C\sqrt{\delta_n},
$$

where $C = \sqrt{2\|v\|_Y}$.                                                    $\square$

## 3.5.2 The L-curve method

We know that if $y \notin D(T^\dagger)$, then $Tx_{\lambda, y} \to \bar{y} = \mathrm{proj}_{\overline{\mathbb{R}(T)}} y$ and $\|x_{\lambda, y}\|_X \to \infty$ as $\lambda \to 0^+$. The L-curve method is a purely *a posteriori* parameter choice method that is based on the following empirical observations: as $\lambda$ is decreased, the residual $\|Tx_{\lambda, y} - y\|_Y$ is initially reduced significantly without a large increase in $\|x_{\lambda, y}\|_X$. However, there comes a point at which any further decrease in $\|Tx_{\lambda, y} - y\|_Y$ comes only with a large increase in $\|x_{\lambda, y}\|_X$. It therefore makes sense to choose a value of $\lambda$ that corresponds to this transition point, which corresponds to a corner on the parametrized curve $(\|Tx_{\lambda, y} - y\|_Y, \|x_{\lambda, y}\|_X)$. Because of the typical shape of this curve, it is referred to as the *L-curve*. The following example is typical.

**Example 3.39.** Consider the equation $Tx = y$, where $T : L^2(0, \pi/2) \to L^2(0, \pi/2)$ is the integral operator defined by

$$(Tx)(s) = \int_0^{\pi/2} e^{s \cos(t)} x(t)\, dt, \ 0 \le s \le \frac{\pi}{2}.$$

We take the exact data to be $y^*(s) = \sinh(s)/s$, which means that the exact solution is $x^*(t) = \sin(t)$. We add random noise to $y^*$ to produce $y$ and compute the curve $(\|Tx_{\lambda,y} - y\|_Y, \|x_{\lambda,y}\|_X)$ for $\lambda$ in the interval $[10^{-10}, 1]$. The result is shown in Figure 3.11, where the characteristic "L" shape is evident. A close-up of the corner is also shown in Figure 3.11.

As suggested above, the L-curve parameter choice method is based on choosing $\lambda$ to correspond to the corner of the curve $(\|Tx_{\lambda,y} - y\|_Y, \|x_{\lambda,y}\|_X)$. Although the graph usually has a corner that is visually evident, it is not completely obvious how to define it. Following Hansen [12], it is typical to define the L-curve to be the curve $(\log(\|Tx_{\lambda,y} - y\|_Y), \log(\|x_{\lambda,y}\|_X))$ (that is, the curve $(\|Tx_{\lambda,y} - y\|_Y, \|x_{\lambda,y}\|_X)$ in a log-log scale), and to define the corner to be its point of maximum curvature.

If we apply these concepts to Example 3.39, the corner of the L-curve corresponds to $\lambda \approx 0.0815$; it is illustrated in Figure 3.12, which shows a close-up of the L-curve in the log-log scale. (The precise location of the point of maximum curvature is not obvious in the graph, because the horizontal and vertical scales are not equal.)

It should be emphasized that the L-curve parameter choice method is a heuristic method. There is no question that the L-curve itself contains much information about a problem. When the curve has a pronounced corner, it seems evident that the optimal choice of $\lambda$ corresponds to a point near it, and numerical experiments suggest that the corner itself (as defined to be the point of maximum curvature) usually yields a good choice for $\lambda$ (see, for example, [13], where it is also stated that choosing the value of $\lambda$ corresponding to the corner sometimes leads to slight under-regularization). Nevertheless, there are relatively few theoretical results guaranteeing the performance of the L-curve method for choosing $\lambda$. For this reason, we will not go into any further details here.

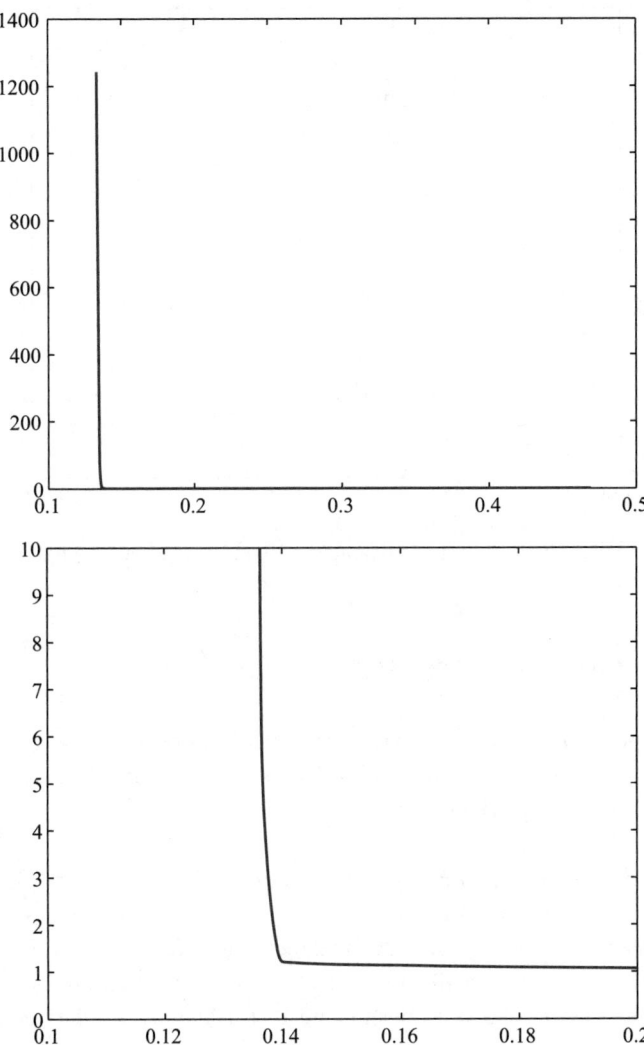

**Figure 3.11.** Top: the L-curve from Example 3.39. Bottom: a close-up of the corner of the L-curve.

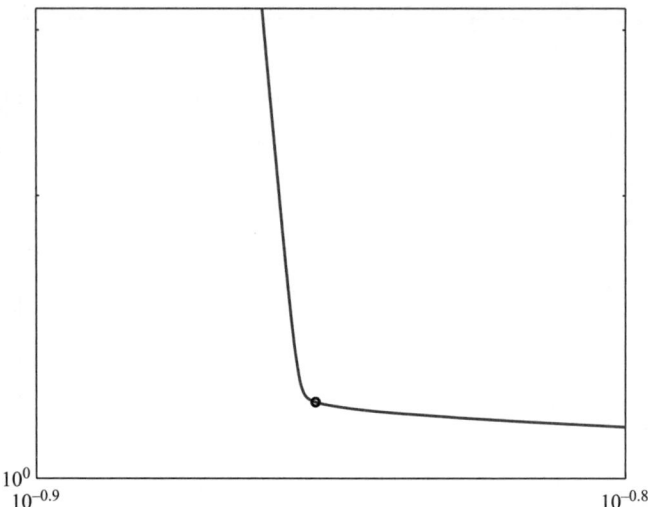

**Figure 3.12.** The L-curve from Example 3.39, in log-log scale, with the corner indicated.

## 3.6   Converse results

In the previous sections, we have derived various results about the convergence of Tikhonov regularization. There are several important conclusions that should not be missed.

*Tikhonov regularization converges when applied to exact data.*   This is expressed in Theorem 3.15: if $y \in D(T^\dagger)$, then $x_{\lambda,y} \to T^\dagger y$ as $\lambda \to 0^+$.

We can express this as a fact about the family $\{R_\lambda\}_{\lambda>0}$ ($R_\lambda = N_\lambda^{-1} T^*$) of operators that approximates $T^\dagger$: $R_\lambda \to T^\dagger$ pointwise on $D(T^\dagger)$ as $\lambda \to 0^+$. This is one way of showing that Tikhonov regularization is effective.

*Tikhonov regularization converges when applied to noisy data, provided the regularization parameter is chosen properly.*   This is expressed by two results. The first, Theorem 3.19, states that if the regularization weight goes to zero, but is not too small compared to the noise level, then $x_{\lambda,y} \to x_{0,y^*}$ strongly: if $y^* \in D(T^\dagger)$, $y_n \to y^*$, $\|y_n - y^*\|_Y \le \delta_n$ for all $n$, where $\delta_n \to 0$, and

$$\lambda_n \to 0^+, \quad \frac{\delta_n^2}{\lambda_n} \to 0 \text{ as } n \to \infty,$$

then $x_{\lambda_n,y_n} \to x_{0,y^*}$ as $n \to \infty$.

The second, Theorem 3.20, identifies the borderline case, in which the regularization weight is just large enough to guarantee weak convergence, but not large enough to imply strong convergence: if $y^* \in D(T^\dagger)$, $y_n \to y^*$, $\|y_n - y^*\|_Y \le \delta_n$ for all $n$, where $\delta_n \to 0$, and there exists $C > 0$ such that

$$\lambda_n \to 0^+, \quad \frac{\delta_n^2}{\lambda_n} \le C \text{ for all } n \in \mathbb{Z}^+,$$

then $x_{\lambda_n,y_n} \to x_{0,y^*}$ weakly as $n \to \infty$.

*With a true inverse problem, there is an inevitable loss of information (amplification of error) when passing from the data to the solution.* This is demonstrated by Theorem 3.24: If $y^* \in D(T^\dagger)$, $x_{0,y^*} \in \mathcal{R}(T^*)$, and $\lambda$ is chosen as $\lambda = k\delta$, where $\|y - y^*\|_Y \le \delta$, then $\|x_{\lambda,y} - x_{0,y^*}\|_X = O(\delta^{1/2})$. If $x_{0,y^*} \in \mathcal{R}(T^*T)$ and $\lambda$ is chosen as $\lambda = k\delta^{2/3}$, then $\|x_{\lambda,y} - x_{0,y^*}\|_X = O(\delta^{2/3})$. These are the optimal rates of convergence under the stated conditions on $x_{0,y^*}$.

The careful reader may have noticed that the interpretations given above are not fully justified by the theorems we have proven. Specifically, our method of proof implies that the regularization weight cannot be too small (compared to the noise level) in order for convergence to occur, and similarly the rates of convergence stated above are only the optimal rates derivable by our method of proof. But it may be the case that these limitations are only limitations of our techniques, not limitations inherent in the inverse problems.

The point of this section is to show that the above interpretations are, in fact, correct by showing that the underlying theorems are sharp. The analysis required to do this is, in places, noticeably more delicate than in the previous sections. The reader may wish to skip the proofs on a first reading.

### 3.6.1   Preliminary results

We begin with several technical facts that are of some independent interest. The first collection of results shows that if any one of $\mathcal{R}(T)$, $\mathcal{R}(T^*)$, $\mathcal{R}(T^*T)$ fails to be closed, then all these subspaces fail to be closed.

We have

$$\mathcal{R}(T^*)^{\perp} = \mathcal{N}(T) \text{ and } \overline{\mathcal{R}(T^*)} = \mathcal{R}(T^*)^{\perp\perp} = \mathcal{N}(T)^{\perp}.$$

It follows that $\mathcal{R}(T^*T) \subset \mathcal{R}(T^*) \subset \mathcal{N}(T)^{\perp}$. Although $\mathcal{R}(T^*T)$ can be a proper subspace of $\mathcal{R}(T^*)$, we do have the following result.

**Lemma 3.40.** $\overline{\mathcal{R}(T^*T)} = \overline{\mathcal{R}(T^*)}$.

*Proof.* Since $\mathcal{R}(T^*T) \subset \mathcal{R}(T^*)$, it follows that $\overline{\mathcal{R}(T^*T)} \subset \overline{\mathcal{R}(T^*)}$. Now suppose that $x \in \overline{\mathcal{R}(T^*)}$; we must show that $x \in \overline{\mathcal{R}(T^*T)}$. Since $x \in \overline{\mathcal{R}(T^*)}$, there exists $\{y_n\} \subset \mathcal{N}(T^*)^{\perp} = \overline{\mathcal{R}(T)}$ such that $T^*y_n \to x$. For each $n$, since $y_n \in \overline{\mathcal{R}(T)}$, there exists $x_n \in X$ such that $\|Tx_n - y_n\|_Y < 1/n$. Then

$$\|T^*Tx_n - x\|_X \leq \|T^*Tx_n - T^*y_n\|_X + \|T^*y_n - x\|_X$$

$$\leq \|T^*\|\|Tx_n - y_n\|_Y + \|T^*y_n - x\|_X$$

$$\leq \frac{1}{n}\|T^*\| + \|T^*y_n - x\|_X \to 0.$$

Thus $T^*Tx_n \to x$, which shows that $x \in \overline{\mathcal{R}(T^*T)}$. $\square$

We can now show that either the sets $\mathcal{R}(T)$, $\mathcal{R}(T^*)$, and $\mathcal{R}(T^*T)$ are all closed or else they all fail to be closed.

**Theorem 3.41.** $\mathcal{R}(T^*T)$ *is closed if and only if* $\mathcal{R}(T)$ *is closed.*

*Proof.* Suppose first that $\mathcal{R}(T)$ is closed and let $u \in \overline{\mathcal{R}(T^*T)}$. Then there exists $\{x_n\} \subset \mathcal{N}(T)^{\perp}$ such that $T^*Tx_n \to u$. Since $\mathcal{R}(T)$ is closed, there exists $\gamma > 0$ such that $\langle T^*Tx, x\rangle_X \geq \gamma \|x\|_X^2$ for all $x \in \mathcal{N}(T)^{\perp}$ (see (2.6) in Section 2.4.2). It follows that

$$\|x_n\|_X^2 \le \gamma^{-1}\langle T^*Tx_n, x_n\rangle_X \le \gamma^{-1}\|T^*Tx_n\|_X\|x_n\|_X \text{ for all } n$$
$$\Rightarrow \|x_n\|_X \le \gamma^{-1}\|T^*Tx_n\|_X \text{ for all } n.$$

Since $\{T^*Tx_n\}$ is convergent, it is bounded and hence $\{x_n\}$ is bounded. Therefore, without loss of generality, there exists $x \in X$ such that $x_n \to x$ weakly in $X$. Since $T^*T$ is bounded, this implies that $T^*Tx_n \to T^*Tx$ weakly, which implies that $u = T^*Tx \in \mathcal{R}(T^*T)$. Thus, $\mathcal{R}(T^*T)$ is closed.

Conversely, suppose $\mathcal{R}(T^*T)$ is closed. Then, with $R = T^*T|_{\mathcal{N}(T)^\perp}$, we know that $R^{-1}$ is bounded (this follows from Theorem 2.18; note that $\mathcal{N}(T^*T) = \mathcal{N}(T)$). Now suppose that $y \in \overline{\mathcal{R}(T)}$. Then there exists $\{x_n\} \subset \mathcal{N}(T)^\perp$ such that $Tx_n \to y$. It follows that $T^*Tx_n \to T^*y$ and hence that $x_n \to R^{-1}T^*y$ because $R^{-1}$ is bounded. But then $Tx_n \to TR^{-1}T^*y$, which implies that $y = TR^{-1}T^*y \in \mathcal{R}(T)$. Therefore, $\mathcal{R}(T)$ is closed. $\qquad\square$

**Corollary 3.42.** $\mathcal{R}(T^*)$ is closed if and only if $\mathcal{R}(T)$ is closed.

*Proof.* Since $T = T^{**}$, it suffices to prove that $\mathcal{R}(T^*)$ is closed if $\mathcal{R}(T)$ is closed. So assume that $\mathcal{R}(T)$ is closed, which implies that $\mathcal{R}(T^*T)$ is closed. But we have

$$\mathcal{R}(T^*T) \subset \mathcal{R}(T^*) \subset \overline{\mathcal{R}(T^*)} = \overline{\mathcal{R}(T^*T)}$$

by Lemma 3.40. Since $\mathcal{R}(T^*T)$ is closed, it follows that $\mathcal{R}(T^*) = \mathcal{R}(T^*T)$ and hence that $\mathcal{R}(T^*)$ is closed. $\qquad\square$

We can use what we just proved to show that $\|N_\lambda^{-1}\|$ is exactly $\lambda^{-1}$, provided $\mathcal{R}(T)$ fails to be closed.

**Theorem 3.43.** *If $\mathcal{R}(T)$ fails to be closed, then there exists a sequence $\{v_n\}$ in $\mathcal{N}(T)^\perp$ such that*

$$\frac{\|N_\lambda^{-1}v_n\|_X}{\|v_n\|_X} \to \lambda^{-1} \text{ as } n \to \infty.$$

*Therefore, $\|N_\lambda^{-1}\| = \lambda^{-1}$; in fact, $\left\|N_\lambda^{-1}|_{\mathcal{N}(T)^\perp}\right\| = \lambda^{-1}$.*

*Proof.* Since $\mathcal{R}(T)$ is not closed, neither is $\mathcal{R}(T^*T)$ and hence there exists $\{x_n\}$ in $\mathcal{N}(T)^\perp$ such that $T^*Tx_n \to v \in \overline{\mathcal{R}(T^*T)} \setminus \mathcal{R}(T^*T)$. It follows that $\|x_n\|_X \to \infty$ as $n \to \infty$. Define $u_n = x_n/\|x_n\|_X$ and $v_n = N_\lambda u_n$; then

$$\|v_n\|_X = \|T^*Tu_n + \lambda u_n\|_X \leq \|T^*Tu_n\|_X + \lambda \to \lambda \text{ as } n \to \infty.$$

It follows that

$$\frac{\|N_\lambda^{-1}v_n\|_X}{\|v_n\|_X} = \frac{\|u_n\|_X}{\|v_n\|_X} \geq \frac{1}{\|T^*Tu_n\|_X + \lambda} \to \frac{1}{\lambda} \text{ as } n \to \infty.$$

$\square$

## 3.6.2   The convergence of Tikhonov regularization can be arbitrarily slow

We now present the first major result of this section. At the end of Section 3.3, we stated that if all we know is that $y^* \in D(T^\dagger)$, then we cannot say anything about how quickly $x_{\lambda,y^*}$ converges to $x_{0,y^*}$; in fact, this convergence can be arbitrarily slow. The following theorem makes this precise.

**Theorem 3.44.** *Suppose $\mathcal{R}(T)$ fails to be closed, $\lambda_0 > 0$, and $f : (0, \lambda_0) \to \mathbb{R}^+$ is any function satisfying $f(\lambda) \to 0$ as $\lambda \to 0$. Then there exists $y^* \in D(T^\dagger)$ such that*

$$\left\{ \frac{\|x_{\lambda,y^*} - x_{0,y^*}\|_X}{f(\lambda)} \; : \; \lambda \in (0, \lambda_0) \right\}$$

*is unbounded.*

*Proof.* We argue by contradiction and suppose that for all $y \in D(T^\dagger)$, there exists $c_y > 0$ such that

$$\|x_{\lambda,y} - x_{0,y}\|_X \leq c_y f(\lambda) \text{ for all } \lambda \in (0, \lambda_0).$$

Given $x \in \mathcal{N}(T)^\perp$ and $y = Tx$, we have $y \in D(T^\dagger)$ and $T^\dagger y = x$. Therefore, for all $x \in \mathcal{N}(T)^\perp$, there exists $\hat{c}_x = c_y > 0$ such that

$$\|N_\lambda^{-1}T^*Tx - x\|_X \leq \hat{c}_x f(\lambda) \text{ for all } \lambda \in (0, \lambda_0).$$

Equivalently, for all $x \in \mathcal{N}(T)^{\perp}$,

$$\|f(\lambda)^{-1}(N_{\lambda}^{-1}T^*T - I)x\|_X \leq \hat{c}_x \text{ for all } \lambda \in (0, \lambda_0).$$

For the remainder of this argument, we regard $f(\lambda)^{-1}(N_{\lambda}^{-1}T^*T - I)$ as an operator mapping $\mathcal{N}(T)^{\perp}$ into itself. It now follows from the uniform boundedness principle (Theorem A.31) that there exists $c > 0$ such that

$$f(\lambda)^{-1}\|N_{\lambda}^{-1}T^*T - I\| \leq c \text{ for all } \lambda \in (0, \lambda_0)$$

or, equivalently,

$$\|N_{\lambda}^{-1}T^*T - I\| \leq cf(\lambda) \text{ for all } \lambda \in (0, \lambda_0).$$

This implies that $\|N_{\lambda}^{-1}T^*T - I\| \to 0$ as $\lambda \to 0$. On the other hand, we have seen that $N_{\lambda}^{-1}T^*T - I = -\lambda N_{\lambda}^{-1}$ and, when $\mathcal{R}(T)$ is not closed, $\left\| N_{\lambda}^{-1}\big|_{\mathcal{N}(T)^{\perp}} \right\| = \lambda^{-1}$ for all $\lambda > 0$. This implies that $\|N_{\lambda}^{-1}T^*T - I\| = 1$ for all $\lambda > 0$ and hence $\|N_{\lambda}^{-1}T^*T - I\|$ does not converge to zero as $\lambda \to 0$. This is a contradiction. $\qquad\square$

Our results about the rate of convergence of $\|x_{\lambda,y} - x_{0,y^*}\|_X$ to zero were based on bounds on how fast $\|x_{\lambda,y^*} - x_{0,y^*}\|_X$ goes to zero for the exact data $y^*$. Theorem 3.44 shows that we should not expect to be able to prove a rate of convergence in the absence of some additional assumption on $x^*$, as in Theorem 3.24.

### 3.6.3   Tikhonov regularization can fail to converge if the regularization parameter is not chosen properly

Our next goal is to show that Theorems 3.19 and 3.20 are tight, in the sense that if $\delta^2/\lambda$ is unbounded as $\delta, \lambda \to 0^+$, then $\|x_{\lambda,y}\|_X$ can be unbounded as well; in particular, then, $x_{\lambda,y}$ does not converge to $x_{0,y^*}$. Although the idea of the proof is not too difficult, the details are rather complicated. For this reason, we give an outline of the proof (albeit a simplified version) first, then explain what we have to do to turn the outline into a real proof, and then give the details.

Let us suppose that we can find a sequence $\{\lambda_n\}$ of values of the regularization parameter such that $\lambda_n \to 0^+$ and, for each $n \in \mathbb{Z}^+$, the bound of Theorem 3.7 is tight:

$$\|N_{\lambda_n}^{-1}T^*\| = \frac{1}{2\sqrt{\lambda_n}}.$$

Moreover, let us suppose that this bound is actually attained; that is, suppose that for each $n \in \mathbb{Z}^+$, there exists $z_n \in Y$ with $\|z_n\|_Y = 1$ and

$$\|N_{\lambda_n}^{-1}T^*z_n\|_X = \frac{1}{2\sqrt{\lambda_n}}. \tag{3.9}$$

Next, we choose any sequence $\{\delta_n\}$ of positive real numbers converging to zero and with the property that $\{\delta_n^2/\lambda_n\}$ is unbounded, say $\delta_n^2/\lambda_n \to \infty$ for simplicity. Then, for any $y^* \in D(T^\dagger)$, we consider $y_n = y^* + \delta_n z_n$. We have

$$
\begin{aligned}
x_{\lambda_n, y_n} - x_{0,y^*} &= N_{\lambda_n}^{-1}T^*y^* + \delta_n N_{\lambda_n}^{-1}T^*z_n - x_{0,y^*} \\
&= \delta_n N_{\lambda_n}^{-1}T^*z_n + N_{\lambda_n}^{-1}\left(T^*y^* - N_{\lambda_n}x_{0,y^*}\right) \\
&= \delta_n N_{\lambda_n}^{-1}T^*z_n + N_{\lambda_n}^{-1}\left(T^*y^* - T^*Tx_{0,y^*} - \lambda_n x_{0,y^*}\right) \\
&= \delta_n N_{\lambda_n}^{-1}T^*z_n - \lambda_n N_{\lambda_n}^{-1}x_{0,y^*},
\end{aligned}
$$

which implies that

$$
\begin{aligned}
\|x_{\lambda_n, y_n} - x_{0,y^*}\|_X &= \left\|\delta_n N_{\lambda_n}^{-1}T^*z_n - \lambda_n N_{\lambda_n}^{-1}x_{0,y^*}\right\|_X \\
&\geq \delta_n \|N_{\lambda_n}^{-1}T^*z_n\|_X - \lambda_n \|N_{\lambda_n}^{-1}x_{0,y^*}\|_X \\
&\geq \frac{d_n}{2\sqrt{\lambda_n}} - \|x_{0,y^*}\|_X.
\end{aligned}
$$

By assumption, this expression goes to infinity as $n \to \infty$, which shows that $\{x_{\lambda_n, y_n}\}$ is unbounded.

How could we show that the bound (3.9) is tight? It turns out that this has to hold if $T^*T$ has an eigenvector corresponding to a positive eigenvalue. If $T^*Tx = \lambda x$, where $\|x\|_X = 1$ and $\lambda > 0$, then

$$N_\lambda x = T^*Tx + \lambda x = \lambda x + \lambda x = 2\lambda x \implies N_\lambda^{-1}x = \frac{1}{2\lambda}x$$

and $y = Tx$ satisfies

$$\|y\|_Y^2 = \langle Tx, Tx \rangle_Y = \langle x, T^*Tx \rangle_X = \langle x, \lambda x \rangle_X = \lambda \implies \|y\|_Y = \sqrt{\lambda}.$$

It follows that

$$N_\lambda^{-1} T^* y = N_\lambda^{-1} T^* T x = \lambda N_\lambda^{-1} x = \frac{1}{2} x$$

$$\implies \|N_\lambda^{-1} T^* y\|_X = \frac{1}{2}$$

$$\implies \frac{\|N_\lambda^{-1} T^* y\|_X}{\|y\|_Y} = \frac{1/2}{\sqrt{\lambda}} = \frac{1}{2\sqrt{\lambda}},$$

as desired.

The above proof is not tenable because the operator $T^*T$ need not have any eigenvalues, much less a sequence of positive eigenvalues converging to zero. However, we can modify the proof to make it work, because it turns out that $T^*T$ must have a sequence of approximate eigenvalues that converges to zero.

**A little spectral theory**

Our analysis requires part of the spectral theory of self-adjoint operators. Rather than quoting it, which is not a prerequisite for reading this book, we develop the results we need directly.

**Definition 3.45.** Let $A : X \to X$ be a self-adjoint bounded linear operator. The *spectrum* of $A$ is the set $\sigma(A)$ of all $\lambda \in \mathbb{R}$ such that $A - \lambda I$ fails to have a bounded inverse.

For general operators, it is necessary to consider complex numbers in defining and analyzing the spectrum. The general theory, though, shows that the spectrum of a self-adjoint operator is a subset of $\mathbb{R}$, and thus we can avoid using complex numbers.

**Definition 3.46.** Let $A : X \to X$ be a self-adjoint bounded linear operator. We say that $\lambda \in \mathbb{R}$ is an *approximate eigenvalue* of $A$ if there exists $\{x_n\} \subset X$ with $\|x_n\|_X = 1$ for all $n \in \mathbb{Z}^+$ and $Ax_n - \lambda x_n \to 0$ as $n \to \infty$. We say that $\lambda \in \mathbb{R}$ is an *eigenvalue* of $A$ if there exists $x \in X, x \neq 0$, such that $Ax = \lambda x$.

It should be noted that every eigenvalue of $A$ is also an approximate eigenvalue of $A$ (as can be shown by taking $x_n = x$ for all $n$ in the definition of approximate eigenvalue). The following theorem shows that the spectrum of a self-adjoint operator consists of its approximate eigenvalues.

**Theorem 3.47.** *Let $A : X \to X$ be a self-adjoint bounded linear operator. Then $\lambda \in \sigma(A)$ if and only if $\lambda$ is an approximate eigenvalue of $A$.*

*Proof.* We first show that if $\lambda \in \sigma(A)$, then $\lambda$ is an approximate eigenvalue of $A$. If $\mathcal{N}(A - \lambda I)$ is nontrivial, then $\lambda$ is an eigenvalue of $A$ and hence an approximate eigenvalue. If $\mathcal{N}(A - \lambda I)$ is trivial, then $A - \lambda I$ is injective and hence $(A - \lambda I)^{-1}$ exists as a mapping from $\mathcal{R}(A - \lambda I)$ to $X$. Moreover,

$$\mathcal{R}(A - \lambda I)^{\perp} = \mathcal{N}((A - \lambda I)^*) = \mathcal{N}(A - \lambda I) = \{0\}$$

(using the fact that $A$ and hence $A - \lambda I$ are self-adjoint), which shows that $\mathcal{R}(A - \lambda I)$ is dense in $X$. We know that $\mathcal{R}(A - \lambda I)$ is a proper subspace of $X$, since otherwise $(A - \lambda I)^{-1}$ would be bounded by the bounded inverse theorem (Theorem A.30), contradicting that $\lambda \in \sigma(A)$. Thus $\mathcal{R}(A - \lambda I)$ fails to be closed and Theorem 2.20 implies that $\lambda$ is an approximate eigenvalue of $A$.

Conversely, suppose $\lambda \notin \sigma(A)$. Then $(A - \lambda I)^{-1}$ exists and is bounded. For any $\{x_n\} \subset X$ with $\|x_n\|_X = 1$ for all $n \in \mathbb{Z}^+$, we have

$$(A - \lambda I)x_n \to 0 \;\Rightarrow\; x_n = (A - \lambda I)^{-1}(A - \lambda I)x_n \to 0,$$

a contradiction. Therefore $\lambda$ is not an approximate eigenvalue.  $\square$

The spectral theory of general (non-self-adjoint) operators is more complicated, not only because complex numbers must be considered, but also because not every point in the spectrum is an approximate eigenvalue. Fortunately, we need only some results for self-adjoint operators to apply to $T^*T$ and $TT^*$.

We will use the following facts about the spectrum.

**Theorem 3.48.** *Let $A : X \to X$ be a self-adjoint bounded linear operator.*

1. *Either $\lambda = \|A\|$ or $\lambda = -\|A\|$ belongs to the spectrum of $A$. (Hence $\sigma(A)$ is always nonempty.)*

2. $\|A\| = \sup\{|\lambda| \ : \ \lambda \in \sigma(A)\}.$

*Proof.*

1. By Lemma 3.9, we have

$$\|A\| = \sup\{|\langle x, Ax \rangle_X| \ : \ x \in X, \|x\|_X = 1\}.$$

Therefore, there exists $\{x_n\} \subset X$, with $\|x_n\|_X = 1$ for all $n \in \mathbb{Z}^+$, such that $|\langle x_n, Ax_n \rangle_X| \to \|A\|$ as $n \to \infty$. We will assume that $\langle x_n, Ax_n \rangle_X \to \|A\|$; the proof in the alternate case is similar. We have

$$\|Ax_n - \|A\|x_n\|_X^2 = \langle Ax_n - \|A\|x_n, Ax_n - \|A\|x_n \rangle_X$$
$$= \|Ax_n\|_X^2 - 2\|A\|\langle x_n, Ax_n \rangle_X + \|A\|^2\|x_n\|_X^2$$
$$\leq \|A\|^2 - 2\|A\|\langle x_n, Ax_n \rangle_X + \|A\|^2,$$

and this last expression converges to zero as $n \to \infty$. This shows that $\|A\|$ is an approximate eigenvalue of $A$ and hence belongs to $\sigma(A)$. If, instead, $\langle x_n, Ax_n \rangle_X \to -\|A\|$, we can show that $-\|A\|$ is an approximate eigenvalue of $A$.

2. By the first part of the theorem, it suffices to prove that if $\lambda \in \mathbb{R}$ satisfies $|\lambda| > \|A\|$, then $\lambda \notin \sigma(A)$. But this is easy: if $\|x_n\|_X = 1$ for all $n \in \mathbb{Z}^+$, then

$$\|Ax_n - \lambda x_n\|_X \geq |\lambda|\|x_n\|_X - \|Ax_n\|_X|$$
$$= ||\lambda| - \|Ax_n\|_X| \geq |\lambda| - \|Ax_n\|_X$$
$$\geq |\lambda| - \|A\|_X > 0.$$

This implies that $\|Ax_n - \lambda x_n\|_X$ does not converge to zero and hence $\lambda$ cannot be an approximate eigenvalue of $A$. $\qquad\square$

## More preliminary results

We are working toward the proof that, if $\mathcal{R}(T)$ fails to be closed, there exists a sequence of positive approximate eigenvalues of $T^*T$ that converges to 0. To prove this, we will show that if it is not the case, then $\|(N_\lambda^{-1}T^*)^*(N_\lambda^{-1}T^*)\| = \|N_\lambda^{-1}T^*\|^2$ is bounded as $\lambda \to 0^+$. The following fact shows that this cannot happen when $\mathcal{R}(T)$ is not closed.

**Theorem 3.49.** *If* $\|N_\lambda^{-1}T^*\|$ *is bounded as* $\lambda \to 0^+$*, then* $\mathcal{R}(T)$ *is closed.*

*Proof.* We know that $N_\lambda^{-1}T^* \to T^\dagger$ as $\lambda \to 0^+$ pointwise on $D(T^\dagger)$. If $\{\|N_\lambda^{-1}T^*\|\}$ is bounded, say $\|N_\lambda^{-1}T^*\| \le M$ for all $\lambda > 0$ sufficiently small, then it follows that

$$\|T^\dagger y\|_X = \lim_{\lambda \to 0^+} \|N_\lambda^{-1}T^*y\|_X \le M\|y\|_Y \text{ for all } y \in D(T^\dagger).$$

Therefore $T^\dagger$ is bounded, and hence $\mathcal{R}(T)$ is closed. $\qquad\square$

We will also use the following facts.

**Theorem 3.50.**

1. *All the elements of* $\sigma(T^*T)$ *are nonnegative.*

2. *The nonzero elements of* $\sigma(T^*T)$ *and* $\sigma(TT^*)$ *are the same.*

3. $(T^*T + \lambda I)^{-1}T^* = T^*(TT^* + \lambda I)^{-1}.$

4. $T(T^*T + \lambda I)^{-1} = (TT^* + \lambda I)^{-1}T.$

*Proof.*

1. We have already proven that $T^*T + \lambda I$ has a bounded inverse if $\lambda > 0$, which shows that no negative number $-\lambda$ can belong to the spectrum of $T^*T$.

2. Suppose $\lambda \ne 0$ belongs to $\sigma(T^*T)$. Then there exists $\{x_n\} \subset X$ such that $\|x_n\|_X = 1$ for all $n \in \mathbb{Z}^+$ and $(T^*T - \lambda I)x_n \to 0$ as $n \to \infty$. Since $T$ is bounded, it follows that

$$T(T^*T - \lambda I)x_n \to 0 \text{ as } n \to \infty.$$

But $T(T^*T - \lambda I) = TT^*T - \lambda T = (TT^* - \lambda I)T$ (notice that in one expression, $I$ is the identity on $X$ and in the other it is the identity on $Y$), and hence we obtain

$$(TT^* - \lambda I)Tx_n \to 0 \text{ as } n \to \infty.$$

Since $\lambda \neq 0$, $(T^*T - \lambda I)x_n \to 0$ implies that $Tx_n$ is bounded away from the zero vector, at least for $n$ sufficiently large. It follows that $\lambda$ is an approximate eigenvalue of $TT^*$, that is, $\lambda \in \sigma(TT^*)$.

The proof that if $\lambda \neq 0$ belongs to $\sigma(TT^*)$, then $\lambda \in \sigma(T^*T)$ follows the same pattern.

3. We start with the desired identity and derive a sequence of equivalent equations until we obtain one that is obviously true:

$$(T^*T + \lambda I)^{-1}T^* = T^*(TT^* + \lambda I)^{-1}$$
$$\Leftrightarrow T^* = (T^*T + \lambda I)T^*(TT^* + \lambda I)^{-1}$$
$$\Leftrightarrow T^*(TT^* + \lambda I) = (T^*T + \lambda I)T^*$$
$$\Leftrightarrow T^*TT^* + \lambda T^* = T^*TT^* + \lambda T^*.$$

This completes the proof.

4. The proof of this part is similar to that of the previous part. $\qquad\square$

As noted above, we are going to prove one of the key results by using the fact that if $\|N_\lambda^{-1}T^*\|$ is bounded as $\lambda \to 0^+$, then $\mathcal{R}(T)$ is closed. Since

$$\|(N_\lambda^{-1}T^*)^*(N_\lambda^{-1}T^*)\| = \|N_\lambda^{-1}T^*\|^2,$$

we can work with the self-adjoint operator $(N_\lambda^{-1}T^*)^*(N_\lambda^{-1}T^*)$.

**Theorem 3.51.** *The spectrum of the self-adjoint operator $(N_\lambda^{-1}T^*)^*(N_\lambda^{-1}T^*)$ is*

$$\left\{ \frac{\omega}{(\omega + \lambda)^2} : \omega \in \sigma(TT^*) \right\}.$$

*Proof.* Using Theorem 3.50, we have

$$(N_\lambda^{-1}T^*)^*(N_\lambda^{-1}T^*) = T(T^*T + \lambda I)^{-1}(T^*T + \lambda I)^{-1}T^*$$
$$= (TT^* + \lambda I)^{-1}T(T^*T + \lambda I)^{-1}T^*$$
$$= (TT^* + \lambda I)^{-1}(TT^* + \lambda I)^{-1}TT^*$$
$$= (TT^* + \lambda I)^{-2}TT^*.$$

We will write $L_\lambda = (TT^* + \lambda I)^{-2}TT^*$ and use this representation throughout the proof.

We first prove that if $\omega \in \sigma(TT^*)$, then $\omega/(\omega + \lambda)^2$ belongs to the spectrum of $L_\lambda$. If $\omega \in \sigma(TT^*)$, then there exists $\{y_n\} \subset Y$ with $\|y_n\|_Y = 1$ for all $n \in \mathbb{Z}^+$ and $(TT^* - \omega I)y_n \to 0$ as $n \to \infty$. We then have

$$L_\lambda y_n - \frac{\omega}{(\omega + \lambda)^2}y_n$$
$$= (TT^* + \lambda I)^{-2}TT^*y_n - \frac{\omega}{(\omega + \lambda)^2}y_n$$
$$= (TT^* + \lambda I)^{-2}\left(TT^*y_n - \frac{\omega}{(\omega + \lambda)^2}(TT^* + \lambda I)^2 y_n\right)$$
$$= (\omega + \lambda)^{-2}(TT^* + \lambda I)^{-2}\left((\omega + \lambda)^2 TT^*y_n - \omega(TT^* + \lambda I)^2 y_n\right)$$
$$= -(\omega + \lambda)^{-2}(TT^* + \lambda I)^{-2}(\omega TT^* - \lambda^2 I)(TT^* - \omega I)y_n.$$

(In the last step, we expanded and then factored the operator

$$(\omega + \lambda)^2 TT^* - \omega(TT^* + \lambda I)^2,$$

similar to expanding and factoring the polynomial $(\omega + \lambda)^2\theta - \omega(\theta + \lambda)^2$ to obtain $-(\omega\theta - \lambda^2)(\theta - \omega)$.) Since $(TT^* - \omega I)y_n \to 0$ as $n \to \infty$, it follows that

$$L_\lambda y_n - \frac{\omega}{(\omega + \lambda)^2}y_n \to 0 \text{ as } n \to \infty$$

also holds. Therefore, $\omega/(\omega + \lambda)^2 \in \sigma(L_\lambda)$.

We now assume that $\theta \in \sigma(L_\lambda)$ and we wish to prove that there exists $\omega$ in $\sigma(TT^*)$ such that $\theta = \omega/(\omega + \lambda)^2$. Since $\|L_\lambda\| = \|N_\lambda^{-1}T^*\|^2$ and

$\theta \leq \|L_\lambda\|$ must hold, we know from our earlier results that $\theta \leq 1/(4\lambda)$. This will be used below. Since $\theta \in \sigma(L_\lambda)$, there exists $\{y_n\} \subset Y$ such that $\|y_n\|_Y = 1$ for all $n \in \mathbb{Z}^+$ and $L_\lambda y_n - \theta y_n \to 0$ as $n \to \infty$. We have

$$
\begin{aligned}
L_\lambda y_n - \theta y_n &= (TT^* + \lambda I)^{-2}TT^* y_n - \theta y_n \\
&= (TT^* + \lambda I)^{-2}\left(TT^* y_n - \theta(TT^* + \lambda I)^2 y_n\right) \\
&= -(TT^* + \lambda I)^{-2}\left(\theta(TT^*)^2 - (1 - 2\theta\lambda)TT^* - \theta\lambda^2 I\right) y_n \\
&= -\theta(TT^* + \lambda I)^{-2}(TT^* - \omega_1 I)(TT^* - \omega_2 I)y_n,
\end{aligned}
$$

where

$$
\omega_1 = \frac{1 - 2\theta\lambda - \sqrt{1 - 4\theta\lambda}}{2\theta}, \quad \omega_2 = \frac{1 - 2\theta\lambda + \sqrt{1 - 4\theta\lambda}}{2\theta}.
$$

Notice that $\omega_1, \omega_2$, the roots of the polynomial $\theta\omega^2 - (1 - 2\theta\lambda)\omega + \theta\lambda^2$, are real numbers because $\theta \leq 1/(4\lambda)$, and we have

$$
\theta\omega^2 - (1 - 2\theta\lambda)\omega + \theta\lambda^2 = 0 \;\Rightarrow\; \theta(\omega + \lambda)^2 - \omega = 0
$$
$$
\Rightarrow\; \frac{\omega}{(\omega + \lambda)^2} = \theta.
$$

Since $(L_\lambda - \theta I)y_n \to 0$ as $n \to \infty$ and $(TT^* + \lambda I)^{-2}$ is nonsingular, it follows that either $(TT^* - \omega_2 I)y_n \to 0$ or $\{\|(TT^* - \omega_2 I)y_n\|_Y\}$ is bounded away from zero and $(TT^* - \omega_1 I)(TT^* - \omega_2 I)y_n \to 0$ (strictly speaking, in the second case we are only guaranteed that a subsequence of $\{\|(TT^* - \omega_2 I)y_n\|_Y\}$ is bounded away from zero, but this is sufficient). In the first case, $\omega_2 \in \sigma(TT^*)$, while in the second, $\omega_1 \in \sigma(TT^*)$. Both satisfy the desired equation.  $\square$

Now we can finally show that the desired sequence of approximate eigenvalues exists.

**Theorem 3.52.** *Suppose $\mathcal{R}(T)$ fails to be closed. Then there exists a sequence $\{\theta_n^2\}$ of positive numbers such that $\theta_n^2 \to 0$ and each $\theta_n^2$ belongs to the spectrum of $T^*T$.*

*Proof.* Suppose that no such sequence $\{\theta_n^2\}$ exists. We will show that $\|N_\lambda^{-1}T^*\|$ must be bounded as $\lambda \to 0^+$ and hence, by Theorem 3.49, that $\mathcal{R}(T)$ is closed. By assumption, there exists $\gamma > 0$ such that $\sigma(T^*T) \subset \{0\} \cup [\gamma, \|T\|^2]$; by Theorem 3.50, it follows that $\sigma(TT^*) \subset \{0\} \cup [\gamma, \|T\|^2]$. From

$$\frac{\omega}{(\omega + \lambda)^2} \leq \frac{\omega}{\omega^2} = \frac{1}{\omega} \leq \frac{1}{\gamma} \text{ for all } \omega \in [\gamma, \|T\|^2]$$

it follows that

$$\|L_\lambda\| = \sup\left\{\frac{\omega}{(\omega + \lambda)^2} : \omega \in \sigma(TT^*)\right\} \leq \frac{1}{\gamma}.$$

This shows that $\|N_\lambda^{-1}T^*\|$ is bounded as $\lambda \to 0^+$.                                  $\square$

Here is the theorem we have been working toward.

**Theorem 3.53.** *If $\mathcal{R}(T)$ fails to be closed, then there exists $\{\lambda_n\} \subset \mathbb{R}^+$ such that $\lambda_n \to 0$ and $\|N_{\lambda_n}^{-1}T^*\| = 1/(2\sqrt{\lambda_n})$ for all n. Thus, in particular, if $\mathcal{R}(T)$ fails to be closed, then $\sqrt{\lambda}\|N_\lambda^{-1}T^*\|$ does not converge to zero as $\lambda \to 0^+$.*

*Proof.* Suppose $\lambda$ is an approximate eigenvalue of $T^*T$. Then there exists a sequence $\{x_n\} \subset X$ such that $\|x_n\|_X = 1$ for all $n$ and $\|T^*Tx_n - \lambda x_n\|_X \to 0$ as $n \to \infty$. We will show that $\sqrt{\lambda}\|N_\lambda^{-1}T^*\| = 1/2$. We have

$$N_\lambda x_n = T^*Tx_n + \lambda x_n = 2\lambda x_n + (T^*T - \lambda I)x_n$$
$$\Rightarrow 2\lambda N_\lambda^{-1}x_n = x_n - N_\lambda^{-1}(T^*T - \lambda I)x_n$$
$$\Rightarrow N_\lambda^{-1}x_n = \frac{1}{2\lambda}x_n - \frac{1}{2\lambda}N_\lambda^{-1}(T^*T - \lambda I)x_n.$$

If we define $y_n = Tx_n$ for all $n$, then we have

$$\|y_n\|_Y^2 = \|Tx_n\|_Y^2 = \langle x_n, T^*Tx_n\rangle_X = \langle x_n, \lambda x_n\rangle_X + \langle x_n, (T^*T - \lambda I)x_n\rangle_X$$
$$\Rightarrow \|y_n\|_Y = \sqrt{\lambda + \langle x_n, (T^*T - \lambda I)x_n\rangle_X} = \sqrt{\lambda}(1 + \epsilon_n),$$

where $\epsilon_n \to 0$ as $n \to \infty$ (because $\|(T^*T - \lambda I)x_n\|_X$ can be made arbitrarily small compared to $\lambda$ by taking $n$ large enough). We then obtain

$$
\begin{aligned}
N_\lambda^{-1} T^* y_n &= N_\lambda^{-1} T^* T x_n \\
&= N_\lambda^{-1}(\lambda x_n + (T^*T - \lambda I)x_n) \\
&= \lambda N_\lambda^{-1} x_n + N_\lambda^{-1}(T^*T - \lambda I)x_n \\
&= \frac{1}{2}x_n - \frac{1}{2}N_\lambda^{-1}(T^*T - \lambda I)x_n + N_\lambda^{-1}(T^*T - \lambda I)x_n \\
&= \frac{1}{2}x_n + \frac{1}{2}N_\lambda^{-1}(T^*T - \lambda I)x_n.
\end{aligned}
$$

It follows that

$$
\|N_\lambda^{-1} T^* y_n\|_X = \frac{1}{2}(1 + \delta_n)
$$

where $\delta_n \to 0$ as $n \to \infty$ (since $\|x_n\|_X = 1$ and $\|(T^*T - \lambda I)x_n\|_X$ can be made arbitrarily small compared to $\lambda$ by taking $n$ large enough). We thus obtain

$$
\frac{\|N_\lambda^{-1} T^* y_n\|_X}{\|y_n\|_Y} = \frac{1 + \delta_n}{2\sqrt{\lambda}(1 + \epsilon_n)} \to \frac{1}{2\sqrt{\lambda}} \text{ as } n \to \infty,
$$

and hence $\|N_\lambda^{-1} T^*\| = 1/(2\sqrt{\lambda})$, as desired.

By Theorem 3.52, there exists a decreasing sequence $\{\lambda_n\}$ converging to zero such that, for each $n$, $\lambda_n$ is an approximate eigenvalue of $T^*T$. It now follows from the above results that $\|N_{\lambda_n}^{-1} T^*\| = 1/(2\sqrt{\lambda_n})$ for each $n$. $\qquad\qquad\square$

### The failure of Tikhonov regularization

Theorems 3.19 and 3.20 guarantee convergence of $x_{\lambda,y}$ to $x_{0,y^*}$, provided $\lambda$ does not go to zero faster than $\delta^2$. The following result shows that, without this assumption, there will always be problems for which convergence does not occur.

**Theorem 3.54.** *Suppose $\mathcal{R}(T)$ fails to be closed and let $y^* \in D(T^\dagger)$. Then there exists a sequence $\{\lambda_n\} \subset \mathbb{R}^+$, $\lambda_n \to 0$, such that for any $\{\delta_n\} \subset \mathbb{R}^+$ such that $\delta_n \to \infty$ and $\delta_n^2/\lambda_n \to \infty$, there exists $\{y_n\} \subset Y$ with $\|y_n - y^*\|_Y = \delta_n$ for all $n \in \mathbb{Z}^+$ and $\|x_{\lambda_n,y_n}\|_X \to \infty$ as $n \to \infty$.*

*Proof.* We saw in Theorem 3.53 that there exists $\{\lambda_n\} \subset \mathbb{R}^+$ such that $\lambda_n \to 0$ and $\|N_\lambda^{-1}T^*\| = 1/(2\sqrt{\lambda_n})$. For each $n$, choose $z_n \in Y$ such that $\|z_n\|_Y = 1$ and

$$\|N_\lambda^{-1}T^*z_n\|_X \geq \frac{1}{2\sqrt{\lambda_n}} - 1.$$

We define $y_n = y^* + \delta_n z_n$ (which implies that $\|y_n - y^*\|_Y = \delta_n$) for all $n$. We then have

$$x_{\lambda_n,y_n} - x_{0,y^*} = x_{\lambda_n,y_n} - x_{\lambda_n,y^*} + x_{\lambda_n,y^*} - x_{0,y^*}$$

$$\Rightarrow \|x_{\lambda_n,y_n} - x_{0,y^*}\|_X \geq \|x_{\lambda_n,y_n} - x_{\lambda_n,y^*}\|_X - \|x_{\lambda_n,y^*} - x_{0,y^*}\|_X$$

$$\Rightarrow \|x_{\lambda_n,y_n} - x_{0,y^*}\|_X \geq \delta_n\|N_\lambda^{-1}T^*z_n\|_X - \|x_{\lambda_n,y^*} - x_{0,y^*}\|_X$$

$$\Rightarrow \|x_{\lambda_n,y_n} - x_{0,y^*}\|_X \geq \frac{\delta_n}{2\sqrt{\lambda_n}} - \delta_n - \|x_{\lambda_n,y^*} - x_{0,y^*}\|_X.$$

Since $\delta_n/\sqrt{\lambda_n} \to \infty$, $\delta_n \to 0$, and $x_{\lambda_n,y^*} \to x_{0,y^*}$, this shows that

$$\|x_{\lambda_n,y_n} - x_{0,y^*}\|_X \to \infty,$$

as desired.                                                                 $\square$

### 3.6.4   Converse results about the rate of convergence

We saw at the end of Section 3.4 that, for $y \in D(T^\dagger)$, the optimal rate for the error $\|x_{\lambda,y} - x_{0,y}\|_X$ to converge to 0 is $O(\lambda)$, and it is achieved if $x_{0,y} \in \mathcal{R}(T^*T)$. It turns out that this sufficient condition is necessary.

**Theorem 3.55.** *Suppose $y^* \in D(T^\dagger)$ and $\|x_{\lambda,y^*} - x_{0,y^*}\|_X = O(\lambda)$ as $\lambda \to 0^+$. Then $x_{0,y^*} \in \mathcal{R}(T^*T)$.*

*Proof.* We begin by deriving an expression for $x_{0,y^*}$:

$$x_{\lambda,y^*} - x_{0,y^*} = -\lambda N_\lambda^{-1}x_{0,y^*}$$

$$\Rightarrow \lambda N_\lambda^{-1}x_{0,y^*} = x_{0,y^*} - x_{\lambda,y^*}$$

$$\Rightarrow x_{0,y^*} = \lambda^{-1}N_\lambda(x_{0,y^*} - x_{\lambda,y^*})$$

$$\Rightarrow x_{0,y^*} = T^*T\left(\lambda^{-1}(x_{0,y^*} - x_{\lambda,y^*})\right) + x_{0,y^*} - x_{\lambda,y^*}.$$

By assumption, $\{\lambda^{-1}(x_{0,y^*} - x_{\lambda,y^*}) : \lambda > 0\}$ is a bounded set. Therefore, there exist $u \in X$ and $\{\lambda_n\} \subset \mathbb{R}^+$ such that $\lambda_n \to 0$ and $\lambda_n^{-1}(x_{0,y^*} - x_{\lambda_n,y^*}) \to u$ weakly. Since $x_{0,y^*} - x_{\lambda_n,y^*} \to 0$, it follows that

$$x_{0,y^*} = T^*T \left(\lambda_n^{-1}(x_{0,y^*} - x_{\lambda_n,y^*})\right) + x_{0,y^*} - x_{\lambda_n,y^*} \to T^*Tu \text{ weakly.}$$

This proves that $x_{0,y^*} = T^*Tu$, as desired. $\qquad\square$

We have already seen that $x_{0,y^*} \in \mathcal{R}(T^*)$ implies that $\|x_{\lambda,y^*} - x_{0,y^*}\|_X = o(\lambda^{1/2})$. We would expect that, conversely, $\|x_{\lambda,y^*} - x_{0,y^*}\|_X = o(\lambda^{1/2})$ should imply that $x_{0,y^*} \in \mathcal{R}(T^*)$. However, this turns out to be false. In the next chapter, we will use the singular value expansion to present an explicit example.

We now consider Theorem 3.24: if $x_{0,y^*} \in \mathcal{R}(T^*T)$ and the regularization parameter $\lambda$ is defined by $\lambda = k\delta^{2/3}$ ($\|y - y^*\|_Y \le \delta$), then $\|x_{\lambda,y^*} - x_{0,y^*}\|_X = O(\delta^{2/3})$. We will show that $x_{0,y^*} \in \mathcal{R}(T^*T)$ is necessary for this to hold, and will also show that $O(\delta^{2/3})$ is the best possible rate of convergence (except in the trivial case that $y^* \in \mathcal{R}(T)^{\perp}$, in which case $x_{0,y^*} = 0$).

First, we prove the necessity of the condition $x_{0,y^*} \in \mathcal{R}(T^*T)$.

**Theorem 3.56.** *Suppose $y^* \in D(T^{\dagger})$ and, given $\{y_n\} \subset Y$ with $y_n \to y$ as $n \to \infty$, $\|x_{\lambda_n,y_n} - x_{0,y^*}\|_X = O(\delta_n^{2/3})$ if $\|y_n - y^*\|_Y \le \delta_n$ for all $n \in \mathbb{Z}^+$ and $\lambda_n = k\delta_n^{2/3}$ for a positive constant $k$. Then $x_{0,y^*} \in \mathcal{R}(T^*T)$.*

*Proof.* We choose a sequence $\{\lambda_n\} \subset (0, \infty)$ with $\lambda_n \to 0$ and define, for each $n \in \mathbb{Z}^+$, $\delta_n = \lambda_n^{3/2}$ and $y_n = (1 + \delta_n)y^*$. By assumption, since $\|y_n - y^*\|_Y = \delta_n \to 0$ and $\lambda_n = \delta_n^{2/3}$, $\|x_{0,y^*} - x_{\lambda_n,y_n}\|_X = O(\delta_n^{2/3})$. Because

$$x_{0,y^*} - x_{\lambda_n,y_n} = x_{0,y^*} - (1 + \delta_n)x_{\lambda_n,y^*} = x_{0,y^*} - x_{\lambda_n,y^*} - \delta_n x_{\lambda_n,y^*},$$

we have

$$x_{0,y^*} - x_{\lambda_n,y^*} = x_{0,y^*} - x_{\lambda_n,y_n} + \delta_n x_{\lambda_n,y^*} = O(\delta_n^{2/3}) + O(\delta_n) = O(\delta_n^{2/3}).$$

Since $\lambda_n = \delta_n^{2/3}$, this shows that

$$x_{0,y^*} - x_{\lambda_n,y^*} = O(\lambda_n).$$

We have shown that this holds for all sequences $\{\lambda_n\}$ such that $\lambda_n \to 0$, and hence Theorem 3.55 implies that $x_{0,y^*} \in \mathcal{R}(T^*T)$, as desired.    $\square$

Now we show that $O(\delta^{2/3})$ is the best possible convergence rate, except in trivial cases.

**Theorem 3.57.** *Assume that $\mathcal{R}(T)$ fails to be closed. Suppose $y^* \in D(T^\dagger)$ and, given $\{y_n\} \subset Y$ with $y_n \to y$ as $n \to \infty$, there exists $\{\lambda_n\} \subset \mathbb{R}^+$ such that*

$$\|x_{\lambda_n, y_n} - x_{0, y^*}\|_X = o(\delta_n^{2/3}),$$

*where $\|y_n - y^*\|_Y \leq \delta_n$ for all $n \in \mathbb{Z}^+$. Then $y^* \in \mathcal{R}(T)^\perp$.*

*Proof.* For $\{y_n\}$, $\{\lambda_n\}$ satisfying the hypotheses of the theorem, we have

$$N_{\lambda_n}(x_{\lambda_n, y_n} - x_{0, y^*}) = N_{\lambda_n} x_{\lambda_n, y_n} - T^*T x_{0, y^*} - \lambda_n x_{0, y^*}$$
$$= T^* y_n - T^* y^* - \lambda_n x_{0, y^*}$$
$$= T^*(y_n - y^*) - \lambda_n x_{0, y^*},$$

which implies

$$\lambda_n x_{0, y^*} = T^*(y_n - y^*) - N_{\lambda_n}(x_{\lambda_n, y_n} - x_{0, y^*}).$$

Since $\|T^*(y_n - y^*)\|_X = O(\delta_n)$ and $\|N_{\lambda_n}(x_{\lambda_n, y_n} - x_{0, y^*})\|_X = o(\delta_n^{2/3})$, it follows that if $x_{0, y^*} \neq 0$, then $\lambda_n = o(\delta_n^{2/3})$ must hold.

Now we assume that $x_{0, y^*} \neq 0$. We will construct a sequence $\{y_n\}$ converging to $y^*$ and use it to obtain a contradiction, which means that $x_{0, y^*} \neq 0$ is impossible, that is, that $x_{0, y^*} = 0$ and $y^* \in \mathcal{R}(T)^\perp$ must hold. Since $\mathcal{R}(T)$ is not closed, it follows from Theorem 3.52 that there exists a sequence $\{\theta_n^2\} \subset \mathbb{R}^+$ of approximate eigenvalues of $T^*T$ with $\theta_n^2 \to 0$ as $n \to \infty$. For each $n$, we can find an approximate eigenvector $\phi_n$ and a vector $\mu_n \in X$ such that

$$T^*T\phi_n = \theta_n^2 \phi_n + \mu_n, \quad \|\mu_n\|_X \leq \theta_n^3.$$

We define $\psi_n = T\phi_n$, $\delta_n = \theta_n^3$, and

$$y_n = y^* + \frac{\delta_n}{\|\psi_n\|_Y} \psi_n.$$

Let $\{\lambda_n\}$ be any sequence satisfying the hypotheses of the theorem. We have

$$x_{\lambda_n, y_n} - x_{0, y^*} = N_{\lambda_n}^{-1} T^* y^* - x_{0, y^*} + \frac{\delta_n}{\|\psi_n\|_Y} N_{\lambda_n}^{-1} T^* \psi_n$$

$$\Rightarrow \frac{\delta_n}{\|\psi_n\|_Y} N_{\lambda_n}^{-1} T^* \psi_n = x_{\lambda_n, y_n} - x_{0, y^*} - \left( N_{\lambda_n}^{-1} T^* y^* - x_{0, y^*} \right).$$

Since $x_{0, y^*} \in \mathcal{R}(T^*T)$, it follows that

$$\left\| N_{\lambda_n}^{-1} T^* y^* - x_{0, y^*} \right\|_X = O(\lambda_n) = o(\delta_n^{2/3}).$$

Since also $\|x_{\lambda_n, y_n} - x_{0, y^*}\|_X = o(\delta^{2/3})$, we see that

$$\frac{\delta_n}{\|\psi_n\|_Y} \left\| N_{\lambda_n}^{-1} T^* \psi_n \right\|_X = o(\delta_n^{2/3}) \tag{3.10}$$

must hold. We will show that this is impossible.

We have

$$\psi_n = T\phi_n \Rightarrow T^* \psi_n = T^*T\phi_n = \theta_n^2 \phi_n + \mu_n$$
$$\Rightarrow N_{\lambda_n}^{-1} T^* \psi_n = \theta_n^2 N_{\lambda_n}^{-1} \phi_n + N_{\lambda_n}^{-1} \mu_n$$

and

$$T^*T\phi_n = \theta_n^2 \phi_n + \mu_n$$
$$\Rightarrow N_{\lambda_n} \phi_n = (\theta_n^2 + \lambda_n)\phi_n + \mu_n$$
$$\Rightarrow \phi_n = (\theta_n^2 + \lambda_n)N_{\lambda_n}^{-1} \phi_n + N_{\lambda_n}^{-1} \mu_n$$
$$\Rightarrow N_{\lambda_n}^{-1} \phi_n = (\theta_n^2 + \lambda_n)^{-1} \phi_n - (\theta_n^2 + \lambda_n)^{-1} N_{\lambda_n}^{-1} \mu_n.$$

Combining these results, we have

$$N_{\lambda_n}^{-1} T^* \psi_n = \frac{\theta_n^2}{\theta_n^2 + \lambda_n} \phi_n - \frac{\theta_n^2}{\theta_n^2 + \lambda_n} N_{\lambda_n}^{-1} \mu_n + N_{\lambda_n}^{-1} \mu_n$$

$$= \frac{\theta_n^2}{\theta_n^2 + \lambda_n} \phi_n + \frac{\lambda_n}{\theta_n^2 + \lambda_n} N_{\lambda_n}^{-1} \mu_n.$$

Recall that $\delta_n = \theta_n^3$, which implies that $\theta_n = \delta_n^{1/3}$. Therefore, $\lambda_n = o(\delta_n^{2/3}) = o(\theta_n^2)$. It follows that

$$\frac{\theta_n^2}{\theta_n^2 + \lambda_n}\|\phi_n\|_X \to 1 \text{ as } n \to \infty.$$

Since

$$\frac{\lambda_n}{\theta_n^2 + \lambda_n}\|N_{\lambda_n}^{-1}\mu_n\|_X \leq \frac{\|\mu_n\|_X}{\theta_n^2 + \lambda_n} \leq \frac{\theta_n^3}{\theta_n^2 + \lambda_n}$$

$$= \frac{\theta_n}{1 + \lambda_n\theta_n^{-2}} \to 0 \text{ as } n \to \infty,$$

we see that

$$\|N_{\lambda_n}^{-1}T^*\psi_n\|_X \to 1 \text{ as } n \to \infty.$$

We have

$$\|\psi_n\|_Y^2 = \|T\phi_n\|_Y^2 = \langle\phi_n, T^*T\phi_n\rangle_X = \theta_n^2\langle\phi_n, \phi_n\rangle_X + \langle\phi_n, \mu_n\rangle_X$$

$$= \theta_n^2 + \langle\phi_n, \mu_n\rangle_X$$

$$\leq \theta_n^2 + \theta_n^3.$$

It now follows easily that

$$\frac{\delta_n}{\|\psi_n\|_Y} \geq \frac{\delta_n}{\theta_n\sqrt{1 + \theta_n}} = \frac{\delta_n^{2/3}}{\sqrt{1 + \delta_n^{1/3}}}$$

and hence

$$\liminf_{n\to\infty} \frac{\frac{\delta_n}{\|\psi_n\|_Y}\left\|N_{\lambda_n}^{-1}T^*\psi_n\right\|_X}{\delta^{2/3}} \geq \liminf_{n\to\infty} \frac{\left\|N_{\lambda_n}^{-1}T^*\psi_n\right\|_X}{\sqrt{1 + \delta^{1/3}}} = 1.$$

This contradicts (3.10).                                                    □

## 3.6.5   The rate of convergence in the discrepancy principle

In Section 3.5.1, we showed that the discrepancy principle parameter choice rule yields $O(\delta^{1/2})$ convergence, provided $T^\dagger y^*$ belongs to $\mathcal{R}(T^*)$.

Based on our other results, it would be natural to conjecture that the rate of convergence would improve to $O(\delta^{2/3})$ if $T^\dagger y^* \in \mathcal{R}(T^*T)$. However, as we now prove, $O(\delta^{1/2})$ is the best we can do when $\lambda$ is chosen by the discrepancy principle, no matter how smooth $T^\dagger y^*$ is, except in trivial cases ($y^* = 0$ or $\mathcal{R}(T)$ is closed). We need the following preliminary result.

**Lemma 3.58.** *Suppose $y^* \in Y$ is given, the sequences $\{y_n\} \subset Y$, $\{\delta_n\} \subset (0, \infty)$ satisfy $\|y_n - y^*\|_Y \leq \delta_n$ and $\|y_n\|_Y > \delta_n$ for all $n \in \mathbb{Z}^+$, and $\lambda_n$ is chosen by the discrepancy principle. Then*

$$\lambda_n \leq \frac{\delta_n \|T\|^2}{\|y_n\|_Y - \delta_n} \text{ for all } n \in \mathbb{Z}^+$$

*and therefore $\lambda_n = O(\delta_n)$.*

*Proof.* We have, by the reverse triangle inequality,

$$\|y_n\|_Y - \delta_n = \|y_n\|_Y - \|Tx_n - y_n\|_Y \leq \|Tx_n\|_Y \text{ for all } n \in \mathbb{Z}^+.$$

We also have

$$T^*(y_n - Tx_n) = \lambda_n x_n \Rightarrow TT^*(y_n - Tx_n) = \lambda_n Tx_n$$

$$\Rightarrow \lambda_n \|Tx_n\|_X \leq \|TT^*\| \|y_n - Tx_n\|_Y = \|T\|^2 \delta_n.$$

Therefore,

$$\lambda_n \leq \frac{\|T\|^2 \delta_n}{\|Tx_n\|_X} \leq \frac{\delta_n \|T\|^2}{\|y_n\|_Y - \delta_n} \text{ for all } n \in \mathbb{Z}^+,$$

as desired.                                                                      □

We can now prove the desired result. We will see that the proof is very similar to the proof of Theorem 3.57.

**Theorem 3.59.** *Suppose $\mathcal{R}(T)$ is not closed, $y^* \in \mathcal{R}(T)$, $x^* = T^\dagger y^* \in \mathcal{R}(T^*T)$, and*

$$\|x_{\lambda(\delta, y), y} - x^*\|_X = o(\delta^{1/2}) \text{ as } \delta \to 0,$$

*where $\|y\|_Y > \delta$, $\|y - y^*\|_Y \leq \delta$, and $\lambda = \lambda(\delta, y)$ is chosen by the discrepancy principle. Then $y^* = 0$.*

*Proof.* Since $x_{0,y^*} \in \mathcal{R}(T^*T)$, Theorem 3.23 implies that $\|x_{\lambda,y^*} - x_{0,y^*}\|_X = O(\lambda)$. Let us assume that $x_{0,y^*} \neq 0$. We will construct a sequence $\{y_n\}$ converging to $y^*$ and use it to obtain a contradiction, which means that $x_{0,y^*} \neq 0$ is impossible, that is, that $x_{0,y^*} = 0$ and $y^* \in \mathcal{R}(T) \cap \mathcal{R}(T)^\perp = \{0\}$ must hold. The construction is exactly as in the proof of Theorem 3.57. Since $\mathcal{R}(T)$ is not closed, it follows from Theorem 3.52 that there exists a sequence $\{\theta_n^2\} \subset \mathbb{R}^+$ of approximate eigenvalues of $T^*T$ with $\theta_n^2 \to 0$ as $n \to \infty$. For each $n$, we can find an approximate eigenvector $\phi_n$ and a vector $\mu_n \in X$ such that

$$T^*T\phi_n = \theta_n^2 \phi_n + \mu_n, \quad \|\mu_n\|_X \leq \theta_n^3.$$

We define $\psi_n = T\phi_n$, $\delta_n = \theta_n^2$, and

$$y_n = y^* + \frac{\delta_n}{\|\psi_n\|_Y}\psi_n.$$

For each $n$, let $\lambda_n$ be defined by the discrepancy principle.

As before,

$$\frac{\delta_n}{\|\psi_n\|_Y}N_{\lambda_n}^{-1}T^*\psi_n = x_{\lambda_n,y_n} - x_{0,y^*} - \left(N_{\lambda_n}^{-1}T^*y^* - x_{0,y^*}\right).$$

Since $x_{0,y^*} \in \mathcal{R}(T^*T)$, it follows from the previous lemma that

$$\left\|N_{\lambda_n}^{-1}T^*y^* - x_{0,y^*}\right\|_X = O(\lambda_n) = O(\delta_n).$$

Since also $\|x_{\lambda_n,y_n} - x_{0,y^*}\|_X = o(\delta^{1/2})$, we see that

$$\frac{\delta_n}{\|\psi_n\|_Y}\left\|N_{\lambda_n}^{-1}T^*\psi_n\right\|_X = o(\delta_n^{1/2}) \qquad (3.11)$$

must hold. We will show that this is impossible.

Exactly as in the proof of Theorem 3.57, we find that

$$N_{\lambda_n}^{-1}T^*\psi_n = \frac{\theta_n^2}{\theta_n^2 + \lambda_n}\phi_n + \frac{\lambda_n}{\theta_n^2 + \lambda_n}N_{\lambda_n}^{-1}\mu_n.$$

Now,

$$\frac{\theta_n^2}{\theta_n^2 + \lambda_n}\|\phi_n\|_X = \frac{1}{1 + \lambda_n\theta_n^{-2}} = \frac{1}{1 + \lambda_n\delta_n^{-1}}$$

and

$$\frac{\lambda_n}{\theta_n^2 + \lambda_n} \|N_{\lambda_n}^{-1} \mu_n\|_X \leq \frac{\|\mu_n\|_X}{\theta_n^2 + \lambda_n} \leq \frac{\theta_n^3}{\theta_n^2 + \lambda_n}$$

$$= \frac{\theta_n}{1 + \lambda_n \theta_n^{-2}} \to 0 \text{ as } n \to \infty.$$

As in the proof of Theorem 3.57,

$$\|\psi_n\|_Y^2 \leq \theta_n^2 + \theta_n^3$$

and it follows that

$$\frac{\delta_n}{\|\psi_n\|_Y} \geq \frac{\delta_n}{\theta_n \sqrt{1 + \theta_n}} = \frac{\delta_n^{1/2}}{\sqrt{1 + \delta_n^{1/2}}}.$$

Therefore,

$$\liminf_{n \to \infty} \frac{\frac{\delta_n}{\|\psi_n\|_Y} \|N_{\lambda_n}^{-1} T^* \psi_n\|_X}{\delta^{1/2}} \geq \liminf_{n \to \infty} \frac{1}{\sqrt{1 + \delta^{1/2}}} \frac{1}{1 + \lambda_n \delta_n^{-1}} > 0,$$

where this last result follows from the fact that $\lambda_n = O(\delta_n)$ (which implies that $\lambda_n \delta_n^{-1}$ is bounded above). This contradicts (3.11). □

## 3.6.6 Summary

We have succeeded in showing that the results of this chapter are sharp: the regularization parameter must be chosen properly (with respect to the noise level) if Tikhonov regularization is to converge, and we have correctly identified the optimal rate of convergence. In an asymptotic sense, as the noise level goes to zero, the error in the solution is larger than the error in the data. We have described this as an inevitable loss of information or amplification of error; it is intrinsic to a true inverse problem. This is one way to characterize precisely the fact that true inverse problems are difficult to solve.

Chapter **4**

# Compact operators and the singular value expansion

The purpose of this chapter is twofold: to introduce a class of operators $T$ for which $Tx = y$ always represents a genuine inverse problem, and to re-examine our results on Tikhonov regularization using a kind of spectral representation called the *singular value expansion*. The operators that we study are called *compact* operators, and they arise frequently in applications, often in the form of a certain kind of integral operator. A compact operator (and only such an operator) can be expressed in a singular value expansion, which uses special orthogonal vectors in the domain and codomain spaces.

## 4.1 Finite-dimensional problems and the singular value decomposition

We begin by introducing the singular value decomposition for a matrix and use it to understand ill-conditioned finite-dimensional problems. This will prepare us to understand the singular value expansion in infinite-dimensional space and also shed more light on the discussion of Section 2.4.

## 4.1.1    An example with a symmetric matrix

The singular value decomposition (SVD) of a matrix is a kind of spectral (eigenvalue/eigenvector) decomposition, and we first describe a simple example for which we can simply use the eigenvalues and eigenvectors of the matrix. The example is obtained by discretizing the inverse problem presented in Section 1.2.

**Example 4.1.** We wish to estimate the function $f$ in the BVP

$$-u''(x) = f(x) \text{ for all } x \in (0, 1),$$
$$u(0) = u(1) = 0 \tag{4.1}$$

from a measurement of $u$. Let us suppose that we establish a grid on the interval $[0, 1]$ by defining $h = 1/m$ and $x_i = ih$, $i = 0, 1, 2, \ldots, m$ ($m$ is a given positive integer). We estimate $-u''(x_i)$ by the central difference approximation

$$-u''(x_i) \approx \frac{-u(x_{i-1}) + 2u(x_i) - u(x_{i+1})}{h^2}, \quad i = 1, 2, \ldots, m - 1.$$

If we now define $\mathbf{u} \in \mathbb{R}^{n-1}$ by $u_i = u(x_i)$, $i = 1, 2, \ldots, m - 1$, then the equation $-u'' = f$ is approximated by the equation $L\mathbf{u} = \mathbf{f}$, where $L$ is the $(m - 1) \times (m - 1)$ matrix

$$L = \frac{1}{h^2} \begin{bmatrix} 2 & -1 & & & \\ -1 & 2 & -1 & & \\ & \ddots & \ddots & \ddots & \\ & & -1 & 2 & -1 \\ & & & -1 & 2 \end{bmatrix}.$$

(The boundary conditions $u(0) = 0$ and $u(1) = 0$ are used in the first and last equations; this is why the first and last rows of $L$ are different from the rest.) If the components of $\mathbf{u}$ are the exact values of $u(x_1), u(x_2), \ldots, u(x_{m-1})$, then $\mathbf{f}$ contains approximations to the values $f(x_1), f(x_2), \ldots, f(x_{m-1})$; the differences $f_i - f(x_i)$ represent the discretization error. If the components of $\mathbf{u}$ are (noisy) measurements of $u(x_1), u(x_2), \ldots, u(x_{m-1})$ and we compute $\mathbf{f} = L\mathbf{u}$, then $f_i - f(x_i)$ is a combination of the discretization error and the propagated measurement error.

Now let us consider a specific example. We assume that the exact values of $u$ and $f$ are $u(x) = x\cos(\pi x/2)$ and $f(x) = -u''(t) = \pi\sin(\pi x/2) + \pi^2 x\cos(\pi t/2)/4$. We then assume that $u$ is measured on the grid $x_1, x_2, \ldots, x_{m-1}$, defined as above, with $m = 40$; to simulate this, we add uniformly distributed random numbers to the components of the exact $\mathbf{u}^*$ to produce a noisy vector $\mathbf{u}$. We scale the vector of errors so that $\|\mathbf{u} - \mathbf{u}^*\|_2 = 0.05\|\mathbf{u}^*\|_2$, where $\|\cdot\|_2$ is the Euclidean norm (that is, the error in the data is 5% in the Euclidean norm). The resulting vectors $\mathbf{u}^*$ and $\mathbf{u}$ are shown in Figure 4.1.

We now compute $\mathbf{f} = L\mathbf{u}$ and compare it to the exact vector $\mathbf{f}^*$ defined by $f_i^* = f(x_i)$, $i = 1, 2, \ldots, m - 1$. The results are also shown in Figure 4.1. The results are comparable to those we have seen earlier in this book (for example, in Section 1.1). The small errors in the data result in large errors in the computed solution, to the extent that the computed solution is completely worthless.

We are going to explain Example 4.1 in a very precise way using the eigenvalues and eigenvectors of the matrix $L$. This matrix is a symmetric matrix with real entries, and we begin by reviewing some facts about a symmetric matrix $A \in \mathbb{R}^{n\times n}$:

- every eigenvalue of $A$ is real;

- eigenvectors of $A$ that correspond to distinct eigenvalues are orthogonal;

- there exists an orthonormal basis $\{\mathbf{x}_1, \mathbf{x}_2, \ldots, \mathbf{x}_n\}$ of $\mathbb{R}^n$ that consists of eigenvectors of $A$;

- if $\lambda_1, \lambda_2, \ldots, \lambda_n$ are the eigenvalues of $A$ corresponding to $\mathbf{x}_1, \mathbf{x}_2, \ldots, \mathbf{x}_n$, then $A = XDX^T$, where $X$ is the matrix whose columns are $\mathbf{x}_1, \mathbf{x}_2, \ldots, \mathbf{x}_n$ and $D$ is the diagonal matrix whose diagonal entries are $\lambda_1, \lambda_2, \ldots, \lambda_n$;

- finally, the representation $A = XDX^T$ can be written as

$$A = \sum_{i=1}^{n} \lambda_i \mathbf{x}_i \mathbf{x}_i^T.$$

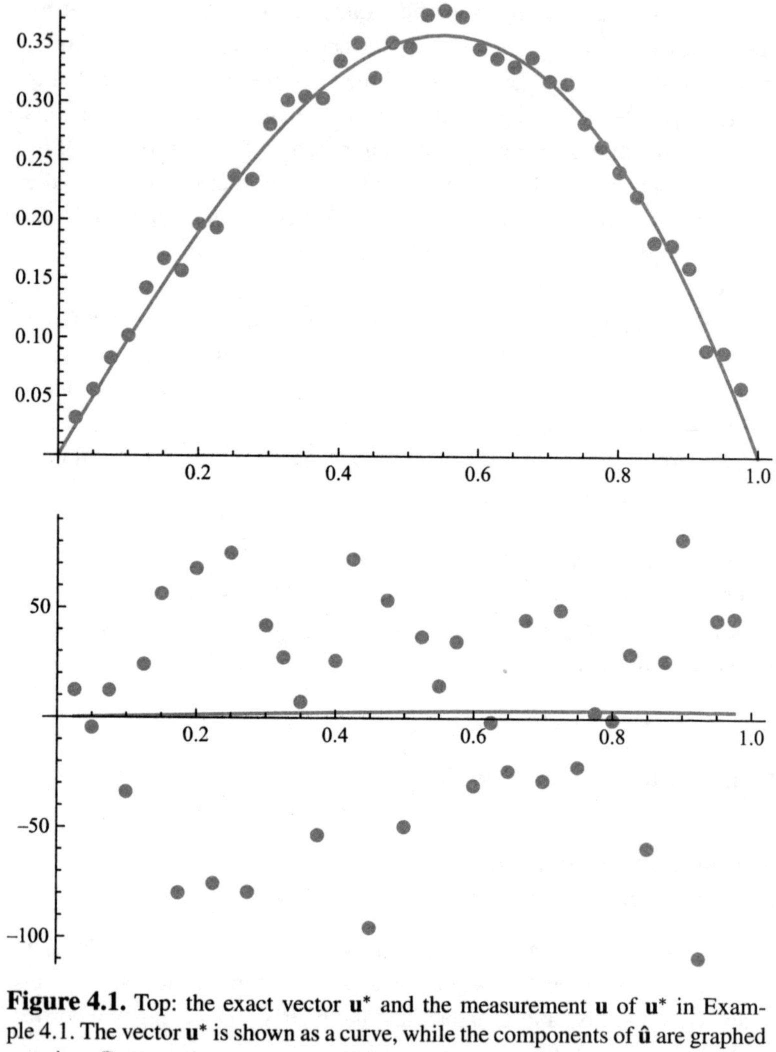

**Figure 4.1.** Top: the exact vector **u**\* and the measurement **u** of **u**\* in Example 4.1. The vector **u**\* is shown as a curve, while the components of **û** are graphed as points. Bottom: the exact vector **f**\* (graphed as a curve) and the computed estimate **f** (with the components graphed as points).

The expression $\mathbf{x}_i\mathbf{x}_i^T$ represents an $n \times n$ matrix whose $k, \ell$ entry is the product of the $k$th and $\ell$th components of the vector $\mathbf{x}_i$. The critical property of $\mathbf{x}_i\mathbf{x}_i^T$ is how it operates on a vector $v \in \mathbb{R}^n$:

$$\left(\mathbf{x}_i\mathbf{x}_i^T\right)\mathbf{v} = (\mathbf{x}_i \cdot \mathbf{v})\mathbf{x}_i.$$

The matrix $A$ is nonsingular if and only if all its eigenvalues are nonzero, in which case

$$A^{-1} = \sum_{i=1}^{n} \lambda_i^{-1}\mathbf{x}_i\mathbf{x}_i^T.$$

It follows that the solution of $A\mathbf{x} = \mathbf{b}$ is

$$\mathbf{x} = A^{-1}\mathbf{b} = \sum_{i=1}^{n} \frac{\mathbf{x}_i \cdot \mathbf{b}}{\lambda_i}\mathbf{x}_i. \tag{4.2}$$

Now let us consider the equation $A\mathbf{x} = \mathbf{b}$ from the point of view of an inverse problem. Actually, as discussed in Section 2.4, we already know that this cannot be a true inverse problem because the column space of $A$, being finite-dimensional, is necessarily closed. Nevertheless, as we pointed out, if the condition number of $A$,

$$\text{cond}(A) = \|A\|\|A^{-1}\| = \frac{\max\left\{\|A\mathbf{x}\|_2 \ : \ \mathbf{x} \in \mathbb{R}^n, \|\mathbf{x}\|_2 = 1\right\}}{\min\left\{\|A\mathbf{x}\|_2 \ : \ \mathbf{x} \in \mathbb{R}^n, \|\mathbf{x}\|_2 = 1\right\}}$$

is large, then the solution $\mathbf{x}$ is very sensitive to errors in the data $\mathbf{b}$. We will now describe this sensitivity in more detail using the eigenvalues and eigenvectors of $A$.

For convenience, let us suppose that the eigenvalues of $A$ are ordered by decreasing magnitude: $|\lambda_1| \geq |\lambda_2| \geq \cdots \geq |\lambda_n|$. Then, for any $\mathbf{x} \in \mathbb{R}^n$,

$$A\mathbf{x} = \sum_{i=1}^{n} \lambda_i(\mathbf{x}_i \cdot \mathbf{x})\mathbf{x}_i$$

$$\Rightarrow \|A\mathbf{x}\|_2^2 = \sum_{i=1}^{n} \lambda_i^2(\mathbf{x}_i \cdot \mathbf{x})^2\|\mathbf{x}_i\|_2^2 = \sum_{i=1}^{n} \lambda_i^2(\mathbf{x}_i \cdot \mathbf{x})^2$$

(using the fact that the basis $\{\mathbf{x}_1, \mathbf{x}_2, \ldots, \mathbf{x}_n\}$ is orthonormal). If we assume that $\|\mathbf{x}\|_2 = 1$, then we have

$$\|A\mathbf{x}\|_2^2 \leq \lambda_1^2 \sum_{i=1}^{n} (\mathbf{x}_i \cdot \mathbf{x})^2 = \lambda_1^2$$

(since $\sum_{i=1}^{n} (\mathbf{x}_i \cdot \mathbf{x})^2 = \|\mathbf{x}\|_2^2 = 1$), which shows that

$$\max\left\{\|A\mathbf{x}\|_2 \: : \: \mathbf{x} \in \mathbb{R}^n, \|\mathbf{x}\|_2 = 1\right\} \leq |\lambda_1|.$$

Since $\|A\mathbf{x}_1\|_2 = |\lambda_1|$, it follows that this inequality is actually an equality.

Similarly, still assuming that $A$ is invertible, we can show that

$$\min\left\{\|A\mathbf{x}\|_2 \: : \: \mathbf{x} \in \mathbb{R}^n, \|\mathbf{x}\|_2 = 1\right\} = |\lambda_n|,$$

and therefore that

$$\mathrm{cond}(A) = \frac{\max\left\{\|A\mathbf{x}\|_2 \: : \: \mathbf{x} \in \mathbb{R}^n, \|\mathbf{x}\|_2 = 1\right\}}{\min\left\{\|A\mathbf{x}\|_2 \: : \: \mathbf{x} \in \mathbb{R}^n, \|\mathbf{x}\|_2 = 1\right\}} = \frac{|\lambda_1|}{|\lambda_n|}.$$

Thus $A$ is ill-conditioned if the ratio of its largest eigenvalue to its smallest is large.

Now let us suppose that $\mathbf{x}^*$ represents the exact solution that we are trying to estimate, and the corresponding exact data is $\mathbf{b}^* = A\mathbf{x}^*$. If $\mathbf{b}$ is a measurement of $\mathbf{b}^*$ and $\mathbf{x}$ is the corresponding solution, then

$$\mathbf{x} - \mathbf{x}^* = A^{-1}\mathbf{b} - A^{-1}\mathbf{b}^* = A^{-1}\left(\mathbf{b} - \mathbf{b}^*\right)$$

$$= \sum_{i=1}^{n} \frac{\mathbf{x}_i \cdot (\mathbf{b} - \mathbf{b}^*)}{\lambda_i} \mathbf{x}_i,$$

which implies that

$$\left\|\mathbf{x} - \mathbf{x}^*\right\|_2^2 = \sum_{i=1}^{n} \frac{|\mathbf{x}_i \cdot (\mathbf{b} - \mathbf{b}^*)|^2}{|\lambda_i|^2}.$$

Since we also have

$$\left\|\mathbf{b} - \mathbf{b}^*\right\|_2^2 = \sum_{i=1}^{n} \left|\mathbf{x}_i \cdot \left(\mathbf{b} - \mathbf{b}^*\right)\right|^2,$$

we see that the components of the error $\mathbf{b} - \mathbf{b}^*$ corresponding to small eigenvalues are magnified when $A\mathbf{x} = \mathbf{b}$ is solved. As a result, it is possible that $\|\mathbf{x} - \mathbf{x}^*\|_2$ is large compared to $\|\mathbf{x}^*\|_2$ even though $\|\mathbf{b} - \mathbf{b}^*\|_2$ is small compared to $\|\mathbf{b}^*\|_2$.

One final note before we apply this analysis to our example: in the example, we are solving $L\mathbf{u} = \mathbf{f}$ for $\mathbf{f}$, given $\mathbf{u}$. This is equivalent to solving $A\mathbf{f} = \mathbf{u}$ for $\mathbf{f}$, where $A = L^{-1}$. If $\lambda_{max}(B)$ and $\lambda_{min}(B)$ denote the eigenvalues of a symmetric matrix $B$ that are largest and smallest in magnitude, then we have

$$\frac{|\lambda_{max}(L^{-1})|}{|\lambda_{min}(L^{-1})|} = \frac{|\lambda_{min}(L)|^{-1}}{|\lambda_{max}(L)|^{-1}} = \frac{|\lambda_{max}(L)|}{|\lambda_{min}(L)|}.$$

This shows that the above analysis applies to computing $\mathbf{f} = L\mathbf{u}$ from $\mathbf{u}$; the only difference is that it is the large eigenvalues of $L$, rather than the small ones, that are problematic.

**Example 4.2.** This is a continuation of Example 4.1. This particular matrix $L$ has the unusual property that we can compute its eigenvalues and eigenvectors exactly. (This fact is convenient, but not essential, for the following discussion.) For each integer $j = 1, 2, \ldots, m - 1, \lambda_j, \mathbf{u}^{(j)}$ form an eigenvalue-eigenvector pair, where

$$\lambda_j = \frac{2(1 - \cos{(j\pi h)})}{h^2}, \ u_k^{(j)} = \sin{(kj\pi h)}, \ k = 1, 2, \ldots, m - 1.$$

This can be verified by computing $L\mathbf{u}^{(j)}$ and simplifying using the angle addition and subtraction identities for the sine function.

It is straightforward to verify that $\lambda_j$ increases with $j$ and, using the approximation $\cos{(\pi h)} \approx 1 - \pi^2 h^2/2$, we see that

$$\lambda_1 = \frac{2(1 - \cos{(\pi h)})}{h^2} \approx \pi^2,$$

$$\lambda_{m-1} = \frac{2(1 - \cos{((m - 1)\pi h)})}{h^2} \approx \frac{4}{h^2} - \pi^2 \approx \frac{4}{h^2}.$$

Therefore, we have $\pi^2 \leq \lambda_j \leq 4/h^2$ for all $j$. In particular, as $h$ becomes small, $\lambda_{m-1}$ grows like $1/h^2$, while $\lambda_1$ is almost constant.

The eigenvectors $u^{(1)}, u^{(2)}, \ldots, u^{(m-1)}$ form a basis for $\mathbb{R}^{m-1}$, and it turns out to be orthogonal:

$$u^{(i)} \cdot u^{(j)} = 0 \text{ for all } i \neq j.$$

Moreover, each $u^{(j)}$ has the same norm:

$$\left\| u^{(j)} \right\|_{\mathbb{R}^{m-1}} = \sqrt{\frac{m}{2}}, \ j = 1, 2, \ldots, m - 1.$$

We now analyze the results of Example 4.1 using the eigenvalues and eigenvectors of $L$.

Recall that the calculations in Example 4.1 were done with $m = 40$, which means that $L$ is $39 \times 39$. The largest and smallest eigenvalues of $L$ are approximately 6390 and 9.9, yielding $\text{cond}(L) \approx 648$. This is the upper bound on the amplification of the error; that is, we know that

$$\frac{\|\mathbf{f} - \mathbf{f}^*\|_2}{\|\mathbf{f}^*\|_2} \leq 648 \frac{\|\mathbf{u} - \mathbf{u}^*\|_2}{\|\mathbf{u}^*\|_2} n$$

must hold. In fact, in the example shown in Figure 4.1, we have

$$\frac{\|\mathbf{f} - \mathbf{f}^*\|_2}{\|\mathbf{f}^*\|_2} \approx 355 \frac{\|\mathbf{u} - \mathbf{u}^*\|_2}{\|\mathbf{u}^*\|_2},$$

which shows that a significant amplification of the error occurred (as is obvious from Figure 4.1), although the amplification was not quite as large as possible.

It is instructive to look at the components of $\mathbf{u} - \mathbf{u}^*$ and $\mathbf{f} - \mathbf{f}^*$ in the directions of the eigenvectors of $L$. They are shown in Figure 4.2, which shows that the error $\mathbf{u} - \mathbf{u}^*$ is small and of approximately the same magnitude in each eigenvector direction. On the other hand, the components of the error $\mathbf{f} - \mathbf{f}^*$ grow with increasing eigenvalue and are significant (ruining the computed solution) except for the components corresponding to the smallest few eigenvalues.

Finally, it should be noted that the eigenvectors of $L$ that correspond to the large eigenvalues—the problematic directions for this problem—are sinusoidal functions of increasing frequency. Since random measurement errors are equally likely to be positive or negative, the noise in the data

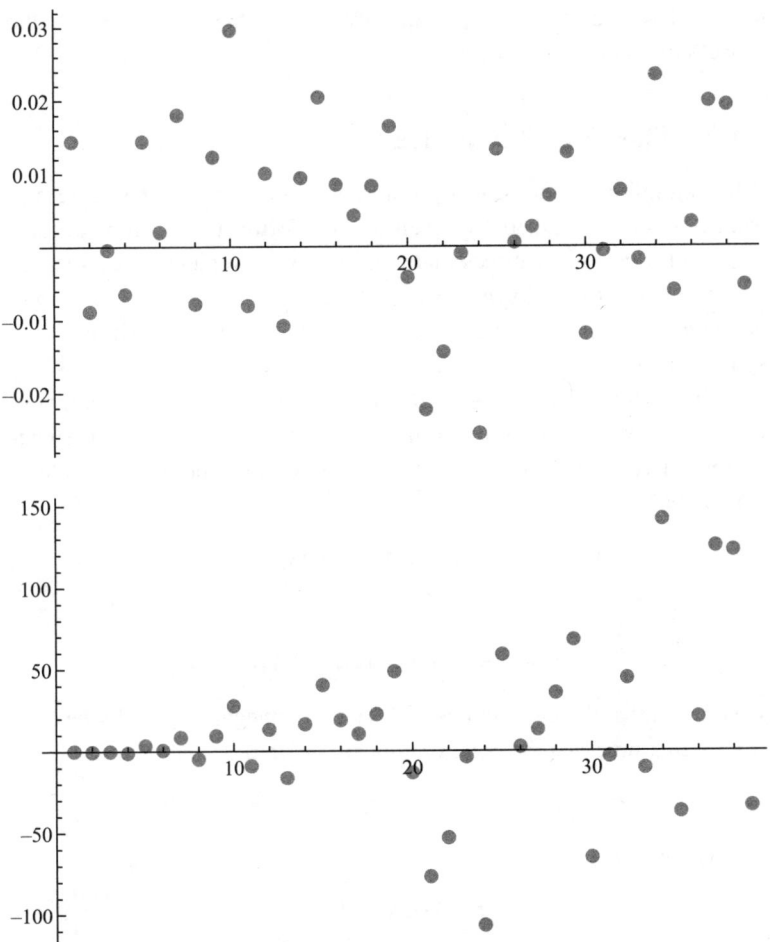

**Figure 4.2.** Top: the components of $\mathbf{u} - \mathbf{u}^*$ in the directions of the eigenvectors of $L$ for Example 4.1. Bottom: the components of $\mathbf{f} - \mathbf{f}^*$ in the same directions. In both graphs, the eigenvector components are ordered by increasing eigenvalue.

tends to have significant high frequency content that is magnified by the process of computing $\mathbf{f} = L\mathbf{u}$.

## 4.1.2    The SVD of a matrix

When an infinite-dimensional problem is discretized, the resulting matrix need not be symmetric or even square. Fortunately, there is decomposition for general matrices that is nearly as convenient as the spectral decomposition for real symmetric matrices. The *singular value decomposition* of $A \in \mathbb{R}^{m \times n}$ is derived from the spectral decomposition of the symmetric matrix $A^T A \in \mathbb{R}^{n \times n}$.

We assume for now that $A \in \mathbb{R}^{m \times n}$ with $m \geq n$. Suppose $\{\mathbf{v}_1, \mathbf{v}_2, \ldots, \mathbf{v}_n\}$ is an orthonormal basis of $\mathbb{R}^n$ consisting of eigenvectors of $A^T A$ and $\lambda_1, \lambda_2, \ldots, \lambda_n$ are the corresponding eigenvalues. Notice that

$$\mathbf{v}_i \cdot A^T A \mathbf{v}_i = \mathbf{v}_i \cdot (\lambda_i \mathbf{v}_i) = \lambda_i (\mathbf{v}_i \cdot \mathbf{v}_i) = \lambda_i$$

and

$$\mathbf{v}_i \cdot A^T A \mathbf{v}_i = (A\mathbf{v}_i) \cdot (A\mathbf{v}_i) = \|A\mathbf{v}_i\|_2^2 \geq 0.$$

This shows that the eigenvalues of $A^T A$ are nonnegative. We define $\sigma_i = \sqrt{\lambda_i}$. We assume that the eigenvectors are ordered so that

$$\sigma_1 \geq \sigma_2 \geq \cdots \geq \sigma_r > \sigma_{r+1} = \cdots = \sigma_n = 0.$$

We then define

$$\mathbf{u}_i = \sigma_i^{-1} A \mathbf{v}_i, \ i = 1, 2, \ldots, r.$$

Then, for $j \neq i$, $1 \leq i, j \leq r$, we have

$$\begin{aligned}
\mathbf{u}_j \cdot \mathbf{u}_i &= \sigma_i^{-1} \sigma_j^{-1} (A\mathbf{v}_j) \cdot (A\mathbf{v}_i) = \sigma_i^{-1} \sigma_j^{-1} \mathbf{v}_j \cdot (A^T A \mathbf{v}_i) \\
&= \sigma_i^{-1} \sigma_j^{-1} \mathbf{v}_j \cdot (\sigma_i^2 \mathbf{v}_i) \\
&= \frac{\sigma_i}{\sigma_j} \mathbf{v}_j \cdot \mathbf{v}_i \\
&= 0
\end{aligned}$$

(since the eigenvectors of $A^T A$ are orthogonal). On the other hand, if $j = i$ in the above calculation, we obtain

$$\mathbf{u}_j \cdot \mathbf{u}_i = \frac{\sigma_i}{\sigma_i} \mathbf{v}_i \cdot \mathbf{v}_i = 1.$$

Therefore $\{\mathbf{u}_1, \mathbf{u}_2, \ldots, \mathbf{u}_r\}$ is an orthonormal set. Let us choose $\mathbf{u}_{r+1}, \mathbf{u}_{r+2}, \ldots, \mathbf{u}_m$ (if necessary, that is, if $r < m$) to extend $\{\mathbf{u}_1, \mathbf{u}_2, \ldots, \mathbf{u}_r\}$ to an orthonormal basis $\{\mathbf{u}_1, \mathbf{u}_2, \ldots, \mathbf{u}_m\}$ of $\mathbb{R}^m$.

We now show that

$$A = U \Sigma V^T = \sum_{i=1}^{r} \sigma_i \mathbf{u}_i \mathbf{v}_i^T, \tag{4.3}$$

where $U = [\mathbf{u}_1 | \mathbf{u}_2 | \cdots | \mathbf{u}_m]$, $V = [\mathbf{v}_1 | \mathbf{v}_2 | \cdots | \mathbf{v}_n]$, and $\Sigma$ is the $m \times n$ diagonal matrix with diagonal entries $\sigma_1, \sigma_2, \ldots, \sigma_n$. A direct calculation shows that

$$U \Sigma V^T = \sum_{i=1}^{r} \sigma_i \mathbf{u}_i \mathbf{v}_i^T$$

(bearing in mind that $\sigma_{r+1} = \sigma_{r+2} = \cdots = \sigma_n = 0$). By definition of $\mathbf{u}_i$, we have

$$A \mathbf{v}_i = \sigma_i \mathbf{u}_i, \ i = 1, 2, \ldots, n.$$

Therefore, given $\mathbf{x} \in \mathbb{R}^n$, we have

$$A\mathbf{x} = A \left( \sum_{i=1}^{n} (\mathbf{v}_i \cdot \mathbf{x}) \mathbf{v}_i \right) = \sum_{i=1}^{n} (\mathbf{v}_i \cdot \mathbf{x}) A \mathbf{v}_i = \sum_{i=1}^{r} \sigma_i (\mathbf{v}_i \cdot \mathbf{x}) \mathbf{u}_i$$

$$= \sum_{i=1}^{r} \sigma_i \mathbf{u}_i \mathbf{v}_i^T \mathbf{x}$$

$$= \left( \sum_{i=1}^{r} \sigma_i \mathbf{u}_i \mathbf{v}_i^T \right) \mathbf{x},$$

as desired.

### 4.1.3   Using the SVD to solve a discrete inverse problem

Now let us consider an equation $A\mathbf{x} = \mathbf{b}$, where $A \in \mathbb{R}^{m \times n}$ and $\mathbf{b} \in \mathbb{R}^m$ are given and we wish to find the minimum-norm least-squares solution. Let us assume that the SVD of $A$ is given by (4.3). Then, since $\{\mathbf{u}_1, \mathbf{u}_2, \ldots, \mathbf{u}_m\}$ is an orthonormal basis of $\mathbb{R}^m$, we have

$$\mathbf{b} = \sum_{i=1}^{m} (\mathbf{u}_i \cdot \mathbf{b})\mathbf{u}_i$$

and

$$A\mathbf{x} - \mathbf{b} = \sum_{i=1}^{r} \sigma_i(\mathbf{v}_i \cdot \mathbf{x})\mathbf{u}_i - \sum_{i=1}^{m} (\mathbf{u}_i \cdot \mathbf{b})\mathbf{u}_i$$

$$= \sum_{i=1}^{r} (\sigma_i(\mathbf{v}_i \cdot \mathbf{x}) - \mathbf{u}_i \cdot \mathbf{b})\,\mathbf{u}_i - \sum_{i=r+1}^{m} (\mathbf{u}_i \cdot \mathbf{b})\mathbf{u}_i.$$

Since the vectors $\mathbf{u}_1, \mathbf{u}_2, \ldots, \mathbf{u}_m$ form an orthonormal set, it follows that

$$\|A\mathbf{x} - \mathbf{b}\|_2^2 = \sum_{i=1}^{r} (\sigma_i(\mathbf{v}_i \cdot \mathbf{x}) - \mathbf{u}_i \cdot \mathbf{b})^2 + \sum_{i=r+1}^{m} (\mathbf{u}_i \cdot \mathbf{b})^2.$$

Now, the goal is to choose $\mathbf{x}$ to make this expression as small as possible. Choosing $\mathbf{x}$ is equivalent to choosing its components $\mathbf{v}_i \cdot \mathbf{x}$, $i = 1, 2, \ldots, n$. (That is, we can represent the unknown vector as $\mathbf{x} = \sum_{i=1}^{n} \alpha_i \mathbf{v}_i$, where $\alpha_1, \alpha_2, \ldots, \alpha_n$ can be chosen to have any values, and, for each $i$, $\alpha_i = \mathbf{v}_i \cdot \mathbf{x}$.) The above expression for $\|A\mathbf{x} - \mathbf{b}\|_2^2$ shows that, for all $\mathbf{x} \in \mathbb{R}^n$,

$$\|A\mathbf{x} - \mathbf{b}\|_2^2 \geq \sum_{i=r+1}^{m} (\mathbf{u}_i \cdot \mathbf{b})^2$$

and

$$\|A\mathbf{x} - \mathbf{b}\|_2^2 = \sum_{i=r+1}^{m} (\mathbf{u}_i \cdot \mathbf{b})^2$$

if we choose $\mathbf{x}$ so that $\sigma_i(\mathbf{v}_i \cdot \mathbf{x}) - \mathbf{u}_i \cdot \mathbf{b} = 0$ for $i = 1, 2, \ldots, r$. This implies that we should choose

$$\mathbf{v}_i \cdot \mathbf{x} = \frac{\mathbf{u}_i \cdot \mathbf{b}}{\sigma_i}, \quad i = 1, 2, \ldots, r.$$

The remaining components of $\mathbf{x}$ are not determined by minimizing $\|A\mathbf{x} - \mathbf{b}\|_2^2$; that is, every vector of the form

$$\mathbf{x} = \sum_{i=1}^{r} \frac{\mathbf{u}_i \cdot \mathbf{b}}{\sigma_i} \mathbf{v}_i + \sum_{i=r+1}^{n} \alpha_i \mathbf{v}_i$$

is a least-squares solution of $A\mathbf{x} = \mathbf{b}$. For any such vector, we have

$$\|\mathbf{x}\|_2^2 = \sum_{i=1}^{r} \frac{(\mathbf{u}_i \cdot \mathbf{b})^2}{\sigma_i^2} + \sum_{i=r+1}^{n} \alpha_i^2,$$

and the minimum-norm solution is obtained by choosing $\alpha_{r+1} = \cdots = \alpha_n = 0$. Therefore, the minimum-norm least-squares solution of $A\mathbf{x} = \mathbf{b}$ is

$$\mathbf{x} = \sum_{i=1}^{r} \frac{\mathbf{u}_i \cdot \mathbf{b}}{\sigma_i} \mathbf{v}_i. \tag{4.4}$$

The reader should notice the similarity of (4.4) and (4.2).

If $A$ has small singular values—more precisely, if the smallest singular values of $A$ are much smaller than the largest—then (4.4) has the same crucial property that (4.2) has in the case $A$ is symmetric, invertible, and ill-conditioned, as we now explain. Let us suppose that $\mathbf{b}$ is a noisy measurement of the exact data $\mathbf{b}^*$, let $\mathbf{x}^*$ be the minimum-norm least-squares solution of $A\mathbf{x} = \mathbf{b}^*$, and let $\mathbf{x}$ be the minimum-norm least-squares solution of $A\mathbf{x} = \mathbf{b}$:

$$\mathbf{x}^* = \sum_{i=1}^{r} \frac{\mathbf{u}_i \cdot \mathbf{b}^*}{\sigma_i} \mathbf{v}_i,$$

$$\mathbf{x} = \sum_{i=1}^{r} \frac{\mathbf{u}_i \cdot \mathbf{b}}{\sigma_i} \mathbf{v}_i.$$

The error in the computed solution is then

$$\mathbf{x} - \mathbf{x}^* = \sum_{i=1}^{r} \frac{\mathbf{u}_i \cdot (\mathbf{b} - \mathbf{b}^*)}{\sigma_i} \mathbf{v}_i.$$

Even if $\|\mathbf{b} - \mathbf{b}^*\|_2$ is small compared to $\|\mathbf{b}^*\|_2$, which means that all of its components $\mathbf{u}_i \cdot (\mathbf{b} - \mathbf{b}^*)$ are small, the components

$$\frac{\mathbf{u}_i \cdot (\mathbf{b} - \mathbf{b}^*)}{\sigma_i}$$

of $\mathbf{x} - \mathbf{x}^*$ corresponding to the small singular values can be large. This situation is exactly analogous to the case of solving $A\mathbf{x} = \mathbf{b}$ when $A$ is symmetric and ill-conditioned.

We have

$$\|\mathbf{x} - \mathbf{x}^*\|_2^2 = \sum_{i=1}^{r} \frac{(\mathbf{u}_i \cdot (\mathbf{b} - \mathbf{b}^*))^2}{\sigma_i^2} \leq \sum_{i=1}^{r} \frac{(\mathbf{u}_i \cdot (\mathbf{b} - \mathbf{b}^*))^2}{\sigma_r^2}$$

$$= \frac{1}{\sigma_r^2} \sum_{i=1}^{r} (\mathbf{u}_i \cdot (\mathbf{b} - \mathbf{b}^*))^2 \leq \frac{1}{\sigma_r^2} \|\overline{\mathbf{b}} - \overline{\mathbf{b}^*}\|_2^2,$$

where $\overline{\mathbf{b}}, \overline{\mathbf{b}^*}$ are the projections of $\mathbf{b}, \mathbf{b}^*$, respectively, onto the column space of $A$. We also have

$$\|\mathbf{x}^*\|_2^2 = \sum_{i=1}^{r} \frac{(\mathbf{u}_i \cdot \mathbf{b}^*)^2}{\sigma_i^2} \geq \sum_{i=1}^{r} \frac{(\mathbf{u}_i \cdot \mathbf{b}^*)^2}{\sigma_1^2} = \frac{1}{\sigma_1^2} \sum_{i=1}^{r} (\mathbf{u}_i \cdot \mathbf{b}^*)^2 = \frac{1}{\sigma_1^2} \|\overline{\mathbf{b}^*}\|_2^2.$$

Here we have used the fact that $\sigma_1$ is the largest singular value of $A$ and $\sigma_r$ is the smallest. It follows that

$$\frac{\|\mathbf{x} - \mathbf{x}^*\|_2}{\|\mathbf{x}^*\|_2} \leq \frac{\sigma_1}{\sigma_r} \frac{\|\overline{\mathbf{b}} - \overline{\mathbf{b}^*}\|_2}{\|\overline{\mathbf{b}^*}\|_2}.$$

It should be noted that the components of $\mathbf{b}^*$ and $\mathbf{b}$ that are orthogonal to the column space of $A$ do not affect $\mathbf{x}^*$ and $\mathbf{x}$.

Based on the above result, we define the condition number of $A$ to be

$$\text{cond}(A) = \frac{\sigma_1}{\sigma_r}.$$

(the ratio of the largest singular value of $A$ to the smallest positive singular value), and we say that $A$ is ill-conditioned if its condition number is large. The reader can verify that the singular values of a symmetric matrix are the absolute values of its eigenvalues; therefore, this definition is consistent with our earlier definition of the condition number of an invertible symmetric matrix.

If the condition number of $A$ is large, then small relative changes in **b** can lead to a large relative change in the minimum-norm least-squares solution of $A\mathbf{x} = \mathbf{b}$. (We could construct an example to illustrate this, but we will not do so here because the example would be so similar to Examples 4.1 and 4.2.) Although the solution **x** is very sensitive to changes in the data **b**, it does depend continuously on **b**. In this chapter, we will study a type of operator, called a *compact operator*, for which a type of singular value decomposition exists. In infinite dimensions, a compact operator can have an infinite sequence $\{\sigma_n\}$ of positive singular values that converge to zero; in such a case, the ratio of $\sigma_1/\sigma_n$ is unbounded as $n \to \infty$. This causes instability; that is, it leads to a true inverse problem.

## 4.2 Compact operators

We begin with the definition of a compact operator.

**Definition 4.3.** Let $U$ and $V$ be Hilbert spaces and let $L : U \to V$ be a linear operator. We say that $L$ is *compact* if the image under $L$ of a bounded subset of $U$ is a precompact subset of $V$. (A set is precompact if its closure is compact.)

Under a norm topology (or any metric topology), a set is precompact if and only if every sequence in the set has a convergent subsequence. Therefore, an operator $L$ is compact if and only if $\{Lu_n\}$ has a convergent subsequence for every bounded sequence $\{u_n\}$ of $U$.

### Some examples
Before we study the properties of compact operators, we will present a few examples of compact and noncompact operators. We have seen

several examples of integral operators of the form

$$K : L^2(a, b) \to L^2(a, b),$$

$$(Ku)(s) = \int_a^b k(s, t)u(t)\, dt,$$

where $k \in L^2((a, b) \times (a, b))$. These are called integral operators *of the first kind*. As we will see in Section 4.4.1, an integral operator of the first kind is always compact. In fact, these operators are the prototypical compact operators—the concept arose in an effort to abstract the essential properties of integral operators of the above form.

A simple example of a noncompact operator is the identity $I : X \to X$, $Ix = x$ for all $x \in X$, where $X$ is an infinite-dimensional Hilbert space. In an infinite-dimensional space $X$, we can always construct an infinite orthonormal sequence $\{x_n\}$, and

$$\|Ix_n - Ix_m\|_X = \|x_n - x_m\|_X = \sqrt{2} \text{ for all } m \neq n$$

(the fact that $\|x_n - x_m\|_X = \sqrt{2}$ follows from the Pythagorean theorem). This shows that $\{Ix_n\}$ cannot have a convergent subsequence, and hence that $I$ is not compact.

A less trivial example of a noncompact operator is an operator of the form $T = I + K$, where $K : X \to X$ is compact, $I$ is the identity operator, and $X$ is infinite-dimensional. Theorem 4.27 shows that an operator $T$ of this form is not compact. Hence, in particular, an integral operator of the form

$$T : L^2(a, b) \to L^2(a, b),$$

$$(Tu)(s) = u(s) + \int_a^b k(s, t)u(t)\, dt,$$

which is called an integral operator *of the second kind*, is not a compact operator. Theorem 4.27 also shows that $\mathcal{R}(I + K)$ is always closed, which means that an integral equation of the form

$$u(s) + \int_a^b k(s, t)u(t)\, dt = v(s), \ a < s < b$$

does not define an inverse problem.

One final example: let $T : L^2(0, 1) \to L^2(0, 1)$ be defined by

$$(Tx)(t) = tx(t) \text{ for all } t \in (0, 1).$$

It is straightforward to show that $\mathcal{R}(T)$ fails to be closed. For instance, the constant function $y(t) = 1$ does not lie in $\mathcal{R}(T)$. To see this, notice that $Tx = y$ implies that

$$x(t) = \frac{y(t)}{t} = \frac{1}{t}.$$

However, $x(t) = 1/t$ does not belong to $L^2(0, 1)$, since

$$\int_0^1 \frac{dt}{t^2} = \infty.$$

On the other hand, it is easy to show that $y(t) = 1$ lies in $\overline{\mathcal{R}(T)}$. In fact, if we define

$$x_n(t) = \begin{cases} \frac{1}{t}, & t \in \left(\frac{1}{n}, 1\right), \\ 0, & t \in \left(0, \frac{1}{n}\right), \end{cases}$$

then $x_n \in L^2(0, 1)$ for all $n \in \mathbb{Z}^+$ and $Tx_n \to y$ as $n \to \infty$. This example is of interest because $T$ is not compact, and hence it shows that compact operators are not the only ones that lead to true inverse problems. To show that $T$ is not compact, we can define the sequence $\{x_n\}$ by

$$x_n(t) = \begin{cases} \sqrt{n(n+1)}, & 1 - \frac{1}{n} < t < 1 - \frac{1}{n+1}, \\ 0, & \text{otherwise.} \end{cases}$$

Then $\{x_n\}$ is orthonormal, while $\{Tx_n\}$ is orthogonal, with $\|Tx_n\|_{L^2}$ bounded away from zero. It follows that $\{Tx_n\}$ cannot have a convergent subsequence, and hence $T$ is not compact.

**Properties of compact operators**

We begin with some elementary properties of compact operators.

**Theorem 4.4.** *Let $U$ and $V$ be Hilbert spaces and suppose $L : U \to V$ is a compact operator. Then $L$ is bounded.*

*Proof.* If $L$ fails to be bounded, then there exists $\{u_n\} \subset X$ such that

$$\frac{\|Lu_n\|_V}{\|u_n\|_U} \to \infty \text{ as } n \to \infty.$$

Without loss of generality, we can assume that $\|u_n\|_U = 1$ for all $n$. But then $\{u_n\}$ is bounded and $\|Lu_n\|_V \to \infty$ as $n \to \infty$, which implies that $\{Lu_n\}$ cannot have a convergent subsequence. Therefore, if $L$ is not bounded, then $L$ is not compact. □

**Theorem 4.5.** *Let $U$, $V$, and $W$ be Hilbert spaces and let $L : U \to V$ and $K : V \to W$ be linear operators.*

1. *If $L$ is compact and $K$ is bounded, then $KL$ is compact.*

2. *If $L$ is bounded and $K$ is compact, then $KL$ is compact.*

*Proof.* Suppose $\{u_n\} \subset U$ is bounded. In each case, we must show that $\{(KL)u_n\}$ has a convergent subsequence.

1. If $L$ is compact, then $\{Lu_n\}$ has a convergent subsequence, say $Lu_{n_k} \to v$. Since $K$ is bounded, it follows that $K(Lu_{n_k}) \to Kv$, that is, $(KL)u_{n_k} \to Kv$. This shows that $\{(KL)u_n\}$ has a convergent subsequence.

2. If $L$ is a bounded operator, then the sequence $\{Lu_n\}$ is also bounded. Therefore, since $K$ is compact, $\{K(Lu_n)\} = \{(KL)u_n\}$ has a convergent subsequence.

Thus, in both cases, $KL$ is compact. □

The previous result implies that $L^*L$ is compact if $L$ is compact. It is also possible to prove that $L^*$ is compact if $L$ is compact. We could prove this directly now, but it will follow from the results about the singular value expansion of a compact operator that we derive below.

A particular type of compact operator is a finite-rank operator.

**Definition 4.6.** Let $U$ and $V$ be Hilbert spaces and let $L : U \to V$ be a linear operator. We say that $L$ is a *finite-rank* operator if $\mathcal{R}(L)$ is finite-dimensional.

It is obvious that a finite-rank operator is compact because every bounded subset of a finite-dimensional space is precompact. However, a finite-dimensional subspace of a Hilbert space is necessarily a closed subspace, and therefore $Tx = y$ is not a true inverse problem if $T$ is a finite-rank operator. On the other hand, if an operator is compact and not finite-rank, then its range cannot be closed.

**Theorem 4.7.** *Let $U$ and $V$ be Hilbert spaces, and let $L : U \to V$ be a compact linear operator that is not finite-rank. Then $\mathcal{R}(L)$ is not closed.*

*Proof.* Assume that $L$ is compact. We will prove that if $\mathcal{R}(L)$ is closed, then $L$ is a finite-rank operator. Therefore, suppose $\mathcal{R}(L)$ is closed and define an operator $L_1 : \mathcal{N}(L)^{\perp} \to \mathcal{R}(L)$ by $L_1 = L|_{\mathcal{N}(L)^{\perp}}$. Since we are assuming that $\mathcal{R}(L)$ is closed, Corollary 2.19 implies that $L_1^{-1}$ is bounded. Clearly $L_1$ is compact. Therefore, by Theorem 4.5, the identity operator $I = L_1^{-1}L_1$ is compact on $\mathcal{N}(L)^{\perp}$. But then $\overline{B_1(0; U_1)}$ is compact, since it is the image under a compact operator of a bounded set (itself) and it is closed. By Theorem A.32 of Section A.6, the closed unit ball in a normed vector space is compact if and only if the space is finite-dimensional. It follows that $\mathcal{N}(L)^{\perp}$ and hence $\mathcal{R}(L)$ are finite-dimensional. $\qquad\square$

One of the most useful ways to prove that an operator is compact is to express it as a limit of compact operators and invoke the following theorem.

**Theorem 4.8.** *Let $U$ and $V$ be Hilbert spaces and let $K_n : U \to V$ be a compact operator for each $n \in \mathbb{Z}^+$. Suppose $K : U \to V$ is a bounded linear operator and assume $K_n \to K$ in the operator norm (that is, assume $\|K_n - K\| \to 0$ as $n \to \infty$). Then $K$ is compact.*

*Proof.* Let $\{u_m\}$ be a bounded sequence in $U$. We must show that $\{Ku_m\}$ has a convergent subsequence. Since $K_1$ is compact, there exists a subsequence $\{u_m^{(1)}\}$ of $\{u_m\}$ and an element $v_1$ of $V$ such that $K_1 u_m^{(1)} \to v_1$ as $m \to \infty$. There then exists a subsequence $\{u_m^{(2)}\}$ of $\{u_m^{(1)}\}$ and an element $v_2$ of $V$ such that $K_2 u_m^{(2)} \to v_2$ as $m \to \infty$. Continuing in this fashion, we construct, for each $j \in \mathbb{Z}^+$, a subsequence $\{u_m^{(j)}\}$ of $\{u_m^{(j-1)}\}$ and an element $v_j$ of $V$ such that $K_j u_m^{(j)} \to v_j$ as $m \to \infty$. Now define $\{x_m\} \subset U$

by $x_m = u_m^{(m)}$. Then $\{x_m\}$ is a subsequence of $\{u_m\}$ and, moreover, $\{x_m\}$ is a subsequence of each $\{u_m^{(j)}\}$. It follows that $K_j x_m \to v_j$ as $m \to \infty$ for each $j \in \mathbb{Z}^+$. We are going to show that there exists $v \in V$ such that $v_j \to v$ as $j \to \infty$ and $Kx_m \to v$ as $m \to \infty$. This will prove the result.

We first show that $\{v_j\}$ is Cauchy. Let $\epsilon > 0$ be given. Since $\{x_m\}$ is bounded, there exists $M > 0$ such that $\|x_m\|_X \leq M$ for all $m$. Since $\{K_n\}$ is convergent, it is Cauchy and therefore there exists $j_0$ such that

$$j, \ell \geq j_0 \;\Rightarrow\; \|K_j - K_\ell\| < \frac{\epsilon}{3M}.$$

Then, given $j, \ell \geq j_0$, there exists $n \in \mathbb{Z}^+$ such that

$$\|K_j x_n - v_j\|_V < \frac{\epsilon}{3}, \quad \|K_\ell x_n - v_\ell\|_V < \frac{\epsilon}{3}.$$

It follows that if $j, \ell \geq j_0$, then

$$\begin{aligned}
\|v_j - v_\ell\|_V &\leq \|v_j - K_j x_n\|_V + \|K_j x_n - K_\ell x_n\|_V + \|K_\ell x_n - v_\ell\|_V \\
&\leq \|v_j - K_j x_n\|_V + \|K_j - K_\ell\|_V \|x_n\|_X . + \|K_\ell x_n - v_\ell\|_V \\
&< \frac{\epsilon}{3} + \frac{\epsilon}{3M} M + \frac{\epsilon}{3} = \epsilon.
\end{aligned}$$

This shows that $\{v_j\}$ is Cauchy and hence that there exists $v \in V$ such that $v_j \to v$ as $j \to \infty$.

We now show that $Kx_m \to v$ as $m \to \infty$. Again let $\epsilon > 0$ be given. Choose $j$ sufficiently large that $\|v_j - v\|_V < \epsilon/3$ and $\|K - K_j\| \leq \epsilon/(3M)$. We then choose $m_0 \in \mathbb{Z}^+$ sufficiently large that

$$m \geq m_0 \;\Rightarrow\; \|K_j x_m - v_j\|_V < \frac{\epsilon}{3}.$$

Then, if $m \geq m_0$,

$$\begin{aligned}
\|Kx_m - v\|_V &\leq \|Kx_m - K_j x_m\|_V + \|K_j x_m - v_j\|_V + \|v_j - v\|_V \\
&\leq \|K - K_j\| \|x_m\|_X + \|K_j x_m - v_j\|_V + \|v_j - v\|_V \\
&\leq \frac{\epsilon}{3M} M + \frac{\epsilon}{3} + \frac{\epsilon}{3} = \epsilon.
\end{aligned}$$

This shows that $Kx_m \to v$ as $m \to \infty$. $\qquad\square$

Since finite-rank operators are compact, the previous theorem yields the following result.

**Corollary 4.9.** *The norm limit of finite-rank operators is compact. That is, if $U$ and $V$ are Hilbert spaces, $K_n : U \to V$ is a finite-rank operator for each $n \in \mathbb{Z}^+$, and $K : U \to V$ is a bounded linear operator such that $K_n \to K$ in the operator norm, then $K$ is compact.*

This last result is important; we will use it to show that every operator with a singular value expansion is compact. In particular, we will show that first-kind integral operators are always compact (see Section 4.4.1).

## 4.3   The spectral theorem for a compact self-adjoint operator

We wish to show that every compact operator can be represented as an infinite series that is reminiscent of the singular value decomposition of a matrix. To do this, we begin with the special case of a compact self-adjoint operator, in which case we can apply spectral theory (that is, the theory of eigenvalues and eigenvectors).

Spectral theory is much more complicated in infinite dimensions than it is in finite dimensions; however, it turns out that compact self-adjoint operators admit a relatively simple theory that is not so different from that of symmetric matrices. In particular, as in the case of symmetric matrices, all eigenvalues are real and there is no need to consider complex numbers.

**Definition 4.10.** Let $U$ be a (real) Hilbert space and let $L : U \to U$ be a self-adjoint linear operator. Then $\lambda \in \mathbb{R}$ is an *eigenvalue* of $L$ if there exists a nonzero $u \in U$ such that $Lu = \lambda u$.

We begin by showing that every compact self-adjoint linear operator has at least one eigenvalue.

**Theorem 4.11.** *Let $U$ be a Hilbert space, and let $L : U \to U$ be a compact self-adjoint linear operator. Then either $\|L\|$ or $-\|L\|$ is an eigenvalue of $L$.*

*Proof.* We can assume that $L \neq 0$, since otherwise the result is trivial. By Lemma 3.9, we have

$$\|L\| = \sup\{|\langle Lu, u\rangle_U| \: : \: u \in U, \|u\|_U = 1\}.$$

It follows that there is a sequence $\{u_n\}$ in $U$ with $\|u_n\| = 1$ for all $n$, and

$$|\langle Lu_n, u_n\rangle_U| \to \|L\|.$$

We may assume that $\langle Lu_n, u_n\rangle_U \to \lambda$, with $\lambda$ equal to either $\|L\|$ or $-\|L\|$, since if this does not hold, it must hold for a subsequence of $\{u_n\}$. Now,

$$0 \leq \|Lu_n - \lambda u_n\|^2 = \|Lu_n\|_U^2 - 2\lambda\langle Lu_n, u_n\rangle_U + \lambda^2\|u_n\|_U^2$$
$$\leq 2\lambda^2 - 2\lambda\langle Lu_n, u_n\rangle_U \to 0.$$

It follows that $Lu_n - \lambda u_n \to 0$. Moreover, $\{u_n\}$ is bounded and $L$ is compact, so, without loss of generality, we may assume that $\{Lu_n\}$ converges, say $Lu_n \to v$. We then have

$$\|\lambda u_n - v\| \leq \|\lambda u_n - Lu_n\| + \|Lu_n - v\| \to 0,$$

so $\lambda u_n \to v$, which implies that $u_n \to \lambda^{-1}v$ ($\lambda \neq 0$ since $L \neq 0$). We then have

$$Lu_n - \lambda u_n \to L\left(\lambda^{-1}v\right) - v,$$

and, since $Lu_n - \lambda u_n \to 0$, we obtain $L\left(\lambda^{-1}v\right) - v = 0$. Thus $Lv = \lambda v$. It remains only to show that $v \neq 0$. But

$$\lambda u_n \to v \Rightarrow \lambda = \|\lambda u_n\| \to \|v\|.$$

Thus $\|v\| = \lambda \neq 0$.  □

The reader may recall that eigenvectors of a symmetric matrix that correspond to distinct eigenvalues are orthogonal. This property generalizes to self-adjoint operators in Hilbert space (and the proof is essentially the same as in the case of a matrix).

**Theorem 4.12.** *Let $U$ be a Hilbert space and let $L : U \to U$ be a self-adjoint bounded linear operator. If $u_1$ and $u_2$ are eigenvectors corresponding to distinct eigenvalues $\lambda_1, \lambda_2 \in \mathbb{R}$, then $u_1$ and $u_2$ are orthogonal.*

*Proof.* We have

$$\lambda_1 \langle u_1, u_2 \rangle_U = \langle \lambda_1 u_1, u_2 \rangle_U = \langle L u_1, u_2 \rangle_U = \langle u_1, L u_2 \rangle_U = \langle u_1, \lambda_2 u_2 \rangle_U$$
$$= \lambda_2 \langle u_1, u_2 \rangle_U,$$

so $(\lambda_1 - \lambda_2) \langle u_1, u_2 \rangle_U = 0$. Since $\lambda_1 \neq \lambda_2$, this implies $\langle u_1, u_2 \rangle_U = 0$. $\qquad\square$

### 4.3.1   Orthogonal sets in Hilbert space

Since the eigenvectors of a self-adjoint compact operator are orthogonal, we will be dealing with orthogonal sets in infinite-dimensional spaces. We need some preliminary results about such sets, beginning with the following definition.

**Definition 4.13.** Let $U$ be a Hilbert space and let $\{\phi_n\}$ be an infinite sequence in $U$. We say that $\{\phi_n\}$ is an *orthonormal sequence* if $\langle \phi_i, \phi_j \rangle_U = 0$ for all $i, j \in \mathbb{Z}^+$, $i \neq j$, and $\langle \phi_i, \phi_i \rangle_U = 1$ for all $i \in \mathbb{Z}^+$.

Given an orthonormal sequence $\{\phi_n\}$, we define the *span* of $\{\phi_n\}$ to be the set of all finite linear combinations of elements of $\{\phi_n\}$. In other words,

$$\mathrm{sp}\{\phi_n\} = \bigcup_{N=1}^{\infty} \mathrm{sp}\{\phi_1, \phi_2, \ldots, \phi_N\}.$$

If every element $u \in U$ can be represented as

$$u = \sum_{n=1}^{\infty} \alpha_n \phi_n,$$

where $\alpha_n \in \mathbb{R}$ for all $n \in \mathbb{Z}^+$, then $\{\phi_n\}$ is called a *complete orthonormal set* for $U$.

We will also need the following standard result from finite-dimensional linear algebra. Given a finite-dimensional subspace $V$ of an inner product space $U$ and a vector $u \in U$, the *projection* of $u$ onto $V$ is the unique vector in $V$ closest to $u$, that is, the vector $v \in V$ satisfying $\|v - u\|_U \le \|w - u\|_U$ for all $w \in V$. The existence of $v$ is guaranteed by Theorem A.16.

**Lemma 4.14.** *If $U$ is an inner product vector space, $V$ is a finite-dimensional subspace of $U$, and $\{x_1, x_2, \ldots, x_n\}$ is an orthonormal basis for $V$, then, for all $u \in U$, the projection of $u$ onto $V$ is given by*

$$\mathrm{proj}_V u = \sum_{i=1}^{n} \langle u, x_i \rangle_U x_i.$$

*Proof.* Since $w = \mathrm{proj}_V u \in V$, it can be represented in the form $w = \sum_{i=1}^{n} \alpha_i x_i$, and $w$ is characterized by the condition that $u - w \in V^\perp$ (see Theorem A.16 in Section A.2). Therefore $\langle u - w, x_j \rangle_U = 0$ for all $j = 1, 2, \ldots, n$. Substituting $w = \sum_{i=1}^{n} \alpha_i x_i$ into this last equation and using the orthogonality of the basis shows that $\alpha_j = \langle u, x_j \rangle_U$. $\qquad\square$

The following lemma collects some basic facts about orthonormal sequences in Hilbert space.

**Lemma 4.15.** *Let $U$ be a Hilbert space, let $\{\phi_n\}$ be an orthonormal sequence in $U$, and define $S = \mathrm{sp}\{\phi_n\}$.*

1. *$\sum_{n=1}^{\infty} \alpha_n \phi_n$ converges to an element $u$ of $U$ if and only if $\sum_{n=1}^{\infty} \alpha_n^2 < \infty$. In this case, $\alpha_n = \langle u, \phi_n \rangle_U$ for all $n \in \mathbb{Z}^+$ and $\|u\|_U^2 = \sum_{n=1}^{\infty} \alpha_n^2$.*

2. *For all $u \in U$, $\sum_{n=1}^{\infty} |\langle u, \phi_n \rangle_U|^2 \le \|u\|_U^2$, and hence $\sum_{n=1}^{\infty} \langle u, \phi_n \rangle_U \phi_n$ converges.*

3. *$\overline{S} = \left\{ \sum_{n=1}^{\infty} \alpha_n \phi_n : \sum_{n=1}^{\infty} \alpha_n^2 < \infty \right\}$.*

4. *For all $u \in U$,*

$$\mathrm{proj}_{\overline{S}} u = \sum_{n=1}^{\infty} \langle u, \phi_n \rangle_U \phi_n.$$

5. *Given $u \in U$, $u \in \overline{S}$ if and only if $\sum_{n=1}^{\infty} |\langle u, \phi_n \rangle_U|^2 = \|u\|_U^2$.*

*Proof.*

1. The series $\sum_{n=1}^{\infty} \alpha_n \phi_n$ converges if and only if the sequence of partial sums is Cauchy. For any $M > N$, we have

$$\sum_{n=1}^{M} \alpha_n \phi_n - \sum_{n=1}^{N} \alpha_n \phi_n = \sum_{n=N+1}^{M} \alpha_n \phi_n$$

and hence

$$\left\| \sum_{n=1}^{M} \alpha_n \phi_n - \sum_{n=1}^{N} \alpha_n \phi_n \right\|_U^2 = \left\| \sum_{n=N+1}^{M} \alpha_n \phi_n \right\|_U^2$$

$$= \left\langle \sum_{n=N+1}^{M} \alpha_n \phi_n, \sum_{m=N+1}^{M} \alpha_m \phi_m \right\rangle_U$$

$$= \sum_{n=N+1}^{M} \sum_{m=N+1}^{M} \alpha_n \alpha_m \langle \phi_n, \phi_m \rangle_U$$

$$= \sum_{n=N+1}^{M} \alpha_n^2.$$

The final simplification follows from the fact that $\{\phi_n\}$ is orthonormal; thus $\langle \phi_n, \phi_m \rangle_U$ is 0 unless $n = m$, in which case the inner product is 1. This equation shows that

$$\left\{ \sum_{n=1}^{N} \alpha_n \phi_n \right\}$$

is Cauchy in $U$ if and only if $\{\sum_{n=1}^{N} \alpha_n^2\}$ is Cauchy in $\mathbb{R}$, and hence $\sum_{n=1}^{\infty} \alpha_n \phi_n$ converges to an element of $U$ if and only if $\sum_{n=1}^{\infty} \alpha_n^2 < \infty$.

Now suppose that $u = \sum_{n=1}^{\infty} \alpha_n \phi_n$. This means that

$$u = \lim_{N \to \infty} \sum_{n=1}^{N} \alpha_n \phi_n,$$

where the limit converges in the $U$-norm. It follows that

$$\langle \phi_m, u \rangle_U = \left\langle \phi_m, \lim_{N \to \infty} \sum_{n=1}^{N} \alpha_n \phi_n \right\rangle_U = \lim_{N \to \infty} \left\langle \phi_m, \sum_{n=1}^{N} \alpha_n \phi_n \right\rangle_U$$

$$= \lim_{N \to \infty} \sum_{n=1}^{N} \alpha_n \langle \phi_m, \phi_n \rangle_U$$

$$= \lim_{N \to \infty} \alpha_m = \alpha_m$$

(because $\{\phi_n\}$ is orthonormal, $\sum_{n=1}^{N} \alpha_n \langle \phi_m, \phi_n \rangle_U$ is zero for $N < m$ and equals $\alpha_m$ for $N \geq m$). Thus $\alpha_m = \langle \phi_m, u \rangle_U$ for each $m$, as desired.

Finally, $u = \lim_{N \to \infty} \sum_{n=1}^{N} \alpha_n \phi_n$ implies that

$$\|u\|_U^2 = \lim_{N \to \infty} \left\| \sum_{n=1}^{N} \alpha_n \phi_n \right\|_U^2 = \lim_{N \to \infty} \sum_{n=1}^{N} \alpha_n^2,$$

where the simplification is accomplished using the orthonormality of $\{\phi_n\}$, as above. Therefore,

$$\|u\|_U^2 = \sum_{n=1}^{\infty} \alpha_n^2,$$

and this completes the proof of the first part of the lemma.

2. Let $S_N$ be the finite-dimensional subspace spanned by $\phi_1, \phi_2, \ldots, \phi_N$. Then $\text{proj}_{S_N} u$ exists and satisfies

$$\|\text{proj}_{S_N} u\|_U^2 + \|u - \text{proj}_{S_N} u\|_U^2 = \|u\|_U^2,$$

which implies that $\|\text{proj}_{S_N} u\|_U^2 \leq \|u\|_U^2$. But

$$\text{proj}_{S_N} u = \sum_{n=1}^{N} \langle u, \phi_n \rangle_U \phi_n$$

and

$$\|\mathrm{proj}_{S_N} u\|_U^2 = \left\|\sum_{n=1}^{N} \langle u, \phi_n \rangle_U \phi_n\right\|_U^2 = \sum_{n=1}^{N} |\langle u, \phi_n \rangle_U|^2.$$

Therefore, for all $N$, $\sum_{n=1}^{N} |\langle u, \phi_n \rangle_U|^2 \le \|u\|_U^2$, and it follows that

$$\sum_{n=1}^{\infty} |\langle u, \phi_n \rangle_U|^2 < \infty,$$

as desired.

3. Clearly each vector of the form $\sum_{n=1}^{\infty} \alpha_n \phi_n$, where the series is convergent, is the limit of vectors lying in $S$ and hence belongs to $\overline{S}$. Conversely, suppose $u \in \overline{S}$. We will show that $u = \sum_{n=1}^{\infty} \langle u, \phi_n \rangle_U \phi_n$, which has the desired form. Since $u \in \overline{S}$, there exists $\{u_j\} \subset S$ such that $u = \lim_{j \to \infty} u_j$. Then there exists an increasing sequence $\{N_j\}$ of positive integers such that $u_j \in S_{N_j}$ for all $u \in \mathbb{Z}^+$, where $S_N = \mathrm{sp}\{\phi_1, \phi_2, \ldots, \phi_N\}$ for all $N \in \mathbb{Z}^+$. Now define, for each $j \in \mathbb{Z}^+$, $\hat{u}_j = \sum_{n=1}^{N_j} \langle u, \phi_n \rangle_U \phi_n$. Since $\hat{u}_j$ is the projection onto $S_{N_j}$ and $u_j \in S_{N_j}$, it follows that $\|u - \hat{u}_j\|_U \le \|u - u_j\|_U$ for all $j$ and therefore

$$u_j \to u \;\Rightarrow\; \hat{u}_j \to u.$$

Since we know that $\sum_{n=1}^{\infty} \langle u, \phi_n \rangle_U \phi_n$ converges, it is now easy (because $\{\hat{u}_j\}$ is a subsequence of the sequence of partial sums) to prove that $\sum_{n=1}^{\infty} \langle u, \phi_n \rangle_U \phi_n$ converges to $u$.

4. Let $u \in U$. Since $w = \mathrm{proj}_{\overline{S}} u \in \overline{S}$, the first and third results imply that

$$w = \sum_{n=1}^{\infty} \langle w, \phi_n \rangle_U \phi_n.$$

But we also know that $w - u \in S^{\perp}$, which yields

$$\langle w - u, \phi_n \rangle_U = 0 \;\Rightarrow\; \langle w, \phi_n \rangle_U = \langle u, \phi_n \rangle_U$$

for all $n$. Therefore, $\mathrm{proj}_{\overline{S}} u = \sum_{n=1}^{\infty} \langle u, \phi_n \rangle_U \phi_n$, as desired.

5. Suppose first that $u \in \bar{S}$. Then

$$u = \text{proj}_{\bar{S}} u = \sum_{n=1}^{\infty} \langle u, \phi_n \rangle_U \phi_n \implies \|u\|_U^2 = \sum_{n=1}^{\infty} |\langle u, \phi_n \rangle_U|^2.$$

Conversely, if $u \notin \bar{S}$, then, defining $\bar{u} = \text{proj}_{\overline{\text{sp}\{S}} u$, we have $u - \bar{u} \neq 0$ and

$$\|u\|_U^2 = \|\bar{u}\|_U^2 + \|u - \bar{u}\|_U^2 \implies \|u\|_U^2 > \|\bar{u}\|_U^2 = \sum_{n=1}^{\infty} |\langle u, \phi_n \rangle_U|^2$$

$$\implies \|u\|_U^2 \neq \sum_{n=1}^{\infty} |\langle u, \phi_n \rangle_U|^2. \qquad \Box$$

We wish to point out two manipulations that are always valid. If $\{\phi_n\}$ is an orthonormal sequence in $U$ and $u = \sum_{n=1}^{\infty} \alpha_n \phi_n$, then for all $w \in U$,

$$\langle u, w \rangle_U = \left\langle \sum_{n=1}^{\infty} \alpha_n \phi_n, w \right\rangle_U = \sum_{n=1}^{\infty} \alpha_n \langle \phi_n, w \rangle_U.$$

This is because the series converges in the $U$-norm and $\langle \cdot, w \rangle_U$ is continuous with respect to the norm topology:

$$\left\langle \sum_{n=1}^{\infty} \alpha_n \phi_n, w \right\rangle_U = \left\langle \lim_{N \to \infty} \sum_{n=1}^{N} \alpha_n \phi_n, w \right\rangle_U$$

$$= \lim_{N \to \infty} \left\langle \sum_{n=1}^{N} \alpha_n \phi_n, w \right\rangle_U \quad \text{(by continuity)}$$

$$= \lim_{N \to \infty} \sum_{n=1}^{N} \alpha_n \langle \phi_n, w \rangle_U \quad \text{(by bilinearity of the inner product)}$$

$$= \sum_{n=1}^{\infty} \alpha_n \langle \phi_n, w \rangle_U.$$

Similarly, if $A$ is a bounded linear operator mapping $U$ to any other Hilbert space, then

$$Au = A\left(\sum_{n=1}^{\infty} \alpha_n \phi_n\right) = \sum_{n=1}^{\infty} \alpha_n A\phi_n.$$

The reasoning is the same: the series converges in the $U$-norm and the operator is continuous with respect to the $U$-norm.

### 4.3.2   Examples of complete orthonormal sets

We will now present some examples of orthonormal sets and complete orthonormal sets in Hilbert spaces.

1. Let us define, for each $n \in \mathbb{Z}^+$, $\hat{\phi}_n(t) = \sin(n\pi t)$. Every continuous function $\phi : [0, 1] \to \mathbb{R}$ belongs to $L^2(0, 1)$, and hence $\hat{\phi}_n \in L^2(0, 1)$ for each $n$. Moreover, a direct calculation shows that

$$\langle \hat{\phi}_n, \hat{\phi}_m \rangle_{L^2} = \int_0^1 \sin(n\pi t)\sin(m\pi t)\, dt = 0 \text{ for all } m \neq n$$

and therefore $\{\hat{\phi}_n : n \in \mathbb{Z}^+\}$ is an orthogonal set. Moreover, another direct calculation shows that

$$\|\hat{\phi}_n\|_{L^2}^2 = \frac{1}{2} \text{ for all } n \in \mathbb{Z}^+.$$

Therefore, if we define $\phi_n = \sqrt{2}\hat{\phi}_n$ for each $n$, then $\{\phi_n : n \in \mathbb{Z}^+\}$ is an orthonormal set in $L^2(0, 1)$.

In fact, $\{\phi_n : n \in \mathbb{Z}^+\}$ is a complete orthonormal set in $L^2(0, 1)$, although it takes some work to prove this. The series

$$\sum_{n=1}^{\infty} \langle x, \phi_n \rangle_{L^2} \phi_n$$

is the *Fourier sine series* of the function $x$. It can be shown (fact number one) that if $x$ is continuous and piecewise smooth (that is, $x'$ is piecewise continuous) on $[0, 1]$ and satisfies $x(0) = x(1) = 0$, then the Fourier sine series converges uniformly to $x$. Let us write $S$

for the subspace of all such $x$. Therefore, if $x \in S$, then

$$\lim_{N \to \infty} \sum_{n=1}^{N} \langle x, \phi_n \rangle_{L^2} \phi_n(t) = x(t) \text{ for all } t \in [0, 1]$$

and this convergence is uniform in $t$. It is easy to verify that uniform convergence implies convergence in the $L^2$-norm. It can also be proven (fact number two) that $S$ is dense in $L^2(0, 1)$; this means that given $x \in L^2(0, 1)$, there exists a sequence $\{x_n\} \subset S$ such that $\|x_n - x\|_{L^2} \to 0$ as $n \to \infty$. Neither of these facts is easy to prove, and we will take them for granted. (For a proof of the first fact, the reader can consult [5] or [7]. For the second, we refer the reader to [4] or [22].) It is straightforward, however, to use these two facts to prove that the Fourier sine series of $x$ converges to $x$ (in the $L^2$-norm) for every $x \in L^2(0, 1)$. Therefore, $\{\phi_n : n \in \mathbb{Z}^+\}$ is a complete orthonormal set for $L^2(0, 1)$.

Every $\phi_n$ satisfies the boundary conditions $\phi_n(0) = \phi_n(1) = 0$. Nonetheless, every function in $L^2(0, 1)$ can be represented by this complete orthonormal set, whether it satisfies the boundary conditions or not. We give two examples.

If $x_1(t) = t(1 - t)$, then $\sum_{n=1}^{N} \langle x_1, \phi_n \rangle_{L^2} \phi_n$ converges rapidly to $x$ as $N \to \infty$. Figure 4.3 shows $x_1$ together with the partial Fourier series $\sum_{n=1}^{N} \langle x_1, \phi_n \rangle_{L^2} \phi_n$ for $N = 1$ and $N = 5$. It should be noted that the approximation is already indistinguishable from the function $x_1$ (on this scale) with only $N = 5$.

As a second example, we take $x_2(t) = t$. Figure 4.4 shows $x_2$ together with $\sum_{n=1}^{N} \langle x_2, \phi_n \rangle_{L^2} \phi_n$ for $N = 10$ and $N = 20$. It should be believable that $\sum_{n=1}^{N} \langle x_2, \phi_n \rangle_{L^2} \phi_n$ converges to $x_2$ in the $L^2$-norm as $N \to \infty$, but now the convergence is not so fast (and in particular, the convergence is not uniform).

The difference in the two examples is that $x_1$ satisfies the boundary conditions $x_1(0) = x_1(1) = 0$, while $x_2(1) \neq 0$.

2. Other complete orthonormal sets for $L^2(0, 1)$ can be derived from Fourier series representations. For instance, if $\psi_n(t) = \sqrt{2} \cos(n\pi t)$ for $n \geq 1$ and $\psi_0(t) = 1$, then $\{\psi_n : n = 0, 1, 2, \ldots\}$ is a complete

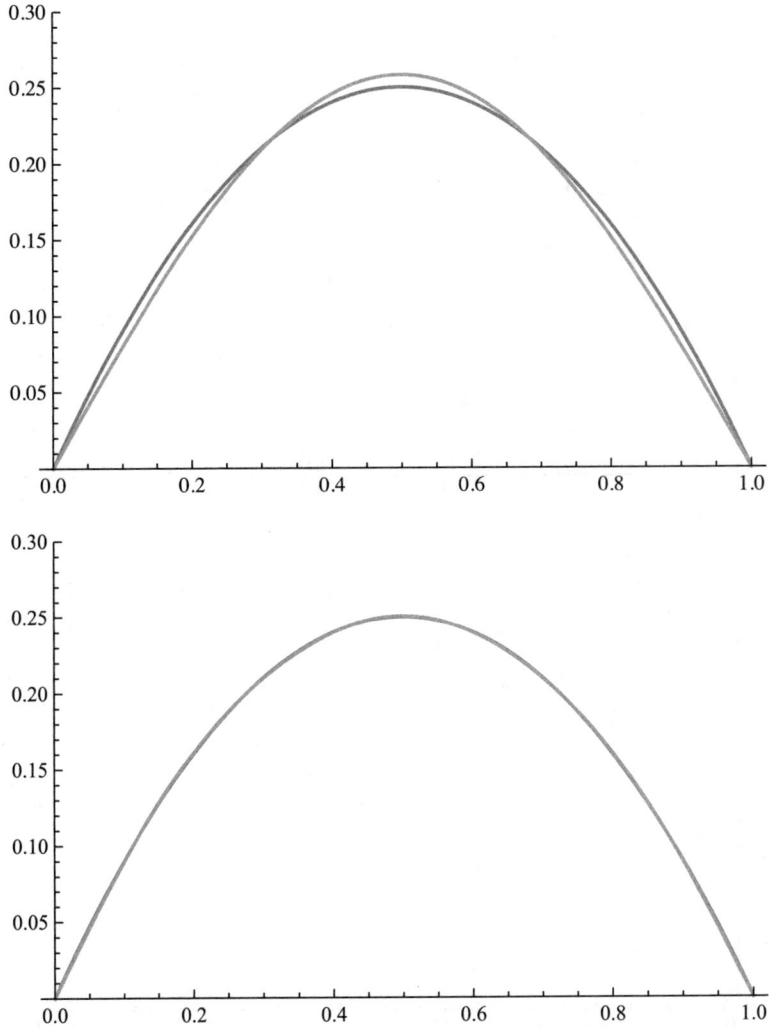

**Figure 4.3.** The function $x_1(t) = t(1-t)$ and the approximation $\sum_{n=1}^{N} \langle x_1, \phi_n \rangle_{L^2} \phi_n$ for $N = 1$ (top) and $N = 5$ (bottom).

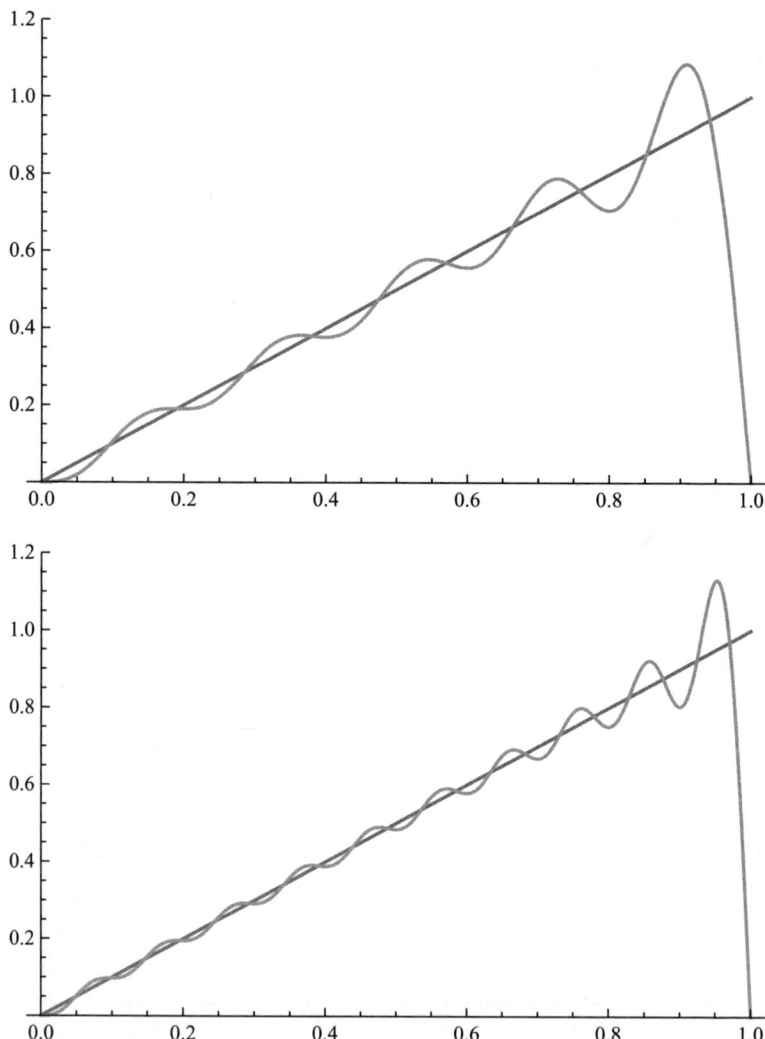

**Figure 4.4.** The function $x_2(t) = t$ and the approximation $\sum_{n=1}^{N} \langle x_2, \phi_n \rangle_{L^2} \phi_n$ for $N = 10$ (top) and $N = 20$ (bottom).

orthonormal set for $L^2(0, 1)$. The corresponding series representation of $x \in L^2(0, 1)$ is called the *Fourier cosine series* of $x$. Another example is $\{\hat{\phi}_n : n \in \mathbb{Z}^+\}$, where

$$\hat{\phi}_n(t) = \sqrt{2} \sin \left( \frac{(2n-1)\pi}{2} t \right), \quad n \in \mathbb{Z}^+.$$

The corresponding series representation of $x \in L^2(0, 1)$ is called the Fourier *quarter-wave sine series* of $x$. Replacing the sine function by cosine yields the *quarter-wave cosine series* of $x$. These orthonormal sets can each be used to represent any function in $L^2(0, 1)$, and each leads to a rapidly converging series when representing a function $x$ satisfying appropriate boundary conditions.

3. Among the many other examples of complete orthonormal sets that could be named, we describe the *Chebyshev polynomials*, which are defined by $T_n(t) = \cos(n \cos^{-1}(t))$, $n = 0, 1, 2, \ldots$. It is not at all obvious that $T_n$ is a polynomial for all $n$. However, it is clear that $T_0(t) = 1$ and $T_1(t) = t$, and the trigonometric identity

$$\cos((n+1)\theta) = 2\cos(\theta)\cos(n\theta) - \cos((n-1)\theta)$$

shows that $T_{n+1}(t) = 2tT_n(t) - T_{n-1}(t)$ for $n \geq 1$. It follows that each $T_n$ is a polynomial, with

$$T_2(t) = 2t^2 - 1, \quad T_3(t) = 4t^3 - 3t, \quad T_4(t) = 8t^4 - 8t^2 + 1, \ldots.$$

Let us define $X$ to be the space of all measurable, real-valued functions $x$ defined on $[-1, 1]$ with the property that

$$\int_{-1}^{1} \frac{x(t)^2}{\sqrt{1-t^2}} \, dt < \infty.$$

The inner product on $X$ is

$$\langle x, y \rangle_X = \int_{-1}^{1} \frac{x(t)y(t)}{\sqrt{1-t^2}} \, dt.$$

It can be verified that $\{T_n : n = 0, 1, 2, \ldots\}$ forms an orthogonal set in $X$; the identity

$$\langle x, y \rangle_X = \int_0^\pi x(\cos(\theta)) y(\cos(\theta)) \, d\theta$$

is useful in this regard. Also, $\|T_n\|_X^2 = \pi/2$ for all $n \geq 1$, with $\|T_0\|_X^2 = 2$. Therefore, the Chebyshev polynomials can easily be normalized to create an orthonormal set. Finally, it can be shown that this set is complete (the corresponding representation of $x \in X$ is the Fourier cosine series of $x(\cos(\theta))$).

### 4.3.3    The spectral theorem

The next theorem describes the eigenvalues of a compact self-adjoint linear operator.

**Theorem 4.16.** *Let $U$ be a Hilbert space and let $L : U \to U$ be a compact self-adjoint linear operator.*

1. *If $\lambda$ is a nonzero eigenvalue of $L$, then $\dim\{u \in U : Lu = \lambda u\}$ is finite.*

2. *If $L$ has infinitely many nonzero eigenvalues, they form a sequence tending to zero.*

*Proof.* We argue by contradiction and suppose that the theorem fails. There are three ways in which the theorem can fail, and we argue that in each case, there must exist a positive number $\epsilon$, a sequence $\{\lambda_n\}$ of eigenvalues with $|\lambda_n| \geq \epsilon$ for all $n$, and a corresponding orthonormal sequence $\{u_n\}$ of eigenvectors. We now justify this.

If conclusion 1 fails, there exists $\lambda \neq 0$ with $\{u \in U : Lu = \lambda u\}$ having infinite dimension. This eigenspace must then contain an infinite orthonormal sequence $\{u_n\}$, and we define the sequence $\{\lambda_n\}$ by $\lambda_n = \lambda$ for all $n$. We also define $\epsilon = |\lambda|$.

If conclusion 2 fails because the set of eigenvalues is uncountable, then there must exist a positive integer $N$ such that the number of (distinct) eigenvalues $\lambda$ with $|\lambda| \geq 1/N$ is infinite. Otherwise, the set of all nonzero eigenvalues could be written as a countable union of finite sets, contradicting that the set is uncountable. Set $\epsilon = 1/N$, choose any sequence of distinct eigenvalues $\{\lambda_n\}$ with $|\lambda_n| \geq \epsilon$ for all $n$, and let $\{u_n\}$ be the sequence of corresponding eigenvectors, each normalized to have unit length. Then, by Theorem 4.12, $\{u_n\}$ is an orthonormal sequence.

Finally, if (2) fails because the nonzero eigenvalues of $L$ form a countable set that includes a sequence not converging to zero, then we can certainly find $\epsilon > 0$ and a sequence of distinct nonzero eigenvalues $\{\lambda_n\}$ with $|\lambda_n| \geq \epsilon$ for all $n$. Again, the sequence $\{u_n\}$ of corresponding (normalized) eigenvectors forms an orthonormal sequence.

Now, we have $\langle u_n, u_m \rangle_U = 0$ for all $n \neq m$, and so

$$
\begin{aligned}
\|Lu_n - Lu_m\|^2 &= \|\lambda_n u_n - \lambda_m Lu_m\|^2 \\
&= \langle \lambda_n u_n - \lambda_m u_m, \lambda_n u_n - \lambda_m u_m \rangle_U \\
&= \lambda_n^2 \|u_n\|_U^2 + \lambda_m^2 \|u_m\|^2 \geq 2\epsilon^2,
\end{aligned}
$$

that is, $\|Lu_n - Lu_m\| \geq \sqrt{2}\epsilon$. But then no subsequence of $\{Lu_n\}$ can possibly be Cauchy, which contradicts the compactness of $L$.    □

The following two lemmas will be needed to derive the spectral decomposition of a compact, self-adjoint linear operator.

**Lemma 4.17.** *Let $H$ be a Hilbert space and let $L : H \to H$ be a bounded linear operator. Suppose $M$ is a closed subspace of $H$. If $M$ is invariant under $L$ (that is, $L(M) \subseteq M$), then $M^\perp$ is invariant under $L^*$.*

*Proof.* We wish to show that $L^*u \in M^\perp$ whenever $u \in M^\perp$. So suppose that $u \in M^\perp$. We have, for each $v \in M$,

$$(L^*u, v) = (u, Lv) = 0,$$

since $Lv \in M$ ($M$ is invariant under $L$) and $u \in M^\perp$. Therefore, $L^*u \in M^\perp$.    □

The operator defined in the following lemma is the analogue of the matrix $\mathbf{vu}^T \in \mathbb{R}^{m \times n}$ defined by $\mathbf{u} \in \mathbb{R}^n$, $\mathbf{v} \in \mathbb{R}^m$.

**Lemma 4.18.** *Let $U$, $V$ be Hilbert spaces and let $u \in U$, $v \in V$. The operator*

$$v \otimes u : U \to V$$

*defined by*

$$(v \otimes u)w = \langle w, u \rangle_U v$$

*is linear and bounded, with $\|v \otimes u\| = \|u\|_U \|v\|_V$.*

*Proof.* The linearity of the operator follows immediately from the linearity of $\langle \cdot, u \rangle_U$. We have

$$\|\langle w, u \rangle_U v\|_V = |\langle w, u \rangle_U| \|v\|_V \le \|u\|_U \|v\|_V \|w\|_U$$

(by the Cauchy-Schwarz inequality), which shows that $v \otimes u$ is bounded and that

$$\|v \otimes u\| \le \|u\|_U \|v\|_V.$$

Moreover, if we take $w = u$, then equality holds in the Cauchy-Schwarz inequality, yielding $\|\langle w, u \rangle_U v\|_V = \|u\|_U \|v\|_V \|w\|_U$. This shows that $\|v \otimes u\| = \|u\|_U \|v\|_V$. $\qquad\square$

The operator $u \otimes v$ is called the *outer product* of $u$ and $v$.

We now prove the spectral theorem for a compact self-adjoint linear operator.

**Theorem 4.19.** *Let $U$ be a Hilbert space and let $L : U \to U$ be a compact self-adjoint linear operator. Then there exists an orthonormal sequence $\{u_n\}$, either finite or infinite, of eigenvectors of $L$ and a corresponding sequence $\{\lambda_n\}$ of eigenvalues such that*

$$L = \sum_n \lambda_n u_n \otimes u_n,$$

*where the series (if infinite) converges to $L$ in the operator norm. In particular,*

$$Lv = \sum_n \lambda_n \langle v, u_n \rangle_U u_n \text{ for all } v \in U.$$

*If the sequence $\{\lambda_n\}$ is infinite, then $\lambda_n \to 0$ as $n \to \infty$.*

*Proof.* We construct the eigenpairs by induction. We know that $L$ has an eigenvalue $\lambda_1 = \pm\|L\|$ and a corresponding unit eigenvector $u_1$. Suppose, by way of induction, that $\lambda_1, \ldots, \lambda_k$ are eigenvalues of $L$ with orthonormal eigenvectors $u_1, \ldots, u_k$. The subspace $M_k = \text{sp}\{u_1, \ldots, u_k\}$ is clearly invariant under $L$, so $M_k^\perp$ is invariant under $L^* = L$. Define $L_k$ to be $L$ restricted to $M_k^\perp$. Then $L_k : M_k^\perp \to M_k^\perp$, $L_k$ is obviously linear, bounded, and compact, and

$$\langle L_k u, v \rangle_U = \langle Lu, v \rangle_U = \langle u, L^* v \rangle_U$$
$$= \langle u, Lv \rangle_U = \langle u, L_k v \rangle_U \quad \text{for all } u, v \in M_k^\perp.$$

Because $L_k v \in M_k^\perp$, this calculation shows that $L_k^* = L_k$ and hence that $L_k$ is self-adjoint. Therefore $L_k$ has an eigenpair $\lambda_{k+1} = \pm\|L_k\|$, $u_{k+1}$ ($\|u_{k+1}\| = 1$), which is also an eigenpair of $L$. Note that $u_{k+1}$ is necessarily orthogonal to $u_1, \ldots, u_k$ (since $u_{k+1} \in M_k^\perp$). For future reference, note that, since $M_1^\perp \supset M_2^\perp \supset \cdots$, we have $|\lambda_1| \geq |\lambda_2| \geq \cdots$.

Suppose that, for some $k$, $L_k$ is the zero operator (that is, that $L|_{M_k^\perp}$ is the zero operator). We can write $U = M_k \oplus M_k^\perp$ and we have

$$Lu = L\left(\text{proj}_{M_k} u + \text{proj}_{M_k^\perp} u\right) = L\left(\text{proj}_{M_k} u\right) + L\left(\text{proj}_{M_k^\perp} u\right)$$
$$= L\left(\text{proj}_{M_k} u\right).$$

But $\text{proj}_{M_k} u = \sum_{n=1}^k \langle u, u_n \rangle_U u_n$, and so

$$Lu = L\left(\sum_{n=1}^k \langle u, u_n \rangle_U u_n\right) = \sum_{n=1}^k \langle u, u_n \rangle_U L u_n = \sum_{n=1}^k \lambda_n \langle u, u_n \rangle_U u_n$$
$$= \left(\sum_{n=1}^k \lambda_n u_n \otimes u_n\right) u.$$

Thus, in this case,

$$L = \sum_{n=1}^k \lambda_n u_n \otimes u_n,$$

and the result is proved.

Now suppose that $L_k \neq 0$ for all $k$. Then we obtain an infinite sequence $\{\lambda_n\}$ of nonzero eigenvalues and a corresponding orthonormal sequence $\{u_n\}$ of eigenvectors. Moreover, by Theorem 4.16, we must have $\lambda_n \to 0$. We wish to show that

$$L = \sum_{n=1}^{\infty} \lambda_n u_n \otimes u_n,$$

that is, that

$$\left\| L - \sum_{n=1}^{k} \lambda_n u_n \otimes u_n \right\| \to 0 \text{ as } k \to \infty. \tag{4.5}$$

For each $x \in U$ with $\|x\| \leq 1$, write

$$x = x_k + \sum_{n=1}^{k} \langle x, u_n \rangle_U u_n,$$

so that $x_k = \operatorname{proj}_{M_k^\perp} x$. Then

$$Lx = Lx_k + L\left( \sum_{n=1}^{k} \langle x, u_n \rangle_U u_n \right) = Lx_k + \sum_{n=1}^{k} \lambda_n \langle x, u_n \rangle_U u_n$$

$$= Lx_k + \left( \sum_{n=1}^{k} \lambda_n u_n \otimes u_n \right) x.$$

Thus

$$\left( L - \sum_{n=1}^{k} \lambda_n u_n \otimes u_n \right) x = Lx_k$$

and, since $x_k \in M_k^\perp$, we have $Lx_k = L_k x_k$. But $\|L_k\| = |\lambda_{k+1}|$, so

$$\left\| \left( L - \sum_{n=1}^{k} \lambda_n u_n \otimes u_n \right) x \right\| = \|Lx_k\| \leq |\lambda_{k+1}|.$$

Since this holds for all $x$ with $\|x\| \leq 1$ and $|\lambda_{k+1}| \to 0$, we have proved (4.5). $\qquad\square$

The spectral representation

$$L = \sum_n \lambda_n u_n \otimes u_n$$

exhibits all the eigenvalues of $L$ except possibly 0. If $\mathcal{N}(L)$ is nontrivial, then 0 is an eigenvalue of $L$. Moreover, we have the following result about $\mathcal{N}(L)$.

**Theorem 4.20.** *Let $U$ be a Hilbert space, and let $L : U \to U$ be a compact self-adjoint linear operator with spectral representation*

$$L = \sum_n \lambda_n u_n \otimes u_n.$$

*Then*

$$\mathrm{sp}\{u_1, u_2, \ldots\}^\perp = \mathcal{N}(L).$$

*Proof.* For any $x \in U$, we have

$$Lx = \sum_n \lambda_n \langle x, u_n \rangle_U u_n.$$

Therefore, $Lx = 0$ if and only if $\langle x, u_n \rangle_U = 0$ for all eigenvectors $u_n$. In other words, $Lx = 0$ if and only if $x \in \mathrm{sp}\{u_1, u_2, \ldots\}^\perp$, which is what we wanted to prove. $\qquad\square$

**Example 4.21.** To illustrate Theorem 4.19, we define $T : L^2(0, 1) \to L^2(0, 1)$ by

$$(Tx)(t) = \int_0^1 k(t, s) x(s) \, ds,$$

where

$$k(t, s) = \begin{cases} s(1 - t) & 0 \le s \le t, \\ t(1 - s) & t \le s \le 1. \end{cases}$$

It can be verified that $T$ is self-adjoint; this follows from the fact that $k(s, t) = k(t, s)$ for all $s, t$. For computational purposes, it is convenient to notice that

$$(Tx)(t) = (1 - t) \int_0^t s x(s) \, ds + t \int_t^1 (1 - s) x(s) \, ds.$$

Let $\phi_n(t) = \sqrt{2}\sin(n\pi t)$ for $n \in \mathbb{Z}^+$. Then, as noted above, $\{\phi_n : n \in \mathbb{Z}^+\}$ is a complete orthonormal set for $L^2(0, 1)$, and a direct calculation shows that, for each $n \in \mathbb{Z}^+$,

$$T\phi_n = \lambda_n\phi_n, \ \lambda_n = \frac{1}{n^2\pi^2}.$$

It follows that

$$T = \sum_{n=1}^{\infty}\lambda_n\phi_n \otimes \phi_n;$$

in other words, for all $x \in L^2(0, 1)$,

$$(Tx)(t) = \sum_{n=1}^{\infty}\frac{a_n}{n^2\pi^2}\sin(n\pi t), \ a_n = 2\int_0^1 x(s)\sin(n\pi s)\,ds.$$

The operator $T$ in this example is the solution operator of the two-point boundary value problem

$$-u''(t) = x(t) \text{ for all } t \in (0, 1),$$
$$u(0) = u(1) = 0 \tag{4.6}$$

that was considered in Section 1.2. The operator $T$ can be derived by solving (4.6) by integrating twice, interchanging the order of integration, and simplifying. The eigenvalues and eigenfunctions represented here can be obtained by solving (4.6) with $f(t) = \lambda u(t)$ using standard techniques for second-order ordinary differential equations. The eigenvalues of $T$ are the reciprocals of the eigenvalues found by solving (4.6).

The reader should note that it is unusual to be able to compute the eigenvalues and eigenvectors of a self-adjoint compact operator $T : X \to X$ explicitly. In a typical case, we know the eigenpairs exist, and we can approximate a finite number of them if needed, but we cannot derive explicit formulas for them.

## 4.4 The singular value expansion of a compact operator

By applying the spectral theorem for compact self-adjoint operators to $L^*L$, we can derive a singular value expansion for $L$ itself, where $L$ is compact (but not necessarily self-adjoint). This is analogous to how, in Section 4.1, we used the spectral representation of a symmetric matrix to derive the singular value decomposition of a general matrix.

**Theorem 4.22.** *Let $U$, $V$ be Hilbert spaces, and let $L : U \to V$ be a compact linear operator. Then there exist (finite or infinite) orthonormal sequences $\{u_n\} \subset U$ and $\{v_n\} \subset V$ and positive numbers $\sigma_1 \geq \sigma_2 \geq \cdots$ such that*

$$L = \sum_n \sigma_n v_n \otimes u_n.$$

*If the series is infinite, it converges in the operator norm to $L$. We have*

$$Lu_n = \sigma_n v_n \text{ for all } n$$

*and*

$$Lu = \sum_n \sigma_n \langle u_n, u \rangle_U v_n \text{ for all } u \in U.$$

*Proof.* Because $L^*L$ is compact and self-adjoint, we can write

$$L^*L = \sum_n \lambda_n u_n \otimes u_n,$$

where $\{u_1, u_2, \ldots\}$ is an orthonormal sequence and $|\lambda_1| \geq |\lambda_2| \geq \cdots > 0$. Moreover,

$$\lambda_n = \lambda_n \langle u_n, u_n \rangle_U = \langle \lambda_n u_n, u_n \rangle_U = \langle L^*Lu_n, u_n \rangle_U = \langle Lu_n, Lu_n \rangle_U \geq 0.$$

Thus we can write $\lambda_n = \sigma_n^2$, with $\sigma_1 \geq \sigma_2 \geq \cdots > 0$.

Define $v_n = \sigma_n^{-1} L u_n$. Note that

$$
\begin{aligned}
\langle v_m, v_n \rangle_U = \langle \sigma_m^{-1} L u_m, \sigma_n^{-1} L u_n \rangle_U &= \sigma_m^{-1} \sigma_n^{-1} \langle u_m, L^* L u_n \rangle_U \\
&= \sigma_m^{-1} \sigma_n^{-1} \langle u_m, \lambda_n u_n \rangle_U \\
&= \sigma_m^{-1} \sigma_n^{-1} \lambda_n \langle u_m, u_n \rangle_U \\
&= \begin{cases} 0, & m \neq n, \\ \frac{\lambda_n}{\sigma_n^2}, & m = n \end{cases} \\
&= \begin{cases} 0, & m \neq n, \\ 1, & m = n. \end{cases}
\end{aligned}
$$

Thus $\{v_n\}$ is an orthonormal sequence.

We wish to show that

$$
L = \sum_n \sigma_n v_n \otimes u_n.
$$

Since $\mathcal{N}(L) = \mathcal{N}(L^* L) = \mathrm{sp}\{u_1, u_2, \ldots\}^{\perp}$ (see Theorem 4.20), we have

$$
\begin{aligned}
Lu = L\left(\mathrm{proj}_{\mathcal{N}(L)^{\perp}} u\right) = L\left(\sum_n \langle u, u_n \rangle_U u_n\right) &= \sum_n \langle u, u_n \rangle_U L u_n \\
&= \sum_n \sigma_n \langle u, u_n \rangle_U v_n \\
&= \sum_n \sigma_n \left(v_n \otimes u_n\right) u,
\end{aligned}
$$

which shows that

$$
L = \sum_n \sigma_n v_n \otimes u_n
$$

is valid in the sense of pointwise convergence. It remains only to show, in the case of infinitely many positive singular values, that the series converges to $L$ in the operator norm. Let $u \in U$ have norm one. Then

$$
Lu - \left(\sum_{n=1}^N \sigma_n v_n \otimes u_n\right) u = \sum_{n=N+1}^{\infty} \sigma_n \langle u, u_n \rangle_U v_n,
$$

which implies that

$$\left\|\left(L - \sum_{n=1}^{N} \sigma_n v_n \otimes u_n\right) u\right\|_V^2 = \left\|\sum_{n=N+1}^{\infty} \sigma_n \langle u, u_n \rangle_U v_n\right\|_V^2$$

$$= \sum_{n=N+1}^{\infty} \sigma_n^2 |\langle u, u_n \rangle_U|^2$$

$$\leq \sigma_{N+1}^2 \sum_{n=N+1}^{\infty} |\langle u, u_n \rangle_U|^2$$

$$\leq \sigma_{N+1}^2 \|u\|_U^2 = \sigma_{N+1}^2.$$

It follows that

$$\left\|L - \sum_{n=1}^{N} \sigma_n v_n \otimes u_n\right\|_V \leq \sigma_{N+1} \to 0 \text{ as } N \to \infty. \qquad \square$$

The representation of Theorem 4.22 is called the *singular value expansion* of the operator $L$. We wish to emphasize several points about it. If $L$ has only finitely many positive singular values, say $N$, then $L$ is clearly of finite rank, since every element of $\mathcal{R}(L)$ belongs to $\mathrm{sp}\{v_1, v_2, \ldots, v_N\}$. Therefore, the interesting case (the case that leads to a true inverse problem) occurs if $L$ has infinitely many positive singular values:

$$L = \sum_{n=1}^{\infty} \sigma_n v_n \otimes u_n.$$

As noted above, $Lu_n = \sigma_n v_n$ for all $n$. Thus, if $x = \sum_{n=1}^{\infty} \alpha_n u_n \in U$, where the series is convergent, then

$$Lx = \sum_{n=1}^{\infty} \alpha_n L u_n = \sum_{n=1}^{\infty} \sigma_n \alpha_n v_n.$$

However, it should be noted that if $\mathcal{N}(L)$ is nontrivial, then not every $x \in U$ can be represented in the form $x = \sum_{n=1}^{\infty} \alpha_n u_n$. In fact, every vector $x$ of this form belongs to $\mathcal{N}(T)^{\perp}$.

One of the consequences of these comments (together with Lemma 4.15) is that

$$\text{sp}\{v_1, v_2, \ldots\} \subset \mathcal{R}(L) \subset \overline{\text{sp}\{v_1, v_2, \ldots\}}.$$

We can now give a second proof of the result that if $L$ is compact and does not have finite rank, then $\mathcal{R}(L)$ is not closed (cf. Theorem 4.7). If $L$ does not have finite rank, then it has an infinite sequence $\{\sigma_n\}$ of positive singular values, and corresponding infinite sequences $\{u_n\}$, $\{v_n\}$ of singular vectors. We have $\|u_n\|_U = 1$, $\|v_n\|_V = 1$ for all $n$, and $Lu_n = \sigma_n v_n$. But then

$$\langle L^*Lu_n, u_n \rangle_U = \|Lu_n\|_V^2 = \sigma_n^2 \to 0 \text{ as } n \to \infty.$$

It follows from Theorem 2.20 that $\mathcal{R}(L)$ is not closed. The converse of this is obvious: if $T$ has only finitely many positive singular values, then $T$ has finite rank and hence $\mathcal{R}(T)$ is closed.

We mentioned in Section 4.2 that $L^*$ is always compact if $L$ is a compact operator. We can prove this using the singular value expansion and the following lemma.

**Lemma 4.23.** *Let $U$ and $V$ be Hilbert spaces, let $\{\phi_n\} \subset U$ and $\{\psi_n\} \subset V$ be orthogonal sequences, and let $\{\sigma_n\}$ be a nonincreasing sequence of positive real numbers that converges to zero. Then*

$$L = \sum_{n=1}^{\infty} \sigma_n \psi_n \otimes \phi_n$$

*defines a compact linear operator mapping $U$ into $V$.*

*Proof.* Since, for each $N \in \mathbb{Z}^+$, $\sum_{n=1}^{N} \sigma_n \psi_n \otimes \phi_n$ is a finite-rank operator, it suffices to show that the series converges in the operator norm; then the sum is the norm-limit of finite-rank operators and hence is compact by Corollary 4.9. To show that the series converges, it suffices to prove that the sequence of partial sums is Cauchy. For all $M, N \in \mathbb{Z}^+$, $M > N$, we have

$$\sum_{n=1}^{M} \sigma_n \psi_n \otimes \phi_n - \sum_{n=1}^{N} \sigma_n \psi_n \otimes \phi_n = \sum_{n=N+1}^{M} \sigma_n \psi_n \otimes \phi_n.$$

Moreover, for all $x \in U$, $\|x\|_U = 1$, we have

$$\left\| \left( \sum_{n=N+1}^{M} \sigma_n \psi_n \otimes \phi_n \right) x \right\|_V^2 = \left\| \sum_{n=N+1}^{M} \sigma_n \langle \phi_n, x \rangle_U \psi_n \right\|_V^2$$

$$= \sum_{n=N+1}^{M} \sigma_n^2 |\langle x, \phi_n \rangle_U|^2$$

$$\leq \sum_{n=N+1}^{\infty} \sigma_n^2 |\langle x, \phi_n \rangle_U|^2$$

$$\leq \sigma_{N+1}^2 \sum_{n=N+1}^{\infty} |\langle x, \phi_n \rangle_U|^2$$

$$\leq \sigma_{N+1}^2 \sum_{n=1}^{\infty} |\langle x, \phi_n \rangle_U|^2$$

$$\leq \sigma_{N+1}^2 \|x\|_U^2 = \sigma_{N+1}^2.$$

Since $\sigma_{N+1} \to 0$ as $N \to \infty$, this shows that the sequence of partial sums is Cauchy. $\qquad \square$

**Theorem 4.24.** *Let $U$ and $V$ be Hilbert spaces and let $L : U \to V$ be a compact operator with singular value expansion*

$$L = \sum_{n=1}^{\infty} \sigma_n \psi_n \otimes \phi_n.$$

*Then $L^*$ is also compact, with SVE*

$$L^* = \sum_{n=1}^{\infty} \sigma_n \phi_n \otimes \psi_n.$$

*Proof.* For all $u \in U$ and $v \in V$, we have

$$Lu = \left( \sum_{n=1}^{\infty} \sigma_n \psi_n \otimes \phi_n \right) u = \sum_{n=1}^{\infty} \sigma_n \langle u, \phi_n \rangle_U \psi_n$$

and hence

$$\langle Lu, v \rangle_V = \left\langle \sum_{n=1}^{\infty} \sigma_n \langle u, \phi_n \rangle_U \psi_n, v \right\rangle_V = \sum_{n=1}^{\infty} \sigma_n \langle u, \phi_n \rangle_U \langle \psi_n, v \rangle_V$$

$$= \sum_{n=1}^{\infty} \langle u, \sigma_n \langle \psi_n, v \rangle_V \phi_n \rangle_U$$

$$= \sum_{n=1}^{\infty} \langle u, \sigma_n (\phi_n \otimes \psi_n) v \rangle_U$$

$$= \left\langle u, \left( \sum_{n=1}^{\infty} \sigma_n \phi_n \otimes \psi_n \right) v \right\rangle_U.$$

By the previous lemma, $\sum_{n=1}^{\infty} \sigma_n \phi_n \otimes \psi_n$ converges to a compact linear operator, and the above calculation shows that the limit is $L^*$.    $\square$

**Example 4.25.** Let us consider the operator $T : L^2(0, 1) \to L^2(0, 1)$ defined by

$$(Tx)(t) = \int_0^t x(s)\, ds.$$

To find the singular values and singular functions of $T$, we follow the proof of Theorem 4.22 and find the spectral representation of the operator $T^*T$. For any $x, y \in L^2(0, 1)$, we have

$$\langle Tx, y \rangle_{L^2(0,1)} = \int_0^1 (Tx)(t)y(t)\, dt = \int_0^1 \left( \int_0^t x(s)\, ds \right) y(t)\, dt$$

$$= \int_0^1 \int_0^t x(s)y(t)\, ds\, dt$$

$$= \int_0^1 \int_s^1 x(s)y(t)\, dt\, ds$$

$$= \int_0^1 x(s) \left( \int_s^1 y(t)\, dt \right) ds$$

$$= \langle x, T^*y \rangle_{L^2(0,1)},$$

where

$$(T^*y)(s) = \int_s^1 y(t)\,dt.$$

The critical step in the calculation was interchanging the order of integration.

If we define $u = T^*Tx$, then it is straightforward to verify that $u'' = -x$ and $u(1) = u'(0) = 0$. Therefore, $T^*T$ is the solution operator of the BVP

$$-u''(t) = x(t) \text{ for all } t \in (0, 1),$$

$$u(1) = 0,$$

$$u'(0) = 0.$$

We can replace $x$ by $\lambda u$ in this BVP and use standard methods of solving second-order ordinary differential equations to find an infinite sequence of eigenpairs,

$$\hat{\lambda}_n = \frac{(2n-1)^2\pi^2}{4}, \quad \hat{\phi}_n(t) = \cos\left(\frac{(2n-1)\pi}{2}t\right), \quad n \in \mathbb{Z}^+.$$

A direct calculation shows that $\|\hat{\phi}_n\|_{L^2} = 1/\sqrt{2}$ for all $n \in \mathbb{Z}^+$, and therefore we define $\phi_n = \sqrt{2}\hat{\phi}_n$ for each $n$. Since $T^*T$ is the solution operator of the BVP (that is, the inverse of the differential operator $-d^2/dt^2$ subject to the given boundary conditions), its eigenvalues are the reciprocals of $\hat{\lambda}_1, \hat{\lambda}_2, \hat{\lambda}_3, \ldots$. Therefore,

$$\lambda_n = \frac{4}{(2n-1)^2\pi^2}, \quad \phi_n(t) = \sqrt{2}\cos\left(\frac{(2n-1)\pi}{2}t\right), \quad n \in \mathbb{Z}^+$$

are eigenpairs of $T^*T$. We noted in Section 4.3.2 that $\{\phi_n : n \in \mathbb{Z}^+\}$ is a complete orthonormal set for $L^2(0, 1)$; therefore, these calculations show that

$$T^*T = \sum_{n=1}^\infty \lambda_n \phi_n \otimes \phi_n.$$

Since $\lambda_n \to 0$ as $n \to \infty$, this representation suffices to show that $T^*T$ is a compact operator (see Lemma 4.23).

Following the proof of Theorem 4.22, we define $\sigma_n = \sqrt{\lambda_n}$ and $\psi_n = \sigma_n^{-1} T \phi_n$; a direct calculation shows that

$$\psi_n(t) = \sqrt{2} \sin\left(\frac{(2n-1)\pi}{2}t\right) \text{ for all } n \in \mathbb{Z}^+.$$

Again using the fact that $\{\phi_n : n \in \mathbb{Z}^+\}$ is a complete orthonormal set for $L^2(0,1)$, we see that $T$ has a singular value expansion, namely

$$T = \sum_{n=1}^{\infty} \sigma_n \psi_n \otimes \phi_n,$$

which shows that $T$ is compact. We can represent $T$ more explicitly: for each $x \in L^2(0,1)$,

$$(Tx)(t) = \sum_{n=1}^{\infty} \frac{2a_n}{(2n-1)\pi} \sin\left(\frac{(2n-1)\pi}{2}t\right),$$

$$a_n = 2 \int_0^1 x(t) \cos\left(\frac{(2n-1)\pi}{2}t\right) dt.$$

## 4.4.1    Integral operators of the first kind

We have presented several examples of operators defined by integration; since the most important inverse problems involve integral operators, we will now consider these operators in more detail. Let $\Omega$ be an open subset of $\mathbb{R}^n$ (in applications, $n$ is typically 1, 2, or 3). Let $k \in L^2(\Omega \times \Omega)$ and define $K : L^2(\Omega) \to L^2(\Omega)$ by

$$(Ku)(x) = \int_\Omega k(x,y)u(y)\,dy. \tag{4.7}$$

The operator $K$ called an *integral operator of the first kind*, and the function $k$ is called the *kernel* of the integral operator (an unfortunate choice of words, since the null space of a general linear operator is also called the kernel of the operator; the two uses of the word kernel must not be confused). By standard Lebesgue integration theory, $k(x, \cdot)$ belongs to $L^2(\Omega)$ for almost every $x \in \Omega$ and therefore $(Ku)(x)$ (which is the $L^2$ inner product of $k(x, \cdot)$ and $u$) is a well-defined real number for almost every $x \in \Omega$.

Moreover, for all $u \in L^2(\Omega)$,

$$(Ku)(x) = \int_\Omega k(x, y)u(y)\, dy = \langle k(x, \cdot), u \rangle_{L^2} \leq \|k(x, \cdot)\|_{L^2} \|u\|_{L^2}$$

$$\Rightarrow \int_\Omega (Ku)^2(x)\, dx \leq \int_\Omega \|k(x, \cdot)\|_{L^2}^2 \|u\|_{L^2}^2\, dx$$

$$= \left( \int_\Omega \int_\Omega k(x, y)^2\, dy\, dx \right) \|u\|_{L^2}^2$$

$$\Rightarrow \int_\Omega (Ku)^2(x)\, dx \leq \|k\|_{L^2(\Omega \times \Omega)}^2 \|u\|_{L^2(\Omega)}^2.$$

This shows that $Ku \in L^2(\Omega)$ and also that $\|Ku\|_{L^2(\Omega)} \leq \|k\|_{L^2(\Omega \times \Omega)}$ $\|u\|_{L^2(\Omega)}$ for each $u \in L^2(\Omega)$. It follows that $K$ is bounded, with $\|K\| \leq \|k\|_{L^2(\Omega \times \Omega)}$.

We will now show that, in fact, $K$ is compact for any $k \in L^2(\Omega \times \Omega)$. It follows that, except in the case that $K$ has finite rank, $\mathcal{R}(K)$ fails to be closed and $Kx = y$ is a genuine inverse problem. The case of a finite-rank integral operator is analyzed below.

To show that the operator $K$ is compact, we will use the fact that there exists a complete orthonormal set $\{\phi_n\}$ for $L^2(\Omega)$. Since $k(x, \cdot) \in L^2(\Omega)$ for almost every $x \in \Omega$, we can find (again, for almost every $x \in \Omega$) a sequence $\{a_n(x)\}$ of real numbers such that

$$k(x, \cdot) = \sum_{n=1}^\infty a_n(x)\phi_n,$$

where the series converges in the $L^2$ norm. The sequence $\{a_n(x)\}$ is defined by

$$a_n(x) = \langle k(x, \cdot), \phi_n \rangle_{L^2} = \int_\Omega k(x, y)\phi_n(y)\, dy = (K\phi_n)(x) \text{ for all } n \in \mathbb{Z}^+.$$

Since $a_n = K\phi_n$, it follows that $a_n \in L^2(\Omega)$ for all $n$. Moreover, since the series for $k(x, \cdot)$ converges, we know that

$$\sum_{n=1}^\infty a_n(x)^2 = \|k(x, \cdot)\|_{L^2}^2.$$

Now, since $a_n \in L^2(\Omega)$ for all $n$, there exists, for each $n$, a sequence $\{c_{mn}\}$ such that

$$a_n = \sum_{m=1}^{\infty} c_{mn}\phi_m.$$

Specifically,

$$
\begin{aligned}
c_{mn} = \langle a_m, \phi_m \rangle_{L^2} &= \int_{\Omega} a_m(x)\phi_m(x)\, dx \\
&= \int_{\Omega} \left( \int_{\Omega} k(x, y)\phi_n(y)\, dy \right) \phi_m(x)\, dx \\
&= \int_{\Omega} \int_{\Omega} k(x, y)\phi_n(y)\phi_m(x)\, dy\, dx \\
&= \langle k, \phi_m\phi_n \rangle_{L^2}.
\end{aligned}
$$

A direct calculation shows that $\{\phi_m\phi_n : m, n \in \mathbb{Z}^+\}$ is a (doubly-indexed) orthonormal sequence in $L^2(\Omega \times \Omega)$, that is, that

$$
\langle \phi_m\phi_n, \phi_k\phi_\ell \rangle_{L^2(\Omega \times \Omega)} = \begin{cases} 1 & \text{if } (m, n) = (k, \ell), \\ 0 & \text{otherwise.} \end{cases}
$$

It follows that, with $c_{mn} = \langle k, \phi_m\phi_n \rangle_{L^2(\Omega \times \Omega)}$, we have

$$\sum_{m=1}^{\infty} \sum_{n=1}^{\infty} c_{mn}^2 \le \|k\|_{L^2(\Omega \times \Omega)}$$

and

$$\sum_{m=1}^{\infty} \sum_{n=1}^{\infty} c_{mn}^2 = \|k\|_{L^2(\Omega \times \Omega)} \iff k = \sum_{m=1}^{\infty} \sum_{n=1}^{\infty} c_{mn}\phi_m\phi_n.$$

(Since the series $\sum_{m=1}^{\infty} \sum_{n=1}^{\infty} c_{mn}^2$ is absolutely convergent, order is unimportant in defining the double sum; it follows that the same is true of $\sum_{m=1}^{\infty} \sum_{n=1}^{\infty} c_{mn}\phi_m\phi_n$.)

We already know that

$$\sum_{n=1}^{\infty} a_n(x)^2 = \|k(x, \cdot)\|_{L^2(\Omega)}^2 \text{ a.e. in } \Omega$$

and

$$\|k\|^2_{L^2(\Omega \times \Omega)} = \int_\Omega \|k(x, \cdot)\|^2_{L^2(\Omega)}\, dx = \int_\Omega \sum_{n=1}^{\infty} a_n(x)^2\, dx.$$

Now, $\{\sum_{n=1}^{N} a_n(x)^2\}$ is a nondecreasing sequence that converges to the integrable function $\|k(x, \cdot)\|^2_{L^2}$ a.e. in $\Omega$. Therefore, by the monotone convergence theorem,

$$\int_\Omega \left( \lim_{N \to \infty} \sum_{n=1}^{N} a_n(x)^2 \right) dx = \lim_{N \to \infty} \int_\Omega \sum_{n=1}^{N} a_n(x)^2\, dx$$

$$= \lim_{N \to \infty} \sum_{n=1}^{N} \int_\Omega a_n(x)^2\, dx,$$

that is,

$$\int_\Omega \sum_{n=1}^{\infty} a_n(x)^2\, dx = \sum_{n=1}^{\infty} \int_\Omega a_n(x)^2\, dx.$$

This yields

$$\sum_{n=1}^{\infty} \int_\Omega a_n(x)^2\, dx = \int_\Omega \|k(x, \cdot)\|^2_{L^2}\, dx = \|k\|^2_{L^2(\Omega \times \Omega)}$$

$$\Rightarrow \sum_{n=1}^{\infty} \|a_n\|^2_{L^2} = \|k\|^2_{L^2(\Omega \times \Omega)}.$$

Since $\sum_{m=1}^{\infty} c^2_{mn} = \|a_n\|^2_{L^2}$, we obtain

$$\sum_{m=1}^{\infty} \sum_{n=1}^{\infty} c^2_{mn} = \|k\|^2_{L^2(\Omega \times \Omega)}$$

and hence

$$k = \sum_{m=1}^{\infty} \sum_{n=1}^{\infty} c_{mn} \phi_m \phi_n,$$

as desired.

It should be noticed that, since the function $k$ was taken to be an arbitrary element of $L^2(\Omega \times \Omega)$, we have shown that the doubly-indexed set $\{\phi_m\phi_n\}$ is a complete orthonormal set for $L^2(\Omega \times \Omega)$. We will now use this representation of the kernel $k$ to show that the integral operator $K$ is the norm limit of a sequence of finite-rank operators and hence that $K$ is compact. Let us define $K_{MN} : L^2(\Omega) \to L^2(\Omega)$ by

$$(K_{MN}u)(x) = \int_\Omega \left( \sum_{m=1}^M \sum_{n=1}^N c_{mn}\phi_m(x)\phi_n(y) \right) u(y)\,dy$$

$$= \sum_{m=1}^M \left( \int_\Omega \left( \sum_{n=1}^N c_{mn}\phi_n(y) \right) u(y)\,dy \right) \phi_m(x).$$

This formula shows that $K_{MN}u \in \mathrm{sp}\{\phi_1, \phi_2, \ldots, \phi_M\}$ for all $u \in L^2(\Omega)$ and hence that $K_{MN}$ has finite rank for each $M, N \in \mathbb{Z}^+$. Also, we have

$$((K - K_{MN})u)(x)$$

$$= \int_\Omega \left( k(x, y) - \sum_{m=1}^M \sum_{n=1}^N c_{mn}\phi_m(x)\phi_n(y) \right) u(y)\,dy$$

$$= \int_\Omega \left( \sum_{m=M+1}^\infty \sum_{n=N+1}^\infty c_{mn}\phi_m(x)\phi_n(y) \right) u(y)\,dy \text{ for all } u \in L^2(\Omega),$$

which shows (by our earlier result) that, for all $u \in L^2(\Omega)$,

$$\|(K - K_{MN})u\|_{L^2(\Omega)} \le \left\| \sum_{m=M+1}^\infty \sum_{n=N+1}^\infty c_{mn}\phi_m\phi_n \right\|_{L^2(\Omega \times \Omega)} \|u\|_{L^2(\Omega)},$$

and hence that

$$\|K - K_{MN}\| \le \left\| \sum_{m=M+1}^\infty \sum_{n=N+1}^\infty c_{mn}\phi_m\phi_n \right\|_{L^2(\Omega \times \Omega)} \to 0 \text{ as } M, N \to \infty.$$

Therefore $K$ is the limit, in the operator norm, of the sequence $\{K_{MN}\}$ of finite-rank operators, which implies by Corollary 4.9 that $K$ is compact.

**Degenerate integral operators**
We have now shown that every integral operator of the form (4.7) is a compact operator. A special case occurs when the kernel $k$ is *degenerate*, that is, when it has the form

$$k(x, y) = \sum_{n=1}^{N} f_n(x) g_n(y). \qquad (4.8)$$

In this case, it is easy to see that

$$Ku = \sum_{n=1}^{N} \left( \int_{\Omega} g_n(y) u(y) \, dy \right) f_n$$

and hence that $K$ has finite rank ($Ku \in \text{sp}\{f_1, f_2, \ldots, f_N\}$ for all $u \in L^2(\Omega)$). The reader will recall that $K$ is certainly compact in this case; however, $Ku = f$ is not a genuine inverse problem in this case because $\mathcal{R}(K)$ is closed and hence $K^{\dagger}$ is bounded.

Conversely, we can show that if $K : L^2(\Omega) \to L^2(\Omega)$ is a finite-rank integral operator of the form (4.7), then the kernel is degenerate. Let us assume that $\mathcal{R}(K)$ is finite-dimensional and that $\{\phi_1, \phi_2, \ldots, \phi_N\}$ is an orthonormal basis for $\mathcal{R}(K)$. Then

$$(Ku)(x) = \sum_{n=1}^{N} \langle Ku, \phi_n \rangle_{L^2} \phi_n(x)$$

$$= \sum_{n=1}^{N} \left( \int_{\Omega} (Ku)(z) \phi_n(z) \, dz \right) \phi_n(x)$$

$$= \sum_{n=1}^{N} \left( \int_{\Omega} \left( \int_{\Omega} k(z, y) u(y) \, dy \right) \phi_n(z) \, dz \right) \phi_n(x)$$

$$= \sum_{n=1}^{N} \left( \int_{\Omega} \left( \int_{\Omega} k(z, y) \phi_n(z) \, dz \right) u(y) \, dy \right) \phi_n(x)$$

$$= \sum_{n=1}^{N} \left( \int_{\Omega} (K^* \phi_n) u(y) \, dy \right) \phi_n(x),$$

where we have used the fact that $K^*$ is defined by $(K^*v)(y) = \int_\Omega k(x,y)v(x)\,dx$. We then obtain

$$(Ku)(x) = \int_\Omega \left( \sum_{n=1}^N \phi_n(x)(K^*\phi_n)(y) \right) u(y)\,dy.$$

Since this holds for all $u \in L^2(\Omega)$, it follows that

$$k(x,y) = \sum_{n=1}^N \phi_n(x)(K^*\phi_n)(y) \text{ for almost every } (x,y) \in \Omega \times \Omega,$$

which shows that $k$ is degenerate.

As we have pointed out, a degenerate integral operator $K$ is a finite-rank operator, implying that $\mathcal{R}(K)$ is closed and therefore that $Ku = v$ is not a true inverse problem. Nevertheless, just as a matrix-vector equation $Ax = \mathbf{b}$ can be ill-conditioned (even though it cannot be unstable), an integral equation $Ku = v$, though stable, can be quite ill-conditioned. Therefore, even though a degenerate first-kind integral equation cannot be a true inverse problem, it can still be difficult to solve in the sense that small changes in the data can lead to large changes in the solution.

The following example shows the behavior of a relatively well-conditioned degenerate integral equation, and presents a nondegenerate integral equation for comparison.

**Example 4.26.** In this example, we compare the integral equation $Tx = y$,

$$(Tx)(s) = \int_0^1 k(s,t)x(t)\,dt, \ 0 < s < 1,$$

for two different choices of the kernel $k$:

$$k_1(s,t) = st + \sin(s+t),$$

$$k_2(s,t) = \frac{s}{2} + \frac{t}{2} + \sin(st).$$

Superficially, the two kernels are quite similar, as shown in Figure 4.5.

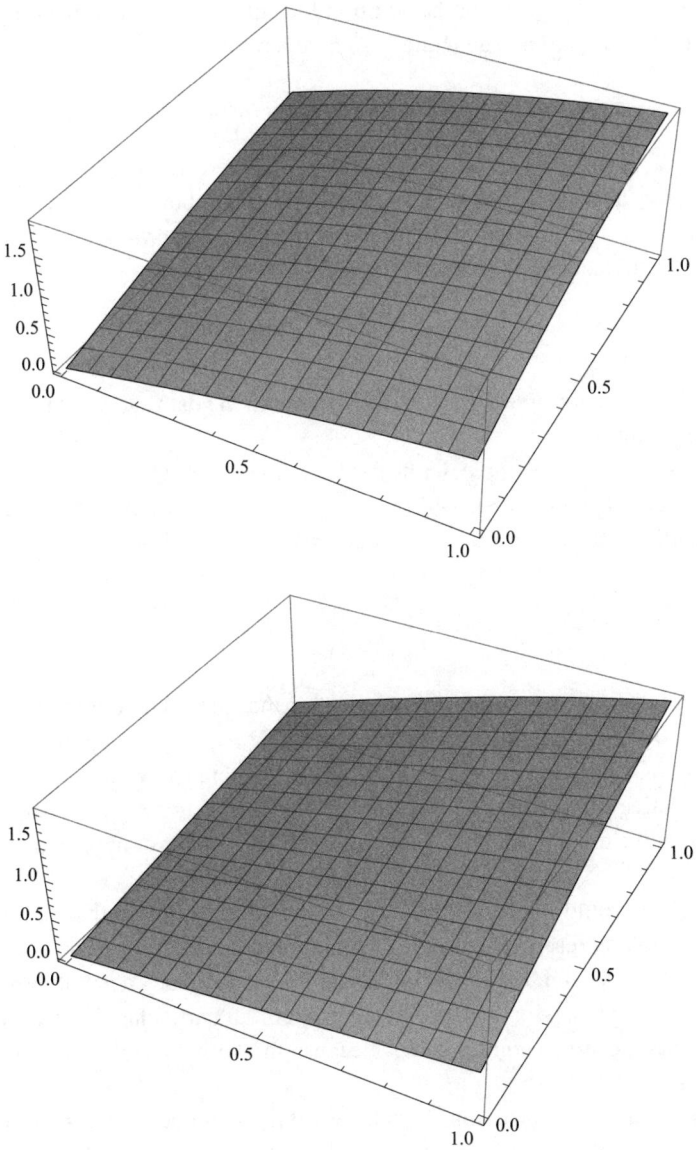

**Figure 4.5.** The two kernels from Example 4.26; $k_1$ is graphed on the top and $k_2$ on the bottom.

However, by applying the addition formula for the sine function, we see that $k_1$ defines a degenerate integral operator:

$$k_1(s, t) = st + \sin(s)\cos(t) + \cos(s)\sin(t).$$

(Thus $k_1$ has the form (4.8), where $N = 3$.) When we discretize $T$, with kernel $k_1$ and use a grid with $n = 50$ points, we obtain a matrix-vector equation $Ax = \mathbf{b}$, where $A$ has three positive singular values, $\sigma_1 \approx 1.1$, $\sigma_2 \approx 0.061$, and $\sigma_3 \approx 0.00040$. Therefore,

$$\text{cond}(A) = \frac{\sigma_1}{\sigma_3} \approx 2724.$$

This is the factor by which the relative error in the data can be magnified in the solution.

As a specific example, let us define $x^*(t) = t$, in which case $y^*(s) = s/3 - \cos(1 + s) - \sin(s) + \sin(1 + s)$. We create a noisy data vector by adding a small multiple of the discontinuous function

$$z(s) = \begin{cases} s, & s < \frac{1}{2}, \\ 2 - 2s, & s > \frac{1}{2} \end{cases}$$

to $y^*$: $y = y^* + \alpha z$, where $\alpha$ is chosen so that the relative error in $y$ is only $10^{-4}$. We then (approximately) solve $Tx = y$ by solving the corresponding discrete equation $Ax = \mathbf{b}$. The results are shown in Figure 4.6; the relative error in the computed solution is approximately 0.12, and we see that the error in the data was magnified by a factor of approximately 1200.

What would happen if we were to refine the grid, so that the discretization represents the original problem more accurately? The answer is that there would be very little change. We have already accurately represented the behavior of the original problem; in particular, the matrix $A$ has three positive singular values because the kernel has the form (4.8) with $N = 3$.

By way of comparison, we perform the analogous experiment using the kernel $k_2$, in which case the integral operator is not degenerate.

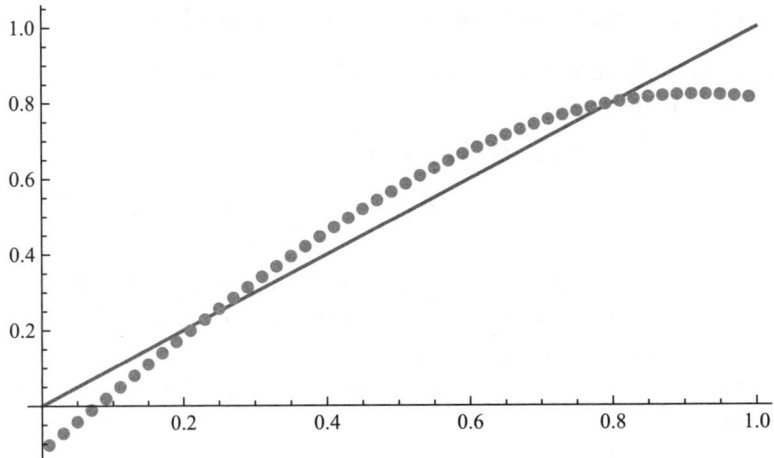

**Figure 4.6.** The exact and computed solution in Example 4.26 with $k_1$ as the kernel for the integral operator. (The integral operator is degenerate.)

Now, when we discretize the equation, the matrix $A$ has six positive singular values, approximately 0.84, 0.029, 0.0017, $8.5 \cdot 10^{-6}$, $1.4 \cdot 10^{-8}$, and $1.3 \cdot 10^{-11}$. We have

$$\text{cond}(A) = \frac{\sigma_1}{\sigma_6} \approx 6.5 \cdot 10^{10},$$

and when we perform the analogous experiement (with $x^*(t) = t$, $y^* = Tx^*$, and $z$ as before), we find that the error is magnified by a factor of approximately $4.3 \cdot 10^9$. (We do not display the computed solution because it is obviously meaningless.) As discussed in Section 2.4, if we were to do the computation more accurately (which in this case requires not only refining the discretization, but also using more precision in the arithmetic— with more precision, we can capture smaller singular values), we would expect the condition number to increase and the amplification of error to get worse.

## 4.4.2   Integral operators of the second kind

In Section 4.2, we briefly mentioned integral operators of the second kind, which have the form

$$T : L^2(\Omega) \to L^2(\Omega),$$

$$(Tu)(x) = u(x) + \int_\Omega k(x, y)u(y)\, dy, \tag{4.9}$$

where $k \in L^2(\Omega \times \Omega)$. We can write such an operator as $T = I + K$, where $K$ is an integral operator of the first kind (and hence $K$ is compact). We wish to prove that $I + K$ always has a closed range, and also that $I + K$ is not compact. The first is true in general, for any compact operator $K$, and the second is true unless $X$ is finite-dimensional.

**Theorem 4.27.** *Let $X$ be a Hilbert space, let $K : X \to X$ be compact, and let $I : X \to X$ be the identity operator. Then*

1. *$\mathcal{R}(I + K)$ is closed;*

2. *if $I + K$ is compact, then $X$ is finite-dimensional.*

*Proof.*

1. Suppose $y \in \overline{\mathcal{R}(I + K)}$, say $y = \lim_{n\to\infty}(I + K)x_n$, where $\{x_n\} \subset X$. We consider two cases. First, let us assume that $\{x_n\}$ is bounded. In this case, the sequence $\{Kx_n\}$ has a convergent subsequence (because $K$ is compact), and there is no loss of generality in assuming that $\{Kx_n\}$ itself converges, say $Kx_n \to z$. But then

$$x_n = (I + K)x_n - Kx_n \to y - z,$$

which implies that

$$(I + K)x_n \to (I + K)(y - z).$$

Since $(I + K)x_n \to y$, it follows that $(I + K)(y - z) = y$, which shows that $y \in \mathcal{R}(I + K)$.

Second, let us assume that $\{x_n\}$ is unbounded. We can assume, without loss of generality, that $\|x_n\|_X \to \infty$, and we can also assume that $x_n \in \mathcal{N}(I + K)^\perp$ for all $n \in \mathbb{Z}^+$. Then

$$(I + K)x_n \to y \ \Rightarrow \ (I + K)\frac{x_n}{\|x_n\|_X} \to 0.$$

Since $\{x_n / \|x_n\|_X\}$ is bounded, we can assume, without loss of generality, that

$$K\frac{x_n}{\|x_n\|_X} \to z$$

for some $z \in X$. But then

$$\frac{x_n}{\|x_n\|_X} = (I + K)\frac{x_n}{\|x_n\|_X} - K\frac{x_n}{\|x_n\|_X} \to 0 - z = -z.$$

It follows that

$$(I + K)z = - \lim_{n \to \infty}(I + K)\frac{x_n}{\|x_n\|_X} = 0,$$

and hence $z \in \mathcal{N}(I + K)$. But $x_n/\|x_n\|_X \in \mathcal{N}(I + K)^\perp$ for all $n$, which implies that $z \in \mathcal{N}(I + K)^\perp$. Therefore, $z \in \mathcal{N}(I + K) \cap \mathcal{N}(I + K)^\perp = \{0\}$, that is $z = 0$. This is impossible, because $z$ is the norm-limit of vectors of length 1. Therefore, this second case cannot occur, and the proof is complete.

2. We now wish to show that if $I + K$ is compact, then $X$ is finite-dimensional. Suppose $\{x_n\}$ is any bounded sequence in $X$. Since $I + K$ and $K$ are both compact, there exists a subsequence $\{x_{n_k}\}$ such that $\{(I + K)x_{n_k}\}$ and $\{Kx_{n_k}\}$ both converge, say

$$(I + K)x_{n_k} \to y, \ Kx_{n_k} \to z.$$

But then

$$x_{n_k} = (I + K)x_{n_k} - Kx_{n_k} \to y - z,$$

and we have shown that $\{x_n\}$ has a convergent subsequence. Since $\{x_n\}$ is an arbitrary bounded sequence, this is possible only if $X$ is finite-dimensional. $\qquad \square$

Applying this theorem to an integral operator $T$ of the second kind (as defined by (4.9)), we see that $Tu = v$ is never an inverse problem, because

$\mathcal{R}(T)$ is necessarily closed. This explains the results presented in Section 1.1, where we compared integral equations of the first and second kind (although we had not introduced this terminology at that point).

## 4.5   The generalized inverse in terms of the SVE

We will now consider a compact operator $T : X \to Y$ and the associated inverse problem $Tx = y$. Let us suppose that the singular value expansion of $T$ is

$$T = \sum_{n=1}^{\infty} \sigma_n \psi_n \otimes \phi_n, \tag{4.10}$$

where $\{\phi_n\}$ is an orthonormal sequence in $X$, $\{\psi_n\}$ is an orthonormal sequence in $Y$, and $\sigma_1 \geq \sigma_2 \geq \sigma_3 \geq \cdots > 0$. As the notation indicates, we will assume, unless we state otherwise, that $T$ is not finite-rank. As we know, this implies that $\mathcal{R}(T)$ is not closed.

We begin by expressing the generalized inverse $T^\dagger$ of $T$ in terms of the SVE and using this representation to rederive the properties of $T^\dagger$. We will need several preliminary results that will allow us to understand the singular value expansion of $T$ and what it tells us about $T$.

The proof of the following simple result will be left to the reader.

**Lemma 4.28.**

1. $\mathcal{N}(T) = \text{sp}\{\phi_n\}^\perp$.

2. $\mathcal{N}(T)^\perp = \overline{\text{sp}\{\phi_n\}}$.

3. $\text{sp}\{\psi_n\} \subset \mathcal{R}(T) \subset \overline{\text{sp}\{\psi_n\}}$.

4. $\overline{\mathcal{R}(T)} = \overline{\text{sp}\{\psi_n\}}$.

The following theorem presents precise descriptions of $\mathcal{R}(T)$ and $D(T^\dagger)$.

**Theorem 4.29.** *Let $y \in Y$. Then $y \in \mathcal{R}(T)$ if and only if $y = \sum_{n=1}^{\infty} \alpha_n \psi_n$, where*

$$\sum_{n=1}^{\infty} \frac{\alpha_n^2}{\sigma_n^2} < \infty. \tag{4.11}$$

*Consequently, $y \in D(T^{\dagger})$ if and only if*

$$\sum_{n=1}^{\infty} \frac{|\langle y, \psi_n \rangle_Y|^2}{\sigma_n^2} < \infty. \tag{4.12}$$

*Proof.* By definition, $y \in \mathcal{R}(T)$ if and only if there exists $x \in X$ such that $Tx = y$; moreover, we might as well assume that $x \in \mathcal{N}(T)^{\perp} = \overline{\mathrm{sp}\{\phi_n\}}$. It follows from Lemma 4.15 that $x = \sum_{n=1}^{\infty} \langle x, \phi_n \rangle_X \phi_n$. Recalling from the definition of the SVE that $T\phi_n = \sigma_n \psi_n$, we have

$$Tx = T\left(\sum_{n=1}^{\infty} \langle x, \phi_n \rangle_X \phi_n\right) = \sum_{n=1}^{\infty} \langle x, \phi_n \rangle_X T\phi_n = \sum_{n=1}^{\infty} \sigma_n \langle x, \phi_n \rangle_X \psi_n.$$

Therefore, if $Tx = y = \sum_{n=1}^{\infty} \langle y, \psi_n \rangle_Y \psi_n$ we have

$$\langle y, \psi_n \rangle_Y = \sigma_n \langle x, \phi_n \rangle_X \implies \frac{\langle y, \psi_n \rangle_Y}{\sigma_n} = \langle x, \phi_n \rangle_X \text{ for all } n \in \mathbb{Z}^+.$$

It follows that

$$\sum_{n=1}^{\infty} \frac{|\langle y, \psi_n \rangle_Y|^2}{\sigma_n^2} = \sum_{n=1}^{\infty} |\langle x, \phi_n \rangle_X|^2 < \infty,$$

as desired.

Conversely, if $y = \sum_{n=1}^{\infty} \alpha_n \psi_n$ and (4.11) holds, then we can define

$$x = \sum_{n=1}^{\infty} \frac{\alpha_n}{\sigma_n} \phi_n.$$

It follows from (4.11) and Lemma 4.15 that $x$ is well-defined (that is, that the series is convergent) and a direct calculation shows that $Tx = y$. Hence $y \in \mathcal{R}(T)$, as desired.

We know that $Y = \overline{\mathcal{R}(T)} \oplus \mathcal{R}(T)^{\perp} = \overline{\mathrm{sp}\{\psi_n\}} \oplus \mathcal{R}(T)^{\perp}$, and hence each $y \in Y$ can be written as $y = \sum_{n=1}^{\infty} \alpha_n \psi_n + z$, where $z \in \mathcal{R}(T)^{\perp}$.

On the other hand, $D(T^\dagger) = \mathcal{R}(T) \oplus \mathcal{R}(T)^\perp$ and therefore $y$ belongs to $D(T^\dagger)$ if and only if, in the given representation of $y$, $\sum_{n=1}^{\infty} \alpha_n^2/\sigma_n^2 < \infty$. Moreover, since $\langle \psi_n, z \rangle_Y = 0$, it follows that $\langle \psi_n, y \rangle_Y = \alpha_n$ for all $n \in \mathbb{Z}^+$. It now follows that $y \in D(T^\dagger)$ if and only if

$$\sum_{n=1}^{\infty} \frac{|\langle y, \psi_n \rangle_Y|^2}{\sigma_n^2} < \infty. \qquad \Box$$

Condition (4.12) is called the *Picard condition*. Since

$$\mathrm{sp}\{\psi_n\} \subset \mathcal{R}(T) \subset \overline{\mathrm{sp}\{\psi_n\}},$$

it follows that $\overline{\mathcal{R}(T)} = \overline{\mathrm{sp}\{\psi_n\}}$. By Lemma 4.15,

$$\overline{\mathrm{sp}\{\psi_n\}} = \left\{ \sum_{n=1}^{\infty} \alpha_n \psi_n \ : \ \sum_{n=1}^{\infty} \alpha_n^2 < \infty \right\}.$$

From this and the fact that $\sigma_n \to 0$ as $n \to \infty$, it is evident that there are elements of $\overline{\mathcal{R}(T)}$ that do not satisfy the Picard condition and hence do not belong to $\mathcal{R}(T)$.[6] This provides another proof that $\mathcal{R}(T)$ is not closed, provided $T$ has infinitely many nonzero singular values.

Next we will derive an explicit formula for $T^\dagger$.

**Theorem 4.30.** *The generalized inverse $T^\dagger$ of $T$ is given by*

$$T^\dagger = \sum_{n=1}^{\infty} \sigma_n^{-1} \phi_n \otimes \psi_n,$$

*where the convergence of the series is to be understood in the pointwise sense. That is, $T^\dagger$ is given by*

$$T^\dagger y = \sum_{n=1}^{\infty} \sigma_n^{-1} (\phi_n \otimes \psi_n) y = \sum_{n=1}^{\infty} \frac{\langle \psi_n, y \rangle_Y}{\sigma_n} \phi_n \text{ for all } y \in D(T^\dagger).$$

---

[6] Given that $\{\sigma_n\}$ is a nonincreasing sequence of positive numbers converging to zero, for each $k \in \mathbb{Z}^+$ there exists $n_k \in \mathbb{Z}^+$ such that $\sigma_{n_k} \leq 1/\sqrt{k}$. We can define $\{\alpha_n\}$ by the conditions that $\alpha_n = \sigma_{n_k}/\sqrt{k}$ if $n = n_k$ for some $k$ and $\alpha_n = 0$ otherwise. Then

$$\sum_{n=1}^{\infty} \alpha_n^2 = \sum_{k=1}^{\infty} \frac{\sigma_{n_k}^2}{k} \leq \sum_{k=1}^{\infty} \frac{1}{k^2} < \infty, \quad \sum_{n=1}^{\infty} \frac{\alpha_n^2}{\sigma_n^2} = \sum_{k=1}^{\infty} \frac{\sigma_{n_k}^2}{k \sigma_{n_k}^2} = \sum_{k=1}^{\infty} \frac{1}{k} = \infty.$$

*Proof.* Let us define $R : \mathcal{R}(T) \oplus \mathcal{R}(T)^{\perp} \to X$ by

$$Ry = \sum_{n=1}^{\infty} \frac{\langle \psi_n, y \rangle_Y}{\sigma_n} \phi_n.$$

We must show that $R$ is well-defined and that $R = T^{\dagger}$. We know from Theorem 2.18 that $T^{\dagger}|_{\mathcal{R}(T)} = S^{-1}$, where $S = T|_{\mathcal{N}(T)^{\perp}}$. Therefore, it suffices to prove that $R(S(x)) = x$ for all $x \in \mathcal{N}(T)^{\perp}$ and that $Ry = 0$ for $y \in \mathcal{R}(T)^{\perp} = \mathcal{N}(T^{\dagger})$.

First, if $y \in \mathcal{R}(T)$, then $y = \sum_{m=1}^{\infty} \alpha_m \psi_m$, where

$$\sum_{m=1}^{\infty} \frac{\alpha_m^2}{\sigma_m^2} < \infty \tag{4.13}$$

and $\langle \psi_n, y \rangle_Y = \alpha_n$. Therefore,

$$Ry = \sum_{n=1}^{\infty} \sigma_n^{-1} \langle \psi_n, y \rangle_Y \phi_n = \sum_{n=1}^{\infty} \sigma_n^{-1} \alpha_n \phi_n.$$

Because (4.13) holds, it follows that $Ry$ is well-defined (that is, the series converges) and $Ry \in \mathcal{N}(T)^{\perp}$ by Lemma 4.28 and Lemma 4.15.

Next, if $x \in \mathcal{N}(T)^{\perp}$, then $x = \sum_{n=1}^{\infty} \alpha_n \phi_n$, where $\sum_{n=1}^{\infty} \alpha_n^2 < \infty$, and

$$Sx = Tx = \sum_{n=1}^{\infty} \sigma_n \alpha_n \psi_n.$$

It then follows immediately that $R(Sx) = x$. Finally, since $\mathcal{R}(T)^{\perp} = \text{sp}\{\psi_n\}^{\perp}$, it follows that $\langle \psi_n, y \rangle_Y = 0$ for all $y$ and hence that $Ry = 0$ for all $y \in \mathcal{R}(T)^{\perp}$.  □

At the risk of pointing out the obvious we remark that the representation of $T^{\dagger}$ is not a singular value expansion, because the numbers $\sigma_n^{-1}$ do not converge to zero as $n \to \infty$. As noted above, the series representation of $T^{\dagger}$ converges in the pointwise sense, not in the operator norm, as has been the case with other series representations of operators that we have seen.

We can now quickly review the properties of $T^\dagger$ using the representation provided by the previous theorem. For example, we have

$$\|T^\dagger \psi_n\|_X = \|\sigma_n^{-1}\phi_n\|_X = \frac{1}{\sigma_n} \to \infty \text{ as } n \to 0.$$

This shows that if $T$ has infinitely many positive singular values then $T^\dagger$ is unbounded. On the other hand, if $T$ has only finitely many nonzero singular values, say

$$T = \sum_{n=1}^{N} \sigma_n \psi_n \otimes \phi_n,$$

then

$$T^\dagger = \sum_{n=1}^{N} \sigma_n^{-1}\phi_n \otimes \psi_n$$

and it follows that $T^\dagger$ is bounded, with $\|T^\dagger\| = \sigma_N^{-1}$ (since $\sigma_1 \geq \sigma_2 \geq \cdots \geq \sigma_N$). Therefore, $T^\dagger$ is bounded if and only if $T$ has only finitely many positive singular values. Since we saw in the previous section that $\mathcal{R}(T)$ is closed if and only $T$ has only finitely many positive singular values, this is consistent with Corollary 2.17: $T^\dagger$ is bounded if and only if $\mathcal{R}(T)$ is closed.

We can directly verify the relationships $T^\dagger T x = \text{proj}_{\mathcal{N}(T)^\perp} x$ for all $x \in X$ and $TT^\dagger T = T$ by using the series representations for $T$ and $T^\dagger$. For all $x \in X$,

$$\text{proj}_{\mathcal{N}(T)^\perp} x = \sum_{n=1}^{\infty} \langle \phi_n, x \rangle_X \phi_n$$

and there exists $\hat{x} \in \mathcal{N}(T)$ such that $x = \hat{x} + \text{proj}_{\mathcal{N}(T)^\perp} x$. Therefore,

$$Tx = T\hat{x} + T\left( \sum_{n=1}^{\infty} \langle \phi_n, x \rangle_X \phi_n \right)$$

$$= 0 + \sum_{n=1}^{\infty} \langle \phi_n, x \rangle_X T\phi_n = \sum_{n=1}^{\infty} \sigma_n \langle \phi_n, x \rangle_X \psi_n$$

and hence

$$T^\dagger T x = \sum_{n=1}^{\infty} \frac{\langle \psi_n, Tx \rangle_Y}{\sigma_n} \phi_n = \sum_{n=1}^{\infty} \frac{\sigma_n \langle \phi_n, x \rangle_X}{\sigma_n} \phi_n$$

$$= \sum_{n=1}^{\infty} \langle \phi_n, x \rangle_X \phi_n = \text{proj}_{\mathcal{N}(T)^\perp} x.$$

It follows immediately that $T T^\dagger T x = T(\text{proj}_{\mathcal{N}(T)^\perp} x) = T x$ because $x - \text{proj}_{\mathcal{N}(T)^\perp} x$ belongs to $\mathcal{N}(T)$.

In a similar fashion, we can use the series representations of $T$ and $T^\dagger$ to prove that $T T^\dagger y = \text{proj}_{\overline{\mathcal{R}(T)}} y$ and $T^\dagger T T^\dagger y = T^\dagger y$ for all $y \in D(T^\dagger)$.

Finally, let us use Theorem 4.30 to prove that $T^\dagger$ is a closed operator. Suppose $\{y_k\} \subset D(T^\dagger)$ and $y_k \to y \in Y$, $T^\dagger y_k \to x \in X$. We have

$$T^\dagger y_k = \sum_{n=1}^{\infty} \frac{\langle \psi_n, y_k \rangle_Y}{\sigma_n} \phi_n$$

and, since $x \in \mathcal{N}(T)^\perp$,

$$x = \sum_{n=1}^{\infty} \langle \phi_n, x \rangle_X \phi_n.$$

It follows that, for each $m \in \mathbb{Z}^+$,

$$\left\| T^\dagger y_k - x \right\|_X^2 = \sum_{n=1}^{\infty} \left| \sigma_n^{-1} \langle \psi_n, y_k \rangle_Y - \langle \phi_n, x \rangle_X \right|^2$$

$$= \sum_{n=1}^{\infty} \frac{\left| \langle \psi_n, y_k \rangle_Y - \sigma_n \langle \phi_n, x \rangle_X \right|^2}{\sigma_n^2}$$

$$\geq \frac{\left| \langle \psi_m, y_k \rangle_Y - \sigma_m \langle \phi_m, x \rangle_X \right|^2}{\sigma_m^2}.$$

Therefore, $\langle \psi_m, y_k \rangle_Y \to \sigma_m \langle \phi_m, x \rangle_X$ for all $m \in \mathbb{Z}^+$ as $k \to \infty$. But, since $y_k \to y$ as $k \to \infty$, it follows that $\langle \psi_m, y_k \rangle_Y \to \langle \psi_m, y \rangle_Y$ as $k \to \infty$ and hence that

$$\langle \psi_m, y \rangle_Y = \sigma_m \langle \phi_m, x \rangle_X \text{ for all } m \in \mathbb{Z}^+.$$

This implies that

$$\sum_{n=1}^{\infty} \frac{|\langle \psi_n, y \rangle_Y|^2}{\sigma_n^2} = \sum_{n=1}^{\infty} |\langle \phi_n, x \rangle_X|^2 < \infty$$

and hence, by the Picard condition, $y \in D(T^{\dagger})$. It also follows, from Theorem 4.30 and the formula for $\langle \psi_n, y \rangle_Y$, that $T^{\dagger} y = x$. We have thus shown that $T^{\dagger}$ is closed.

## 4.5.1   Interpretation of the series representation of $T^{\dagger}$

We discussed in Chapter 1 the typical nature of direct and inverse problems, and the SVE of $T$ allows us state these observations more precisely. In terms of a compact operator $T$, the direct problem is to compute $y = Tx$ from $x$, while the inverse problem is to solve $Tx = y$ for $x$ given $y$. We have reinterpreted the inverse problem as computing the best (minimum-norm) least-squares solution of $Tx = y$, which means computing $x_{0,y} = T^{\dagger} y$ from $y$.

Let us assume that $T : X \to Y$ has a singular value expansion

$$T = \sum_{n=1}^{\infty} \sigma_n \psi_n \otimes \phi_n.$$

The *singular components* of $x \in X$ are the quantities $\langle x, \phi_n \rangle_X \phi_n$, $n = 1, 2, 3, \ldots$, the projections of $x$ onto the singular vectors $\phi_n$, $n = 1, 2, 3, \ldots$. Similarly, the singular components of $y \in Y$ are the quantities $\langle y, \psi_n \rangle_Y \psi_n$, $n = 1, 2, 3, \ldots$.

In a typical application, $X$ and $Y$ are spaces of functions (as in the examples we have presented), and the singular vectors in both sequences ($\{\phi_n\}$ and $\{\psi_n\}$) tend to increase in frequency with increasing $n$. In other words, the high-frequency singular vectors correspond to the small singular values, and any high frequencies contained in $x$ tend to be dampened when $T$ acts on $x$. For this reason, $T$ is a smoothing operator.

Now consider an exact vector $x = \sum_{n=1}^{\infty} \alpha_n \phi_n + w$, where $w \in \mathcal{N}(T)$, and a measurement $\hat{x} = x + \sum_{n=1}^{\infty} \epsilon_n \phi_n + \hat{\epsilon} \hat{w}$, where $\hat{w} \in \mathcal{N}(T)$

and $\|\hat{w}\|_X = 1$. We have

$$\|\hat{x} - x\|_X^2 = \sum_{n=1}^{\infty} \epsilon_n^2 + \hat{\epsilon}^2,$$

and therefore if $\|\hat{x} - x\|_X$ is small, then so is each $\epsilon_n$.

We then have

$$T\hat{x} = \sum_{n=1}^{\infty} \sigma_n \alpha_n \psi_n + \sum_{n=1}^{\infty} \sigma_n \epsilon_n \psi_n.$$

Here is the typical scenario:

- The exact $x$ is typically relatively smooth, that is, the singular components of $x$ corresponding to large values of $n$ are negligible. More specifically, there exists $N \in \mathbb{Z}^+$ that is not too large and such that

$$\sum_{n=1}^{N} \alpha_n^2 \approx \sum_{n=1}^{\infty} \alpha_n^2.$$

- The error $\hat{x} - x$ is small compared to $x$, that is, $\|\hat{x} - x\|_X \ll \|x\|_X$. However, since the error is likely generated by some random process, it is entirely possible that it contains much higher frequencies than does $x$; in other words, in order for

$$\sum_{n=1}^{M} \epsilon_n^2 \approx \sum_{n=1}^{\infty} \epsilon_n^2$$

to hold, it may be necessary to take $M$ much larger than $N$.

- The first two facts imply that $\hat{x}$ may contain large relative errors in the singular components corresponding to $n$ with $N < n \leq M$. In other words, for these values of $n$, it may be that $|\epsilon_n/\alpha_n|$ is not small. The error in $\hat{x}$ is still not large, though, because

$$\sum_{n=N+1}^{M} \epsilon_n^2 \leq \|\hat{x} - x\|_X^2 \ll \|x\|_X^2.$$

- Finally, applying $T$ to $\hat{x}$ does not change these relationships, because the singular components in $\hat{x}$ in which there are large relative errors are multiplied by small singular values; therefore,

$$\sum_{n=N+1}^{M} \sigma_n^2 \epsilon_n^2 \ll \sum_{n=1}^{N} \sigma_n^2 \alpha_n^2,$$

which means that $\|T\hat{x} - Tx\|_Y \ll \|Tx\|_Y$ holds.

On the other hand, we have

$$T^\dagger = \sum_{n=1}^{\infty} \frac{1}{\sigma_n} \phi_n \otimes \psi_n.$$

In applying $T^\dagger$ to $y$, the singular components corresponding to large $n$ are amplified by the factor $1/\sigma_n$, which is large for large $n$. Since large $n$ corresponds to higher frequencies, it follows that $T^\dagger$ amplifies large frequencies.

To illustrate this, suppose the exact data vector is $y = \sum_{n=1}^{\infty} \beta_n \psi_n$ and the noisy data vector is $\hat{y} = y + \sum_{n=1}^{\infty} \delta_n \psi_n + \hat{z}$, where $\hat{z} \in \mathcal{R}(T)^\perp$. Then

$$T^\dagger \hat{y} = \sum_{n=1}^{\infty} \frac{\alpha_n}{\sigma_n} \phi_n + \sum_{n=1}^{\infty} \frac{\delta_n}{\sigma_n} \phi_n.$$

Now we can describe the typical scenario in solving the inverse problem with noisy data:

- The exact $y$ is typically very smooth, because it is the result of applying the smoothing operator $T$ to an $x$ that is relatively smooth. Thus, if we choose $K \in \mathbb{Z}^+$ such that

$$\sum_{n=1}^{K} \beta_n^2 \approx \sum_{n=1}^{\infty} \beta_n^2,$$

then this $K$ is likely to be much smaller than the $N$ defined in analyzing the direct problem.

- The error $\hat{y} - y$ is small compared to $y$, that is, $\|\hat{y} - y\|_Y \ll \|y\|_Y$. As before, since the error is generated by some random process, it probably contains much higher frequencies than does $y$. Thus, if we choose

$L \in \mathbb{Z}^+$ such that

$$\sum_{n=1}^{L} \delta_n^2 \approx \sum_{n=1}^{\infty} \delta_n^2,$$

then $L$ will be much larger than $K$.

- As before, this implies that $\hat{y}$ may contain large relative errors in the singular components corresponding to $n$ with $K < n \le L$. For these values of $n$, $|\delta_n/\beta_n|$ may not be small. The error in $\hat{y}$ is still not large, because

$$\sum_{n=K+1}^{L} \delta_n^2 \le \|\hat{y} - y\|_Y^2 \ll \|y\|_Y^2.$$

- Here is where the situation is completely different from our description of the direct problem. When $T^\dagger$ is applied to $\hat{y}$, the singular components in $\hat{y}$ in which there are large relative errors are multiplied by the large values $1/\sigma_n, K + 1 < n \le L$. It is entirely possible that the resulting errors in components $n = K + 1, K + 2, \dots, L$ of $T^\dagger \hat{y}$ completely overwhelm components $n = 1, 2, \dots, K$ that contain only small relative error.[7] In other words, it may be that

$$\|T^\dagger \hat{y} - T^\dagger y\|_Y^2 \ge \sum_{n=K+1}^{L} \frac{\delta_n^2}{\sigma_n^2} \gg \|T^\dagger y\|_Y^2.$$

This is consistent with various examples we have given. Below we present another example and analyze it in terms of the singular components.

In short, in both the direct and inverse problems, the singular components in the data corresponding to large $n$ are typically completely corrupted by random noise. However, in the direct problem, this is harmless because these components are multiplied by small singular values, and in

---

[7] Here we are not trying to suggest that there is really an abrupt change from "small relative error" to "large relative error" in passing from $n = K$ to $n = K + 1$. The change would be more gradual, from "small relative error" to "noticeable relative error" to "large relative error."

the inverse problem it is catastrophic because they are multiplied by the reciprocals of small singular values.

**Example 4.31.** We now present an example that illustrates the points raised above. Consider the integral operator $T : L^2(0, 1) \to L^2(0, 1)$ defined by

$$(Tx)(s) = \int_0^s x(t) \, dt, \ 0 < s < 1.$$

This operator can be written in the standard form $(Tx)(s) = \int_0^1 k(s, t) x(t) \, dt$ by defining

$$k(s, t) = \begin{cases} 1, & 0 < t < s, \\ 0, & s < t < 1. \end{cases}$$

If $y = Tx$, then $y$ is the solution of the initial value problem $y' = x$, $y(0) = 0$. The functions $x^*(t) = t$ and $y^*(s) = t - t^2/2$ satisfy $Tx^* = y^*$ exactly. We will define a noisy data vector $y$ by $y = y^* + \delta z$, where $z$ is a blip centered at $t = 1/2$:

$$z(t) = \begin{cases} \frac{\sqrt{15}(4\epsilon^2 - (1-2t)^2)}{16\epsilon^{5/2}}, & \frac{1}{2} - \epsilon \leq t \leq \frac{1}{2} + \epsilon, \\ 0, & \text{otherwise.} \end{cases}$$

The function $z$ is shown in Figure 4.7; note that $z$ has been normalized to have $L^2$-norm one. (This perturbation has been chosen because it has significant high-frequency content and yet is simple enough to allow analytic computations.) Figure 4.7 also displays the noisy data.

The solution of the inverse problem $Tx = y$ is simply $x = x^* + \delta z'$, which is a piecewise continuous function with a large error on the interval $(1/2 - \epsilon, 1/2 + \epsilon)$. In fact, we have

$$\|z'\|_{L^2} = \frac{\sqrt{5}}{\epsilon\sqrt{2}};$$

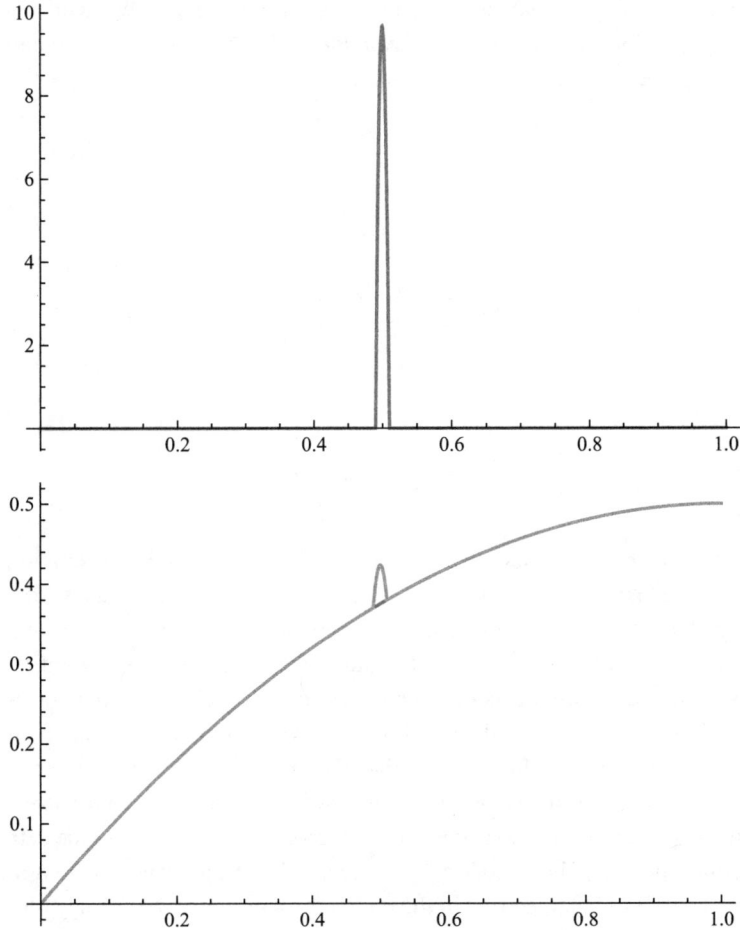

**Figure 4.7.** Top: the perturbation $z$ in Example 4.31 for $\epsilon = 0.01$. Bottom: the exact data $y^*$ and the noisy data $y$.

since we normalized $z$ to have $L^2$-norm one, $\|z'\|_{L^2}$ is simply the ratio of $\|x - x^*\|_{L^2}$ to $\|y - y^*\|_{L^2}$.

The purpose of this example is to examine the above results using the singular value expansion of $T$. This operator is so simple that we can compute its singular values and singular functions analytically (which is rarely possible); this was done in Example 4.25. The singular values and singular vectors are

$$\sigma_n = \frac{2}{(2n-1)\pi},$$

$$\phi_n(t) = \sqrt{2}\cos\left(\frac{(2n-1)\pi t}{2}\right),$$

$$\psi_n(s) = \sqrt{2}\sin\left(\frac{(2n-1)\pi s}{2}\right), \quad n = 1, 2, 3, \ldots,$$

and

$$T = \sum_{n=1}^{\infty} \sigma_n \psi_n \otimes \phi_n.$$

Figure 4.8 shows the first 40 singular values of $T$, graphed on a semilog scale (that is, we graph $\log_{10}\sigma_n$ versus $n$). The same figure shows the magnitude of the first 40 singular components of the exact data $y^*$ and the noise $y - y^* = \delta z$, for $\epsilon = 0.01$ and $\delta = 0.005$. We see that the first three singular components of the error are much smaller than the corresponding components of the exact data. However, the fourth component of the error is only slightly smaller than the exact value, and all the rest of the components of the error are significantly larger than the corresponding components of the exact data. The components of the error in the data are multiplied by the factors $\sigma_n^{-1}$, which are large for large values of $n$. This explains the large error in the computed solution $x$. The exact solution $x^*$, the computed solution $x$, and the singular components of each are shown in Figure 4.9. As expected, the first three singular components of $x$ are quite accurate.

This example suggests one strategy for stabilizing an inverse problem: discard the terms in the series

$$T^\dagger y = \sum_{n=1}^{\infty} \frac{\langle \psi_n, y \rangle_Y}{\sigma_n} \phi_n.$$

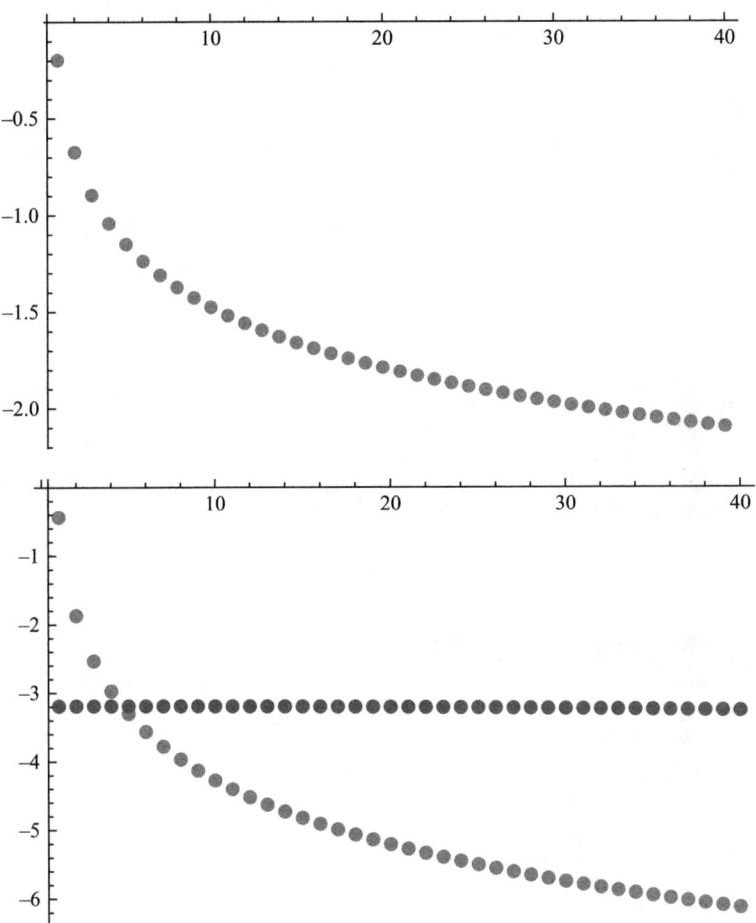

**Figure 4.8.** Top: the first 40 singular values of the operator $T$ in Example 4.31. Bottom: the first 40 singular components of the exact data $y^*$ and the error $y - y^*$ in $y$.

that are expected to be inaccurate. In the above example, this would mean discarding all but the first three terms, and using the approximation

$$x^* \approx \sum_{n=1}^{3} \frac{\langle \psi_n, y \rangle_Y}{\sigma_n} \phi_n.$$

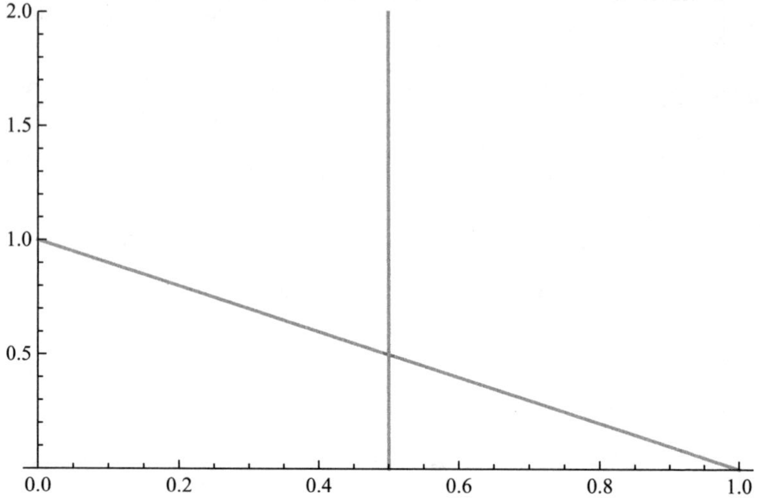

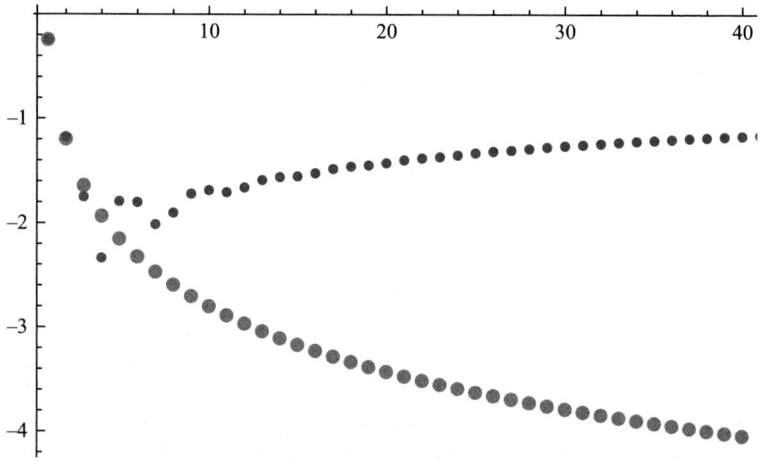

**Figure 4.9.** Top: the exact solution $x^*$ and the computed solution $x$ from Example 4.31. (Note that $x$ is piecewise linear, differing from $x^*$ only on the interval $(1/2 - \epsilon, 1/2 + \epsilon)$, where it has a slope approximately $-970$ when $\epsilon = 0.01$. Therefore, the graph of $x$ appears nearly vertical on $(1/2 - \epsilon, 1/2 + \epsilon)$.) Bottom: the first 40 singular components of the exact solution $x^*$ and the computed solution $x$.

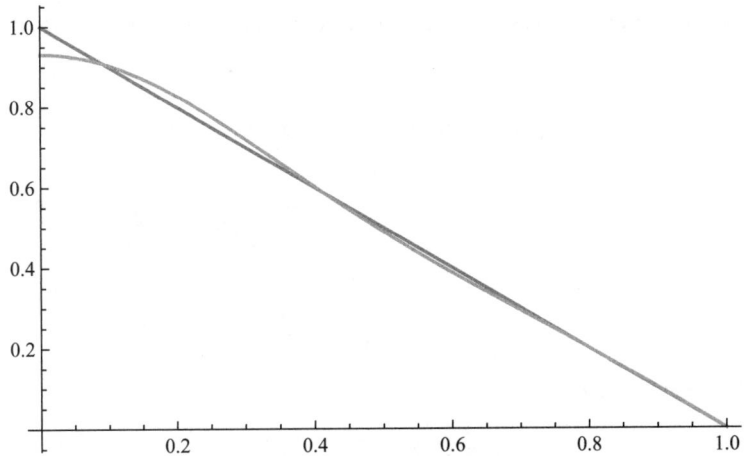

**Figure 4.10.** The exact solution $x^*$ and truncated SVE approximation obtained by discarding all but the first three singular components of $x$.

As Figure 4.10 shows, this produces quite a reasonable approximation of $x^*$. This strategy is called the *truncated SVE* method (or truncated SVD method in finite dimensions), and it can be quite effective. However, it has the drawback that it is necessary to compute the singular values and singular vectors, making the approach impractical for many problems.

## 4.6 Tikhonov regularization in terms of the SVE

We will now express the Tikhonov solution of $Tx = y$ in terms of the singular value expansion of $T$. We continue to assume that the SVE of $T$ is

$$T = \sum_{n=1}^{\infty} \sigma_n \psi_n \otimes \phi_n.$$

We know that $T^*$ is also compact, with

$$T^* = \sum_{n=1}^{\infty} \sigma_n \phi_n \otimes \psi_n.$$

We will need to express the operators $T^*T$ and $N_\lambda = T^*T + \lambda I$ as well. For every $x \in X$, we have

$$T^*Tx = T^* \left( \sum_{n=1}^{\infty} \sigma_n \langle \phi_n, x \rangle_X \psi_n \right) = \sum_{n=1}^{\infty} \sigma_n \langle \phi_n, x \rangle_X T^* \psi_n$$

$$= \sum_{n=1}^{\infty} \sigma_n^2 \langle \phi_n, x \rangle_X \phi_n$$

$$= \sum_{n=1}^{\infty} \sigma_n^2 (\phi_n \otimes \phi_n)x,$$

which shows that

$$T^*T = \sum_{n=1}^{\infty} \sigma_n^2 \phi_n \otimes \phi_n.$$

This confirms that $T^*T$ is compact, which we already proved in Section 4.2, because the series on the right is an SVE. It also shows that $T^*T\phi_n = \sigma_n^2 \phi_n$ for each $n$.

To express $N_\lambda$ in terms of the singular values and singular vectors of $T$, we first note that each $x \in X$ can be written as

$$x = \text{proj}_{N(T)^\perp} x + \text{proj}_{N(T)} x = \sum_{n=1}^{\infty} \langle x, \phi_n \rangle_X \phi_n + \text{proj}_{N(T)} x.$$

Therefore,

$$(T^*T + \lambda I)x = T^*T \left( \sum_{n=1}^{\infty} \langle x, \phi_n \rangle_X \phi_n \right) + T^*T(\text{proj}_{N(T)} x) +$$

$$\lambda \sum_{n=1}^{\infty} \langle x, \phi_n \rangle_X \phi_n + \lambda \text{proj}_{N(T)} x$$

$$= \sum_{n=1}^{\infty} \langle x, \phi_n \rangle_X T^*T\phi_n + \sum_{n=1}^{\infty} \lambda \langle x, \phi_n \rangle_X \phi_n + \lambda \text{proj}_{N(T)} x$$

$$= \sum_{n=1}^{\infty} \left( \sigma_n^2 + \lambda \right) \langle x, \phi_n \rangle_X \phi_n + \lambda \text{proj}_{N(T)} x,$$

which shows that

$$N_\lambda = T^*T + \lambda I = \sum_{n=1}^{\infty} \left( \sigma_n^2 + \lambda \right) \phi_n \otimes \phi_n + \lambda \text{proj}_{\mathcal{N}(T)}.$$

We can now derive the Tikhonov solution $x_{\lambda,y} = N_\lambda^{-1} T^* y$ in terms of the singular values and singular vectors of $T$. We have $(T^*T + \lambda I)x_{\lambda,y} = T^* y$ if and only if

$$\sum_{n=1}^{\infty} \left( \sigma_n^2 + \lambda \right) \langle x_{\lambda,y}, \phi_n \rangle_X \phi_n + \lambda \text{proj}_{\mathcal{N}(T)} x_{\lambda,y} = \sum_{n=1}^{\infty} \sigma_n \langle \psi_n, y \rangle_Y \phi_n.$$

Since the two series represent vectors in $\mathcal{N}(T)^\perp$, it follows that $\text{proj}_{\mathcal{N}(T)} x_{\lambda,y}$ must be zero (which we proved previously—specifically, we proved in Corollary 3.5 that $x_{\lambda,y} \in \mathcal{N}(T)^\perp$). It then follows that

$$\sum_{n=1}^{\infty} \left( \sigma_n^2 + \lambda \right) \langle x_{\lambda,y}, \phi_n \rangle_X \phi_n = \sum_{n=1}^{\infty} \sigma_n \langle \psi_n, y \rangle_Y \phi_n$$

$$\Rightarrow \left( \sigma_n^2 + \lambda \right) \langle x_{\lambda,y}, \phi_n \rangle_X = \sigma_n \langle \psi_n, y \rangle_Y \text{ for all } n \in \mathbb{Z}^+$$

$$\Rightarrow \langle x_{\lambda,y}, \phi_n \rangle_X = \frac{\sigma_n}{\sigma_n^2 + \lambda} \langle \psi_n, y \rangle_Y \text{ for all } n \in \mathbb{Z}^+$$

$$\Rightarrow x_{\lambda,y} = \sum_{n=1}^{\infty} \frac{\sigma_n}{\sigma_n^2 + \lambda} \langle \psi_n, y \rangle_Y \phi_n,$$

and hence that

$$N_\lambda^{-1} T^* = \sum_{n=1}^{\infty} \frac{\sigma_n}{\sigma_n^2 + \lambda} \phi_n \otimes \psi_n.$$

Since

$$\frac{\sigma_n}{\sigma_n^2 + \lambda} \to 0 \text{ as } n \to \infty,$$

Lemma 4.23 implies that $N_\lambda^{-1} T^*$ is not only bounded but compact. Under Tikhonov regularization, $N_\lambda^{-1} T^*$ is the operator approximating the unbounded operator $T^\dagger$. As before, we will write $R_\lambda = N_\lambda^{-1} T^*$.

The advantage of the SVE is that it gives detailed information about how $R_\lambda$ approximates $T^\dagger$. From the analysis in the last chapter, we know that $R_\lambda \to T^\dagger$ pointwise on $D(T^\dagger)$, but analyzing the SVE gives much more information. We have

$$T^\dagger = \sum_{n=1}^\infty \frac{1}{\sigma_n} \phi_n \otimes \psi_n, \ R_\lambda = \sum_{n=1}^\infty \frac{\sigma_n}{\sigma_n^2 + \lambda} \phi_n \otimes \psi_n,$$

and these formulas reveal exactly how each operator acts on each singular component of the data $y$. The generalized inverse $T^\dagger$ maps the $n$th singular component of $y$ to a multiple of $\phi_n$, with the magnitude amplified or dampened by the factor $1/\sigma_n$, while the Tikhonov operator does the same thing, except that the factor is $\sigma_n/(\sigma_n^2 + \lambda)$. If $\sigma_n^2 \gg \lambda$, then

$$\frac{\sigma_n}{\sigma_n^2 + \lambda} \approx \frac{\sigma_n}{\sigma_n^2} = \frac{1}{\sigma_n},$$

and therefore the two operators treat the $n$th singular component approximately the same if $\sigma_n^2 \gg \lambda$. On the other hand, if $\sigma_n^2 \ll \lambda$, then

$$\frac{\sigma_n}{\sigma_n^2 + \lambda} \approx \frac{\sigma_n}{\lambda} \ll \frac{1}{\sigma_n}.$$

This shows that, if $\sigma_n^2 \ll \lambda$, then $R_\lambda$ dampens (or filters out) the $n$th singular component.

The reasoning in the previous paragraph is confirmed by Figure 4.11, which shows graphs of $f(\sigma) = 1/\sigma$ and $g(s) = \sigma/(\sigma^2 + \lambda)$, in both semilog and log-log scales. The following facts, evident in these graphs, can be easily verified.

1. $1/\sigma > \sigma/(\sigma^2 + \lambda)$ for all $\sigma > 0$ and the difference is increasing as $\sigma$ decreases.

2. The graphs suggest that the value of $\sigma$ that maximizes $\sigma/(\sigma^2 + \lambda)$ is of interest (clearly $\sigma/(\sigma^2 + \lambda)$ is not a good approximation for $\sigma$ less than this value). The quantity $\sigma/(\sigma^2 + \lambda)$ is maximized at $\sigma = \sqrt{\lambda}$, and

$$f(\sqrt{\lambda}) - g(\sqrt{\lambda}) = \frac{1}{2\sqrt{\lambda}}.$$

This is consistent with the discussion above.

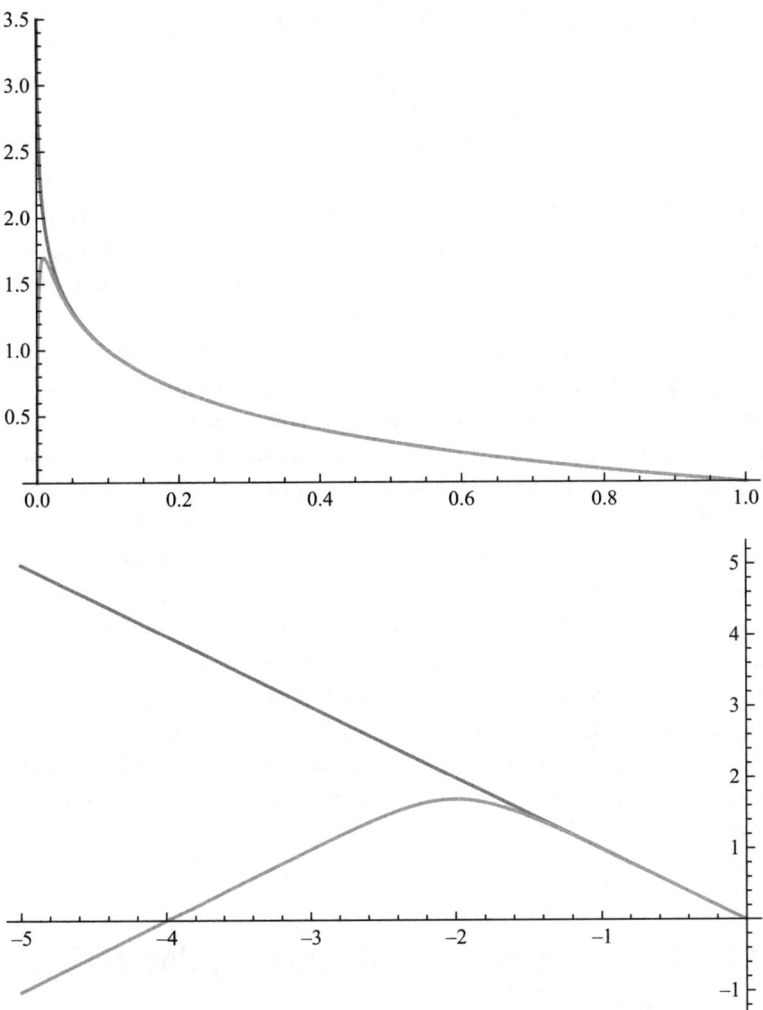

**Figure 4.11.** The functions $f(\sigma) = 1/\sigma$ and $g(\sigma) = \sigma/(\sigma^2 + \lambda)$, in semilog (top) and log-log (bottom) scales. In both plots, the graph of $f$ is the top curve. The value $\lambda = 10^{-4}$ is used.

It is common to write the Tikhonov solution as

$$x_{\lambda,y} = \sum_{n=1}^{\infty} \frac{\sigma_n^2}{\sigma_n^2 + \lambda} \frac{\langle \psi_n, y \rangle_Y}{\sigma_n} \phi_n$$

and to call the values $\sigma_n^2/(\sigma_n^2 + \lambda)^2$, $n = 1, 2, 3, \ldots$, the *filter factors* of the regularization method. As we will see in Section 4.8, we can design new regularization methods by choosing different filter factors.

As discussed at the end of Section 4.5, it is the error in the singular components of the data that correspond to large values of $n$ that lead to difficulties in solving the inverse problem. The discussion above shows that Tikhonov regularization preserves the singular components that correspond to $\sigma_n > \sqrt{\lambda}$ and then (quickly) dampens the singular components as $\sigma_n$ falls below $\sqrt{\lambda}$. In this way, the method is able to preserve the information in the data that has not been corrupted by noise, without allowing the noise in the other singular components to ruin the computed solution. Of course, the last statement depends on our ability to choose $\lambda$ correctly. In many problems, the singular values are not known and we do not know a good cut-off for keeping versus eliminating singular components. Our analysis shows that it is possible in principle to choose a good value of $\lambda$, and we understand exactly how a good value of $\lambda$ leads to a good solution. At the same time, we still need a method for choosing $\lambda$ in a specific problem.

## 4.7   Convergence of Tikhonov regularization via the SVE

We will now rederive our earlier results using the singular value expansion of $T$. We begin with the basic convergence result.

**Theorem 4.32.** *For each $y \in D(T^\dagger)$, $x_{\lambda,y} \to x_{0,y}$ as $\lambda \to 0^+$.*

*Proof.* Given $y \in D(T^{\dagger})$, we have

$$
\begin{aligned}
x_{\lambda,y} - x_{0,y} = x_{\lambda,y} &= \sum_{n=1}^{\infty} \frac{\sigma_n}{\sigma_n^2 + \lambda} \langle \psi_n, y \rangle_Y \phi_n - \sum_{n=1}^{\infty} \frac{1}{\sigma_n} \langle \psi_n, y \rangle_Y \phi_n \\
&= \sum_{n=1}^{\infty} \left( \frac{\sigma_n}{\sigma_n^2 + \lambda} - \frac{1}{\sigma_n} \right) \langle \psi_n, y \rangle_Y \phi_n \\
&= - \sum_{n=1}^{\infty} \frac{\lambda}{\sigma_n (\sigma_n^2 + \lambda)} \langle \psi_n, y \rangle_Y \phi_n.
\end{aligned}
$$

It follows that, for any $N \in \mathbb{Z}^+$,

$$
\begin{aligned}
\| x_{\lambda,y} - x_{0,y} \|_X^2 &= \sum_{n=1}^{\infty} \frac{\lambda^2}{(\sigma_n^2 + \lambda)^2} \frac{|\langle \psi_n, y \rangle_Y|^2}{\sigma_n^2} \\
&= \sum_{n=1}^{N} \frac{\lambda^2}{(\sigma_n^2 + \lambda)^2} \frac{|\langle \psi_n, y \rangle_Y|^2}{\sigma_n^2} + \sum_{n=N+1}^{\infty} \frac{\lambda^2}{(\sigma_n^2 + \lambda)^2} \frac{|\langle \psi_n, y \rangle_Y|^2}{\sigma_n^2} \\
&\leq \sum_{n=1}^{N} \frac{\lambda^2}{(\sigma_n^2 + \lambda)^2} \frac{|\langle \psi_n, y \rangle_Y|^2}{\sigma_n^2} + \sum_{n=N+1}^{\infty} \frac{|\langle \psi_n, y \rangle_Y|^2}{\sigma_n^2}.
\end{aligned}
$$

Let $\epsilon > 0$ be given. Since

$$
\sum_{n=1}^{\infty} \frac{|\langle \psi_n, y \rangle_Y|^2}{\sigma_n^2}
$$

is convergent, we can choose $N$ sufficiently large that

$$
\sum_{n=N+1}^{\infty} \frac{|\langle \psi_n, y \rangle_Y|^2}{\sigma_n^2} < \frac{\epsilon}{2}
$$

holds. We also have, for each fixed $n$, $\lambda^2 / (\sigma_n^2 + \lambda)^2 \to 0$ as $\lambda \to 0$. Therefore, we can find $\lambda_0 > 0$ sufficiently small that

$$
0 < \lambda < \lambda_0 \implies \sum_{n=1}^{N} \frac{\lambda^2}{(\sigma_n^2 + \lambda)^2} \frac{|\langle \psi_n, y \rangle_Y|^2}{\sigma_n^2} < \frac{\epsilon}{2}.
$$

It then follows that if $0 < \lambda < \lambda_0$, then $\|x_{\lambda,y} - x_{0,y}\|_X^2 < \epsilon$, and this proves that $x_{\lambda,y} \to x_{0,y}$ as $\lambda \to 0^+$, as desired. $\qquad\square$

The technique we used in the previous proof to show that the series representing $x_{\lambda,y} - x_{0,y}$ converges to zero as $\lambda \to 0$ will be useful again, so we describe it abstractly in the following lemma.

**Lemma 4.33.** *Suppose $\{\beta_n\}$ is a sequence of nonnegative real numbers and, for each $\lambda > 0$, $\{\alpha_{n,\lambda}\}$ is a sequence of nonnegative real numbers. Suppose further that*

$$\sum_{n=1}^{\infty} \beta_n < \infty,$$

*there exists $M > 0$ such that $\alpha_{n,\lambda} \leq M$ for all $n$ and all $\lambda > 0$, and, for each $n \in \mathbb{Z}^+$, $\alpha_{n,\lambda} \to 0$ as $\lambda \to 0^+$. Then*

$$\sum_{n=1}^{\infty} \alpha_{n,\lambda}\beta_n \to 0 \text{ as } \lambda \to 0^+.$$

*Proof.* (Sketch) We write

$$\sum_{n=1}^{\infty} \alpha_{n,\lambda}\beta_n = \sum_{n=1}^{N} \alpha_{n,\lambda}\beta_n + \sum_{n=N+1}^{\infty} \alpha_{n,\lambda}\beta_n \leq \sum_{n=1}^{N} \alpha_{n,\lambda}\beta_n + M \sum_{n=N+1}^{\infty} \beta_n.$$

Given $\epsilon > 0$, we use the fact that $\sum_{n=1}^{\infty} \beta_n < \infty$ to choose $N$ sufficiently large that $M \sum_{n=N+1}^{\infty} \beta_n < \epsilon/2$, and then choose $\lambda_0 > 0$ sufficiently small that, for all $\lambda \in (0, \lambda_0)$, $\sum_{n=1}^{N} \alpha_{n,\lambda}\beta_n < \epsilon/2$. This second step is possible because we can choose $\lambda_0$ sufficiently small that $\alpha_{n,\lambda}\beta_n < \epsilon/(2N)$ for all $\lambda \in (0, \lambda_0)$ and each $n = 1, 2, \ldots, N$. $\qquad\square$

We now present a proof of Theorem 3.17 using the formula for $x_{\lambda,y}$ derived above.

**Theorem 4.34.** *If $y \notin D(T^\dagger)$, then $\|x_{\lambda,y}\|_X \to \infty$ as $\lambda \to 0^+$.*

*Proof.* We have

$$x_{\lambda,y} = \sum_{n=1}^{\infty} \frac{\sigma_n}{\sigma_n^2 + \lambda} \langle \psi_n, y \rangle_Y \phi_n$$

and therefore, for any $N \in \mathbb{Z}^+$,

$$\|x_{\lambda,y}\|_X^2 = \sum_{n=1}^{\infty} \frac{\sigma_n^2 \, |\langle \psi_n, y \rangle_Y|^2}{(\sigma_n^2 + \lambda)^2} = \sum_{n=1}^{\infty} \frac{\sigma_n^4}{(\sigma_n^2 + \lambda)^2} \frac{|\langle \psi_n, y \rangle_Y|^2}{\sigma_n^2}$$

$$\geq \sum_{n=1}^{N} \frac{\sigma_n^4}{(\sigma_n^2 + \lambda)^2} \frac{|\langle \psi_n, y \rangle_Y|^2}{\sigma_n^2}.$$

Now choose $N = n_\lambda$ such that $\lambda \leq \sigma_n^2$ if and only if $n = 1, 2, \ldots, n_\lambda$. Clearly $n_\lambda$ is well-defined for $\lambda$ sufficiently small, and $n_\lambda \to \infty$ as $\lambda \to 0^+$. It follows that

$$\|x_{\lambda,y}\|_X^2 \geq \sum_{n=1}^{n_\lambda} \frac{\sigma_n^4}{(\sigma_n^2 + \lambda)^2} \frac{|\langle \psi_n, y \rangle_Y|^2}{\sigma_n^2} \geq \sum_{n=1}^{n_\lambda} \frac{\sigma_n^4}{(\sigma_n^2 + \sigma_n^2)^2} \frac{|\langle \psi_n, y \rangle_Y|^2}{\sigma_n^2}$$

$$= \frac{1}{4} \sum_{n=1}^{n_\lambda} \frac{|\langle \psi_n, y \rangle_Y|^2}{\sigma_n^2}.$$

Since $y \notin D(T^\dagger)$, the failure of the Picard condition and the fact that $n_\lambda \to \infty$ as $\lambda \to 0^+$ imply that

$$\|x_{\lambda,y}\|_X^2 \geq \frac{1}{4} \sum_{n=1}^{n_\lambda} \frac{|\langle \psi_n, y \rangle_Y|^2}{\sigma_n^2} \to \infty \text{ as } \lambda \to 0^+. \qquad \square$$

For the next theorem, we need an expression for the operator $TN_\lambda^{-1}T^*$. We have

$$(TN_\lambda^{-1}T^*)y = T\left(N_\lambda^{-1}T^*y\right) = T\left(\sum_{n=1}^{\infty} \frac{\sigma_n}{\sigma_n^2 + \lambda} \langle \psi_n, y \rangle_Y \phi_n\right)$$

$$= \sum_{n=1}^{\infty} \frac{\sigma_n}{\sigma_n^2 + \lambda} \langle \psi_n, y \rangle_Y T\phi_n$$

$$= \sum_{n=1}^{\infty} \frac{\sigma_n^2}{\sigma_n^2 + \lambda} \langle \psi_n, y \rangle_Y \psi_n.$$

Therefore,

$$TN_\lambda^{-1}T^* = \sum_{n=1}^\infty \frac{\sigma_n^2}{\sigma_n^2 + \lambda} \psi_n \otimes \psi_n.$$

We now use Lemma 4.33 to prove the following result.

**Theorem 4.35.** *For all* $y \in Y$, $Tx_{\lambda,y} \to \bar{y} = \mathrm{proj}_{\overline{\mathcal{R}(T)}}$ *as* $\lambda \to 0^+$.

*Proof.* We have $\bar{y} = \sum_{n=1}^\infty \langle \psi_n, y \rangle_Y \psi_n$ and hence

$$\begin{aligned}
Tx_{\lambda,y} - \bar{y} = TN_\lambda^{-1}T^*y - \bar{y} &= \sum_{n=1}^\infty \frac{\sigma_n^2}{\sigma_n^2 + \lambda} \langle \psi_n, y \rangle_Y \psi_n - \sum_{n=1}^\infty \langle \psi_n, y \rangle_Y \psi_n \\
&= \sum_{n=1}^\infty \left( \frac{\sigma_n^2}{\sigma_n^2 + \lambda} - 1 \right) \langle \psi_n, y \rangle_Y \psi_n \\
&= -\sum_{n=1}^\infty \frac{\lambda}{\sigma_n^2 + \lambda} \langle \psi_n, y \rangle_Y \psi_n,
\end{aligned}$$

which implies that

$$\|Tx_{\lambda,y} - \bar{y}\|_Y^2 = \sum_{n=1}^\infty \frac{\lambda^2}{(\sigma_n^2 + \lambda)^2} |\langle \psi_n, y \rangle_Y|^2.$$

If we define $\alpha_{n,\lambda} = \lambda^2/(\sigma_n + \lambda)^2$ and $\beta_n = |\langle \psi_n, y \rangle_Y|^2$, then Lemma 4.33 shows that $\|Tx_{\lambda,y} - \bar{y}\|_Y^2 \to 0$ as $\lambda \to 0^+$, as desired. $\qquad\square$

## 4.7.1   Rates of convergence

To prove the following theorems, we remind the reader of the following basic identity, which was first derived in Section 3.4:

$$\begin{aligned}
N_\lambda(x_{\lambda,y} - x_{0,y}) = T^*y - T^*Tx_{0,y} - \lambda x_{0,y} &\Rightarrow N_\lambda(x_{\lambda,y} - x_{0,y}) = -\lambda x_{0,y} \\
&\Rightarrow x_{\lambda,y} - x_{0,y} = -\lambda N_\lambda^{-1} x_{0,y}.
\end{aligned}$$

We now give another proof of Theorem 3.22.

**Theorem 4.36.** *Suppose* $y \in D(T^\dagger)$ *and* $x_{0,y} \in \mathcal{R}(T^*)$. *Then*

$$\|x_{\lambda,y} - x_{0,y}\|_X = o\left(\lambda^{1/2}\right) \text{ as } \lambda \to 0^+.$$

*Proof.* Assume that $x_{0,y} = T^* w_{0,y}$ for some $w_{0,y} \in Y$. Then

$$x_{0,y} = T^* w_{0,y} = \sum_{n=1}^{\infty} \sigma_n \langle \psi_n, w_{0,y} \rangle_Y \phi_n$$

$$\Rightarrow x_{\lambda,y} - x_{0,y} = -\lambda N_\lambda^{-1} x_{0,y} = -\lambda \sum_{n=1}^{\infty} \frac{\sigma_n}{\sigma_n^2 + \lambda} \langle \psi_n, w_{0,y} \rangle_Y \phi_n.$$

Therefore,

$$\|x_{\lambda,y} - x_{0,y}\|_X^2 = \sum_{n=1}^{\infty} \frac{\sigma_n^2 \lambda^2}{(\sigma_n^2 + \lambda)^2} |\langle \psi_n, w_{0,y} \rangle_Y|^2$$

$$= \lambda \sum_{n=1}^{\infty} \frac{\sigma_n^2 \lambda}{(\sigma_n^2 + \lambda)^2} |\langle \psi_n, w_{0,y} \rangle_Y|^2.$$

Now define $\alpha_{n,\lambda} = \sigma_n^2 \lambda / (\sigma_n^2 + \lambda)^2$, $\beta_n = |\langle \psi_n, w_{0,y} \rangle_Y|^2$. Then $\sum_{n=1}^{\infty} \beta_n < \infty$ and, for each $n$, $\alpha_{n,\lambda} \to 0$ as $\lambda \to 0^+$. Moreover,

$$\frac{\sigma_n^2 \lambda}{(\sigma_n^2 + \lambda)^2} \le \frac{\frac{1}{2}(\sigma_n^4 + \lambda^2)}{\sigma_n^4 + \lambda^2 + 2\sigma_n^2 \lambda} \le \frac{\sigma_n^4 + \lambda^2}{2(\sigma_n^4 + \lambda^2)} = \frac{1}{2} \text{ for all } \sigma_n > 0, \lambda > 0.$$

Therefore Lemma 4.33 applies, yielding

$$\lambda \sum_{n=1}^{\infty} \frac{\sigma_n^2 \lambda}{(\sigma_n^2 + \lambda)^2} |\langle \psi_n, w_{0,y} \rangle_Y|^2 = o(\lambda) \text{ as } \lambda \to 0.$$

It follows that $\|x_{\lambda,y} - x_{0,y}\|_X = o(\lambda^{1/2})$, as desired. $\qquad\square$

Similarly, we can use the singular value expansion to prove Theorem 3.23.

**Theorem 4.37.** *Suppose $y \in D(T^\dagger)$ and $x_{0,y} \in \mathcal{R}(T^*T)$. Then*

$$\|x_{\lambda,y} - x_{0,y}\|_X = O(\lambda) \text{ as } \lambda \to 0^+.$$

*Proof.* Suppose $x_{0,y} = T^* T u_{0,y}$ for some $u_{0,y} \in X$. We then have

$$x_{\lambda,y} - x_{0,y} = -\lambda N_\lambda^{-1} x_{0,y} = -\lambda N_\lambda^{-1} T^* T u_{0,y}.$$

We have

$$T^*Tu_{0,y} = \sum_{n=1}^{\infty} \sigma_n^2 \langle \phi_n, u_{0,y} \rangle_X \phi_n$$

$$\Rightarrow x_{\lambda,y} - x_{0,y} = -\lambda N_\lambda^{-1} T^*Tu_{0,y} = \lambda \sum_{n=1}^{\infty} \frac{\sigma_n^2}{\sigma_n^2 + \lambda} \langle \phi_n, u_{0,y} \rangle_X \phi_n,$$

which implies that

$$\|x_{\lambda,y} - x_{0,y}\|_X^2 = \lambda^2 \sum_{n=1}^{\infty} \frac{\sigma_n^4}{(\sigma_n^2 + \lambda)^2} |\langle \phi_n, u_{0,y} \rangle_X|^2 \le \lambda^2 \sum_{n=1}^{\infty} |\langle \phi_n, u_{0,y} \rangle_X|^2$$

$$\le \lambda^2 \|u_{0,y}\|_X^2. \qquad \square$$

### 4.7.2   Another converse result

In Section 3.6, we presented a number of results to demonstrate that our convergence theorems are sharp. We noted there, however, that there is one exception: if $x_{0,y^*}$ belongs to $\mathcal{R}(T^*)$, then $\|x_{\lambda,y} - x_{0,y^*}\|_X = o(\delta^{1/2})$. However, the converse is not true; it is possible that $\|x_{\lambda,y} - x_{0,y^*}\|_X = o(\delta^{1/2})$ and yet that $x_{0,y^*}$ does not belong to $\mathcal{R}(T^*)$. Using the SVE, this can be demonstrated by an explicit example (this example is taken from Neubauer [21]).

**Example 4.38.** Let $X$, $Y$ be any (infinite-dimensional) Hilbert spaces, let $\{\phi_n\}$, $\{\psi_n\}$ be orthonormal sequences in $X$, $Y$, respectively, and define $T : X \to Y$ by

$$T = \sum_{n=1}^{\infty} \sigma_n \psi_n \otimes \phi_n,$$

where $\sigma_n = n^{-2}$. Define $x^* \in X$ by

$$x^* = \sum_{n=1}^{\infty} \alpha_n \phi_n, \quad \alpha_n = \frac{1}{\sqrt{n^5 \ln(n)}}, \quad n \in \mathbb{Z}^+,$$

and define $y^* = Tx^*$. Then

$$y^* = \sum_{n=1}^{\infty} \sigma_n \alpha_n \psi_n.$$

Recall that $x^* \in \mathcal{R}(T^*)$ if and only if

$$\sum_{n=1}^{\infty} \left(\frac{\alpha_n}{\sigma_n}\right)^2 < \infty.$$

But

$$\left(\frac{\alpha_n}{\sigma_n}\right)^2 = \frac{n^4}{n^5 \ln(n)} = \frac{1}{n \ln(n)}$$

and

$$\sum_{n=1}^{\infty} \frac{1}{n \ln(n)} = \infty.$$

This shows that $x^* \notin \mathcal{R}(T^*)$.

We have

$$x_{\lambda,y^*} = \sum_{n=1}^{\infty} \frac{\sigma_n}{\sigma_n^2 + \lambda} \langle \psi_n, y^* \rangle_Y \phi_n = \sum_{n=1}^{\infty} \frac{\sigma_n^2 \alpha_n}{\sigma_n^2 + \lambda} \phi_n$$

and hence

$$x_{\lambda,y^*} - x_{0,y^*} = \sum_{n=1}^{\infty} \frac{\sigma_n^2 \alpha_n}{\sigma_n^2 + \lambda} \phi_n - \sum_{n=1}^{\infty} \alpha_n \phi_n = -\sum_{n=1}^{\infty} \frac{\lambda \alpha_n}{\sigma_n^2 + \lambda} \phi_n.$$

Therefore,

$$\|x_{\lambda,y^*} - x_{0,y^*}\|_X^2 = \sum_{n=1}^{\infty} \frac{\lambda^2 \alpha_n^2}{(\sigma_n^2 + \lambda)^2} = \sum_{n=1}^{\infty} \frac{\lambda^2}{(1 + \lambda n^4)^2 n \ln(n)}.$$

We wish to show that $\|x_{\lambda,y^*} - x_{0,y^*}\|_X^2 = o(\lambda)$, that is, that

$$\sum_{n=1}^{\infty} \frac{\lambda}{(1 + \lambda n^4)^2 n \ln(n)} \to 0 \text{ as } \lambda \to 0^+. \tag{4.14}$$

This can be shown using Lemma 4.33. We define

$$\alpha_{n,\lambda} = \frac{\lambda n^4}{(1 + \lambda n^4)^2}, \quad \beta_n = \frac{1}{n^5 \ln(n)}.$$

Then, for each $n \in \mathbb{Z}^+$, $\alpha_{n,\lambda} \to 0$ as $\lambda \to 0^+$, and $\alpha_{n,\lambda} \leq 1/2$ for all $n \in \mathbb{Z}^+$, $\lambda \geq 0$. Also, $\sum_{n=1}^{\infty} \beta_n < \infty$. Therefore, by Lemma 4.33, (4.14) holds, and we have shown that $\|x_{\lambda,y^*} - x_{0,y^*}\|_X = o(\lambda^{1/2})$, in spite of the fact that $x^* \notin \mathcal{R}(T^*)$.

# 4.8   General regularization methods

Tikhonov regularization produces the approximate solution

$$x_{\lambda,y} = (T^*T + \lambda I)^{-1} T^* y = \left( \sum_{n=1}^{\infty} \frac{\sigma_n}{\sigma_n^2 + \lambda} \phi_n \otimes \psi_n \right) y.$$

We saw in Section 4.6 that the filter factors $\sigma_n^2/(\sigma_n^2 + \lambda)$ contain the essential information about the method, at least when applied to a compact operator. It is natural to ask if we could design other regularization methods by choosing different filter factors. The key feature of the filter factors for Tikhonov regularization is the fact that the function $\sigma \mapsto \sigma/(\sigma^2 + \lambda)$ approximates $\sigma \mapsto 1/\sigma$, at least for $\sigma$ not too close to 0. This suggests that we might choose different functions that approximate $\sigma \mapsto 1/\sigma$ and obtain other regularization methods.

## 4.8.1   Functions of $T^*T$

To make it convenient to define and analyze alternative regularization methods, we begin by describing a general method for defining functions of $T^*T$, where $T$ is a compact operator with SVE

$$T = \sum_{n=1}^{\infty} \sigma_n \psi_n \otimes \phi_n.$$

Polynomials in $T^*T$ can be defined algebraically, with $(T^*T)^2 = (T^*T)(T^*T)$, $(T^*T)^3 = (T^*T)(T^*T)(T^*T)$, and so forth. If $p$ is a polynomial function, with

$$p(\beta) = a_k \beta^k + a_{k-1} \beta^{k-1} + \cdots + a_1 \beta + a_0,$$

then

$$p(T^*T) = a_k(T^*T)^k + a_{k-1}(T^*T)^{k-1} + \cdots + a_1 T^*T + a_0 I,$$

where $I$ is the identity operator on $X$. Now, it can be verified directly that, for each positive integer $k$,

$$(T^*T)^k = \sum_{n=1}^{\infty} \sigma_n^{2k} \phi_n \otimes \phi_n.$$

This follows from the fact that $T^*T\phi_n = \sigma_n^2 \phi_n$; for instance,

$$
\begin{aligned}
(T^*T)^2 x = T^*T\left(T^*Tx\right) = T^*T &\left( \sum_{n=1}^{\infty} \sigma_n^2 \langle \phi_n, x \rangle_X \phi_n \right) \\
&= \sum_{n=1}^{\infty} \sigma_n^2 \langle \phi_n, x \rangle_X T^*T\phi_n \\
&= \sum_{n=1}^{\infty} \sigma_n^4 \langle \phi_n, x \rangle_X \phi_n \\
&= \left( \sum_{n=1}^{\infty} \sigma_n^4 \phi_n \otimes \phi_n \right) x.
\end{aligned}
$$

It then follows that

$$p(T^*T) = \sum_{n=1}^{\infty} p(\sigma_n^2)\phi_n \otimes \phi_n.$$

This suggests that, if $f$ is a more general function, we also define

$$f(T^*T) = \sum_{n=1}^{\infty} f(\sigma_n^2)\phi_n \otimes \phi_n.$$

For convenience, we will assume that $f$ is continuous, although this is not really necessary. (We could, for example, assume that $f$ is piecewise continuous.)

**Theorem 4.39.** *Let $f : [0, \sigma_1^2] \to \mathbb{R}$ be continuous. Then*

$$f(T^*T) = \sum_{n=1}^{\infty} f(\sigma_n^2)\phi_n \otimes \phi_n$$

*defines a bounded linear operator mapping $X$ into itself.*

*Proof.* Since $f$ is continuous, $M = \max\{|f(\beta)| : \beta \in [0, \sigma_1^2]\}$ exists. For each $x \in X$, we have

$$\sum_{n=1}^{\infty} |f(\sigma_n^2)|^2 |\langle \phi_n, x \rangle_X|^2 \leq M \sum_{n=1}^{\infty} |\langle \phi_n, x \rangle_X|^2 \leq M\|x\|_X^2.$$

Therefore,

$$f(T^*T)x = \sum_{n=1}^{\infty} f(\sigma_n^2)\langle \phi_n, x \rangle_X \phi_n$$

is well-defined (because the series is convergent) and $\|f(T^*T)x\|_X \leq \sqrt{M}\|x\|_X$ for all $x \in X$. This shows that $f(T^*T)$ is bounded with $\|f(T^*T)\| \leq \sqrt{M}$. $\qquad\square$

Here are some examples of $f(T^*T)$ for different functions $f$.

1. $f(\beta) = 1$. In this case,

$$f(T^*T)x = \sum_{n=1}^{\infty} \langle \phi_n, x \rangle_X \phi_n \text{ for all } x \in X.$$

Since $\mathcal{N}(T)^\perp = \overline{\mathrm{sp}\{\phi_n\}}$, it follows that $f(T^*T)x = \mathrm{proj}_{\mathcal{N}(T)^\perp}x$; that is, $f(T^*T)$ is the orthogonal projection operator mapping $X$ onto $\mathcal{N}(T)^\perp$.

2. $f(\beta) = \beta^\nu$, where $\nu > 0$ is a real number (not necessarily an integer). Then

$$f(T^*T) = \sum_{n=1}^{\infty} \sigma_n^{2\nu}\phi_n \otimes \phi_n.$$

This will be an important example below, where we refine our earlier results about rates of convergence of $x_{\lambda,y} \to x_{0,y}$. Since $\sigma_n^{2\nu} \to 0$ as $n \to \infty$, it follows that $(T^*T)^\nu$ is compact for all $\nu > 0$.

3. $f_\lambda(\beta) = 1/(\beta + \lambda)$, where $\lambda > 0$. Then

$$f(T^*T) = \sum_{n=1}^{\infty} \frac{1}{\sigma_n^2 + \lambda} \phi_n \otimes \phi_n = (T^*T + \lambda I)^{-1} = N_\lambda^{-1}.$$

Tikhonov regularization is based on the related operator $R_\lambda = f_\lambda(T^*T)T^*$. Since $f_\lambda(T^*T)\phi_n = f_\lambda(\sigma_n^2)\phi_n$, we have

$$R_\lambda y = f_\lambda(T^*T)(T^*y) = f_\lambda(T^*T)\left(\sum_{n=1}^{\infty} \sigma_n \langle \psi_n, y \rangle_Y \phi_n\right)$$

$$= \sum_{n=1}^{\infty} \sigma_n \langle \psi_n, y \rangle_Y f_\lambda(T^*T)\phi_n$$

$$= \sum_{n=1}^{\infty} f_\lambda(\sigma_n^2)\sigma_n \langle \psi_n, y \rangle_Y \phi_n$$

$$= \sum_{n=1}^{\infty} \frac{\sigma_n}{\sigma_n^2 + \lambda} \langle \psi_n, y \rangle_Y \phi_n,$$

which is the formula we derived before.

### 4.8.2 An abstract regularization method

Now let us suppose $\{s_\lambda : \lambda > 0\}$ is a family of continuous real-valued functions defined on $[0, \sigma_1^2]$ with the property that $s_\lambda(\beta)$ is an increasingly good approximation of $1/\beta$ as $\lambda \to 0^+$. We take this to mean that, at the very least, $s_\lambda(\beta) \to 1/\beta$ as $\lambda \to 0^+$ for each fixed $\beta > 0$. The purpose of the following analysis is to determine what other properties we need in $s_\lambda$ in order to define a regularization method.

Given such a family of functions, we define a family of regularization operators by

$$S_\lambda = s_\lambda(T^*T)T^* = \sum_{n=1}^{\infty} \sigma_n s_\lambda(\sigma_n^2)\phi_n \otimes \psi_n.$$

For this to be a valid regularization method, we need $S_\lambda y \to T^\dagger y$ for all $y \in D(T^\dagger)$. We have

$$S_\lambda y - T^\dagger y = \sum_{n=1}^{\infty} \sigma_n s_\lambda(\sigma_n^2) \langle \psi_n, y \rangle_Y \phi_n - \sum_{n=1}^{\infty} \frac{1}{\sigma_n} \langle \psi_n, y \rangle_Y \phi_n$$

$$= \sum_{n=1}^{\infty} (\sigma_n^2 s_\lambda(\sigma_n^2) - 1) \frac{\langle \psi_n, y \rangle_Y}{\sigma_n} \phi_n,$$

and therefore

$$\|S_\lambda y - T^\dagger y\|_X^2 = \sum_{n=1}^{\infty} |\sigma_n^2 s_\lambda(\sigma_n^2) - 1|^2 \frac{|\langle \psi_n, y \rangle_Y|^2}{|\sigma_n|^2}.$$

Since $y \in D(T^\dagger)$, the Picard condition implies that

$$\sum_{n=1}^{\infty} \frac{|\langle \psi_n, y \rangle_Y|^2}{|\sigma_n|^2} < \infty.$$

Also,

$$\text{for all } \beta > 0, \ s_\lambda(\beta) \to \frac{1}{\beta} \text{ as } \lambda \to 0^+$$

$$\Rightarrow \text{ for all } n \in \mathbb{Z}^+, \ \sigma_n^2 s_\lambda(\sigma_n^2) \to 1 \text{ as } \lambda \to 0^+.$$

Therefore, to apply Lemma 4.33 to prove that $\|S_\lambda y - T^\dagger y\|_Y \to 0$ as $\lambda \to 0^+$ (with $\alpha_{\lambda,n} = |\sigma_n^2 s_\lambda(\sigma_n^2) - 1|^2$, $\beta_n = |\langle \psi_n, y \rangle_Y|^2/|\sigma_n|^2$), we need to assume one additional condition: $\{|\beta s_\lambda(\beta)| : \beta \in [0, \sigma_1^2], \lambda > 0\}$ is bounded (which implies that $\alpha_{\lambda,n} = |\sigma_n^2 s_\lambda(\sigma_n^2) - 1|^2$ is bounded for $\lambda > 0$ and for all $n \in \mathbb{Z}^+$).

We have thus proved the following theorem.

**Theorem 4.40.** *For each $\lambda > 0$, let $s_\lambda : [0, \sigma_1^2] \to \mathbb{R}$ be continuous and assume*

$$\text{for all } \beta \in (0, \sigma_1^2], \ s_\lambda(\beta) \to \frac{1}{\beta} \text{ as } \lambda \to 0^+$$

*and*

*there exists $M > 0$ such that $|\beta s_\lambda(\beta)| \leq M$ for all $\beta \in [0, \sigma_1^2], \lambda > 0$.*

*Suppose $S_\lambda : Y \to X$ is defined by $S_\lambda = s_\lambda(T^*T)T^*$. Then*

$$\text{for all } y \in D(T^\dagger), \ S_\lambda y \to T^\dagger y \text{ as } \lambda \to 0^+.$$

### 4.8.3   Rates of convergence

We now wish to explore the possibility of deriving the rate of convergence of $S_\lambda y$ to $T^\dagger y$ by assuming that the exact solution $x_{0,y} = T^\dagger y$ has some extra property. In the case of Tikhonov regularization, the extra property was either $x_{0,y} \in \mathcal{R}(T^*)$ or $x_{0,y} \in \mathcal{R}(T^*T)$. Since we have now defined $(T^*T)^\nu$ for any $\nu > 0$, we will consider the more general condition $x_{0,y} \in \mathcal{R}((T^*T)^\nu)$. (We will show that $\mathcal{R}(T^*) = \mathcal{R}((T^*T)^{1/2})$; hence the more general condition $x_{0,y} \in \mathcal{R}((T^*T)^\nu)$ includes both of the previous cases.)

We can also express the condition in terms of the exact data rather than the exact solution. If $x_{0,y} \in \mathcal{R}((T^*T)^\nu)$, then $\bar{y} \in \mathcal{R}(T(T^*T)^\nu)$, where $\bar{y} = \text{proj}_{\overline{\mathcal{R}(T)}} y$, or, equivalently, $y \in \mathcal{R}(T(T^*T)^\nu) \oplus \mathcal{R}(T)^\perp$. The converse is also true.

Here is the basic result that we will need.

**Theorem 4.41.**

$$\mathcal{R}((T^*T)^\nu) = \left\{ \sum_{n=1}^\infty \alpha_n \phi_n : \sum_{n=1}^\infty \frac{\alpha_n^2}{\sigma_n^{4\nu}} < \infty \right\}.$$

*Consequently,*

$$\mathcal{R}(T(T^*T)^\nu) \oplus \mathcal{R}(T)^\perp = \left\{ y \in Y : \sum_{n=1}^\infty \frac{|\langle \psi_n, y \rangle_Y|^2}{\sigma_n^{4\nu+2}} < \infty \right\}.$$

*Proof.* We have

$$x \in \mathcal{R}((T^*T)^\nu) \Leftrightarrow \text{ there exists } u \in X, \ x = (T^*T)^\nu u$$

$$\Leftrightarrow \text{ there exists } u \in X, \ x = \sum_{n=1}^\infty \sigma_n^{2\nu} \langle \phi_n, u \rangle_X \phi_n.$$

Therefore, with $\alpha_n = \sigma_n^{2v} \langle \phi_n, u \rangle_X$, we have $x = \sum_{n=1}^{\infty} \alpha_n \phi_n$ and

$$\sum_{n=1}^{\infty} \frac{\alpha_n^2}{\sigma_n^{4v}} = \sum_{n=1}^{\infty} |\langle \phi_n, u \rangle_X|^2 < \infty.$$

Conversely, if $x = \sum_{n=1}^{\infty} \alpha_n \phi_n$ and

$$\sum_{n=1}^{\infty} \frac{\alpha_n^2}{\sigma_n^{4v}} < \infty$$

then we can define

$$u = \sum_{n=1}^{\infty} \frac{\alpha_n}{\sigma_n^{2v}} \psi_n.$$

It follows that $u$ is a well-defined element of $X$ (that is, the series defining $u$ converges) and $(T^*T)^{2v} u = x$.    □

**Corollary 4.42.** $\mathcal{R}(T^*) = \mathcal{R}((T^*T)^{1/2})$.

*Proof.* By the previous theorem, it suffices to prove that

$$\mathcal{R}(T^*) = \left\{ \sum_{n=1}^{\infty} \alpha_n \phi_n : \sum_{n=1}^{\infty} \frac{\alpha_n^2}{\sigma_n^2} < \infty \right\}. \qquad (4.15)$$

We have

$$x \in \mathcal{R}(T^*) \Leftrightarrow \text{ there exists } w \in Y, \ x = T^* w$$

$$\Leftrightarrow \text{ there exists } w \in Y, \ x = \sum_{n=1}^{\infty} \sigma_n \langle \psi_n, w \rangle_Y \phi_n.$$

Therefore, with $\alpha_n = \sigma_n \langle \psi_n, w \rangle_Y$, we have $x = \sum_{n=1}^{\infty} \alpha_n \phi_n$ and

$$\sum_{n=1}^{\infty} \frac{\alpha_n^2}{\sigma_n^2} = \sum_{n=1}^{\infty} |\langle \psi_n, w \rangle_Y|^2 < \infty.$$

Conversely, if $x = \sum_{n=1}^{\infty} \alpha_n \phi_n$ and

$$\sum_{n=1}^{\infty} \frac{\alpha_n^2}{\sigma_n^2} < \infty,$$

then we can define

$$w = \sum_{n=1}^{\infty} \frac{\alpha_n}{\sigma_n} \psi_n.$$

It follows that $w$ is a well-defined element of $Y$ (that is, the series defining $w$ converges) and $T^*w = x$. This proves (4.15). $\qquad\square$

Now let us consider bounding $\|S_\lambda y - T^\dagger y\|_X$ when $y \in \mathcal{R}(T(T^*T)^\nu) \oplus \mathcal{R}(T)^\perp$ for some $\nu > 0$. We have

$$S_\lambda y - T^\dagger y = \sum_{n=1}^{\infty} (\sigma_n^2 s_\lambda(\sigma_n^2) - 1) \frac{\langle \psi_n, y \rangle_Y}{\sigma_n} \phi_n.$$

If $\bar{y} = T(T^*T)^\nu u$, then

$$\langle \psi_n, y \rangle_Y = \sigma_n^{2\nu+1} \langle \phi_n, u \rangle_X \text{ for all } n \in \mathbb{Z}^+$$

and hence

$$S_\lambda y - T^\dagger y = \sum_{n=1}^{\infty} \sigma_n^{2\nu} (\sigma_n^2 s_\lambda(\sigma_n^2) - 1) \langle \phi_n, u \rangle_X \phi_n$$

$$\Rightarrow \|S_\lambda y - T^\dagger y\|_X^2 = \sum_{n=1}^{\infty} \left| \sigma_n^{2\nu} (\sigma_n^2 s_\lambda(\sigma_n^2) - 1) \right|^2 |\langle \phi_n, u \rangle_X|^2.$$

The possibility of deriving a rate of convergence now reduces to the question of deriving a bound on $\left| \sigma_n^{2\nu} (\sigma_n^2 s_\lambda(\sigma_n^2) - 1) \right|$ in terms of $\lambda$ and $\nu$ that is independent of $\sigma_n$ (equivalently, we need a bound on $|\beta^\nu(\beta s_\lambda(\beta) - 1)|$ that is independent of $\beta \in [0, \sigma_1^2]$). This is a question that cannot be answered in the abstract but depends on the function $s_\lambda$ under consideration.

In the case of Tikhonov regularization, we have $s_\lambda(\beta) = 1/(\beta + \lambda)$ and

$$f(\beta) = |\beta^\nu(\beta s_\lambda(\beta) - 1)| = \frac{\lambda \beta^\nu}{\beta + \lambda}.$$

A simple exercise in single-variable calculus will verify the following:

- For $0 < \nu < 1$, $f$ has a maximizer at $\beta = (\nu/(1 - \nu))\lambda$ and

  for all $\beta \in [0, \sigma_1^2]$, $f(\beta) \le f\left(\dfrac{\nu}{1 - \nu}\lambda\right) = \nu^\nu(1 - \nu)^{1-\nu}\lambda^\nu = O(\lambda^\nu)$.

  Notice that $\nu^\nu(1 - \nu)^{1-\nu} \in [1/2, 1)$ for all $\nu \in (0, 1)$.

- For $\nu \ge 1$, $f$ is an increasing function of $\beta$ and hence

  for all $\beta \in [0, \sigma_1^2]$, $f(\beta) \le f(\sigma_1^2) = \dfrac{\sigma_1^{2\nu}}{\sigma_1^2 + \lambda}\lambda = O(\lambda)$.

This shows that increasing smoothness beyond $\nu = 1$ does not result in an increased rate of convergence in Tikhonov regularization, which is consistent with our results in Section 3.4 (see Theorem 3.25).

The above results lead to the following theorem.

**Theorem 4.43.** *Suppose $T$ is compact and $y \in \mathcal{R}(T(T^*T)^\nu) \oplus \mathcal{R}(T)^\perp$ (that is, $T^\dagger y \in \mathcal{R}((T^*T)^\nu))$ for some $\nu > 0$. Let $x_{\lambda,y}$ be the Tikhonov solution for each $\lambda > 0$ and let $x_{0,y} = T^\dagger y$.*

- *If $0 < \nu \le 1$, then $\|x_{\lambda,y} - x_{0,y}\|_X = O(\lambda^\nu)$ as $\lambda \to 0^+$.*

- *If $\nu > 1$, then $\|x_{\lambda,y} - x_{0,y}\|_X = O(\lambda)$ as $\lambda \to 0^+$.*

This result is included in the following theorem, so we will not give the proof.

As we noted above, for an abstract regularization operator $S_\lambda = s_\lambda(T^*T)T^*$, the rate of convergence depends on a bound on $|\beta^\nu(\beta s_\lambda(\beta) - 1)|$. We have the following result.

**Theorem 4.44.** *Suppose that $\{s_\lambda\}$ satisfies the hypotheses of Theorem 4.40. Suppose that, in addition,*

$$|\beta^\nu(\beta s_\lambda(\beta) - 1)| \le b(\nu, \lambda) \text{ for all } \beta \in [0, \sigma_1^2], \text{ for all } \nu > 0,$$

$$\lambda > 0 \text{ sufficiently small,}$$

*where, for each $\nu > 0$, $b(\nu, \lambda) \to 0$ as $\lambda \to 0^+$. Define $S_\lambda = s_\lambda(T^*T)T^*$. Then, for each $\nu > 0$ and each $y \in \mathcal{R}(T(T^*T)^\nu) \oplus \mathcal{R}(T)^\perp$, there exists* *a constant $C > 0$ such that*

$$\|S_\lambda y - T^\dagger y\|_X \leq C b(v, \lambda) \text{ for all } \lambda > 0 \text{ sufficiently small.}$$

*Proof.* We have

$$\|S_\lambda y - T^\dagger y\|_X^2 = \sum_{n=1}^\infty \left| \sigma_n^{2v}(\sigma_n^2 s_\lambda(\sigma_n^2) - 1) \right|^2 |\langle \phi_n, u \rangle_X|^2.$$

By assumption,

$$\left| \sigma_n^{2v}(\sigma_n^2 s_\lambda(\sigma_n^2) - 1) \right|^2 \leq b(v, \lambda)^2 \text{ for all } n \in \mathbb{Z}^+$$

and therefore

$$\|S_\lambda y - T^\dagger y\|_X^2 \leq \sum_{n=1}^\infty b(v, \lambda)^2 |\langle \phi_n, u \rangle_X|^2 = b(v, \lambda)^2 \sum_{n=1}^\infty |\langle \phi_n, u \rangle_X|^2$$

$$\leq b(v, \lambda)^2 \|u\|_X^2.$$

Thus, $\|S_\lambda y - T^\dagger y\|_X \leq \|u\|_X b(v, \lambda)$. $\qquad \square$

## 4.8.4 Asymptotic regularization

We now present a regularization method that can be analyzed using the results derived above. In principle, we can choose any family $\{s_\lambda : \lambda > 0\}$ of functions that satisfies the properties described above and define the regularization operator $S_\lambda$ by $S_\lambda = s_\lambda(T^*T)T^*$. In practice, however, the SVE of $T$ will not be known and a regularization operator based on $s_\lambda$ is of interest only if there is some direct way of computing $S_\lambda$. This is the case with Tikhonov regularization, where we can compute $x_{\lambda,y} = R_\lambda y = (T^*T + \lambda I)^{-1} T^* y$ by solving $(T^*T + \lambda I)x_{\lambda,y} = T^* y$.

*Asymptotic regularization* is defined as follows: given $y \in Y$, define $u = u(t)$ to be the solution of the initial value problem (IVP)

$$\begin{aligned} u' + T^*Tu &= T^*y, \\ u(0) &= 0. \end{aligned} \qquad (4.16)$$

We will show that $u(t) \to T^\dagger y$ as $t \to \infty$. We can then define $S_\lambda y = u(1/\lambda)$, so that $S_\lambda y \to T^\dagger y$ as $\lambda \to 0^+$. Moreover, we will show that $S_\lambda = s_\lambda(T^*T)T^*$ for a certain $s_\lambda$.

We will assume, in the following derivation, that the IVP (4.16) has a unique solution $u = u(t)$ that exists for all $t \geq 0$ (thus $u : [0, \infty) \to X$). Since the differential equation is linear, this would be a standard result if $X$ were finite-dimensional. Below we compute the solution explicitly in terms of the SVE of $T$, allowing a direct verification of the existence of the solution, whereas uniqueness could be verified by an appropriate version of Gronwall's inequality.

Given that $T = \sum_{n=1}^\infty \sigma_n \psi_n \otimes \phi_n$ is the SVE of $T$, we assume that (4.16) has a solution of the form

$$u(t) = \sum_{n=1}^\infty \alpha_n(t)\phi_n$$

and also assume that term-by-term differentiation is valid:

$$u'(t) = \sum_{n=1}^\infty \alpha_n'(t)\phi_n.$$

(Our assumptions mean that both series converge in the norm topology of $X$.) We then have

$$T^*Tu(t) = \sum_{n=1}^\infty \alpha_n(t)T^*T\phi_n = \sum_{n=1}^\infty \sigma_n^2 \alpha_n(t)\phi_n,$$

$$T^*y = \sum_{n=1}^\infty \sigma_n \langle \psi_n, y \rangle_Y \phi_n.$$

Substituting into the IVP (4.16) yields

$$\sum_{n=1}^\infty \alpha_n'(t)\phi_n + \sum_{n=1}^\infty \sigma_n^2 \alpha_n(t)\phi_n = \sum_{n=1}^\infty \sigma_n \langle \psi_n, y \rangle_Y \phi_n$$

$$\Rightarrow \sum_{n=1}^\infty \left( \alpha_n'(t) + \sigma_n^2 \alpha_n(t) - \sigma_n \langle \psi_n, y \rangle_Y \right) \phi_n = 0$$

$$\Rightarrow \alpha_n'(t) + \sigma_n^2 \alpha_n(t) - \sigma_n \langle \psi_n, y \rangle_Y = 0, \quad n = 1, 2, 3, \ldots.$$

Moreover, the initial condition $u(0) = 0$ implies that $\alpha_n(0) = 0$ must hold for $n = 1, 2, 3, \ldots$.

The solution of the scalar IVP

$$\alpha_n'(t) + \sigma_n^2 \alpha_n(t) - \sigma_n \langle \psi_n, y \rangle_Y = 0,$$
$$\alpha_n(0) = 0$$

is

$$\alpha_n(t) = \frac{1 - e^{-\sigma_n^2 t}}{\sigma_n} \langle \psi_n, y \rangle_Y,$$

and hence our calculations suggest that the solution of (4.16) is

$$u(t) = \sum_{n=1}^{\infty} \frac{1 - e^{-\sigma_n^2 t}}{\sigma_n} \langle \psi_n, y \rangle_Y \phi_n.$$

We leave it as an exercise for the reader to prove that this formula defines a continuously differentiable function $u$, that term-by-term differentiation is valid, and $u$ satisfies the IVP (4.16).

We now define

$$s_\lambda(\beta) = \begin{cases} \frac{1 - e^{-\beta/\lambda}}{\beta}, & \beta > 0, \\ \frac{1}{\lambda}, & \beta = 0. \end{cases}$$

Then

$$S_\lambda y = u(1/\lambda) = \sum_{n=1}^{\infty} \sigma_n s_\lambda(\sigma_n^2) \langle \psi_n, y \rangle_Y \phi_n = s_\lambda(T^*T)T^*y,$$

as desired. It is immediately obvious that $s_\lambda(\beta) \to 1/\beta$ as $\lambda \to 0^+$ for each $\beta > 0$ and also that $|\beta s_\lambda(\beta)| \leq 1$ for all $b \in [0, \sigma_1^2]$ and all $\lambda > 0$. Therefore, it follows from Theorem 4.40 that $S_\lambda y \to T^\dagger y$ as $\lambda \to 0^+$ for all $y \in D(T^\dagger)$. Moreover, a simple exercise in calculus shows that, for all $\nu > 0$,

$$|\beta^\nu(\beta s_\lambda(\beta) - 1)| \leq \nu^\nu e^{-\nu} \lambda^\nu.$$

Therefore, by Theorem 4.44, if $T^\dagger y \in \mathcal{R}((T^*T)^\nu)$, then

$$\|S_\lambda y - T^\dagger y\|_X = O(\lambda^\nu).$$

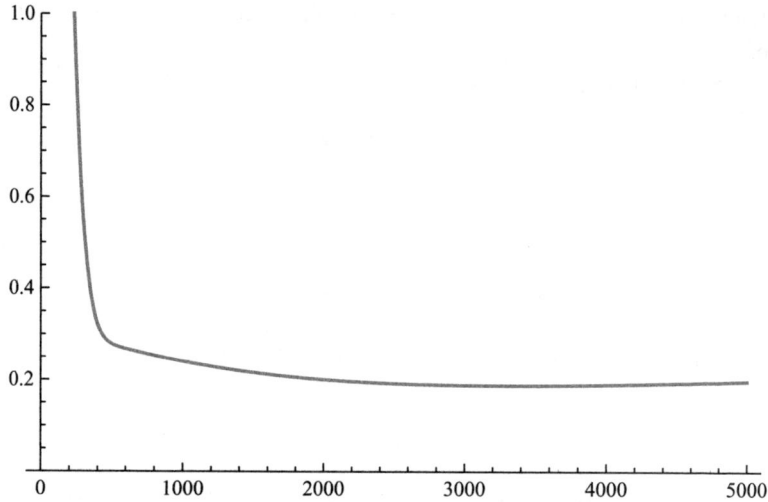

**Figure 4.12.** The error $\|S_\lambda y - x^*\|_{L^2}$ versus $t = 1/\lambda$ in Example 4.45.

In contrast with Tikhonov regularization, this is valid for all $\nu > 0$, including $\nu > 1$, and hence asymptotic regularization is able to take advantage of a smoother solution to obtain faster convergence.

**Example 4.45.** Let $T : L^2(0, 1) \to L^2(0, 1)$ be the operator of Example 3.16 in Section 3.2:

$$(Tx)(s) = \int_0^1 k(s, t)x(t)\, dt, \ 0 < s < 1,$$

where

$$k(s, t) = \begin{cases} s(1 - t), & s < t, \\ t(1 - s), & s \geq t. \end{cases}$$

We take $x^*(t) = 7t - 10t^3 + 3t^5$ and $y^*(t) = (31t - 49t^3 + 21t^5 - 3t^7)/42$; then $Tx^* = y^*$ and the exact solution $x^*$ lies in $\mathcal{R}(T^*T)$. We generate a noisy data vector by adding uniformly distributed random noise to (the discrete version of) $y^*$. The error amounts to 1% in the $L^2$-norm. We then solve (4.16) and compute the error $\|S_\lambda y - x^*\|_{L^2}$ for various values

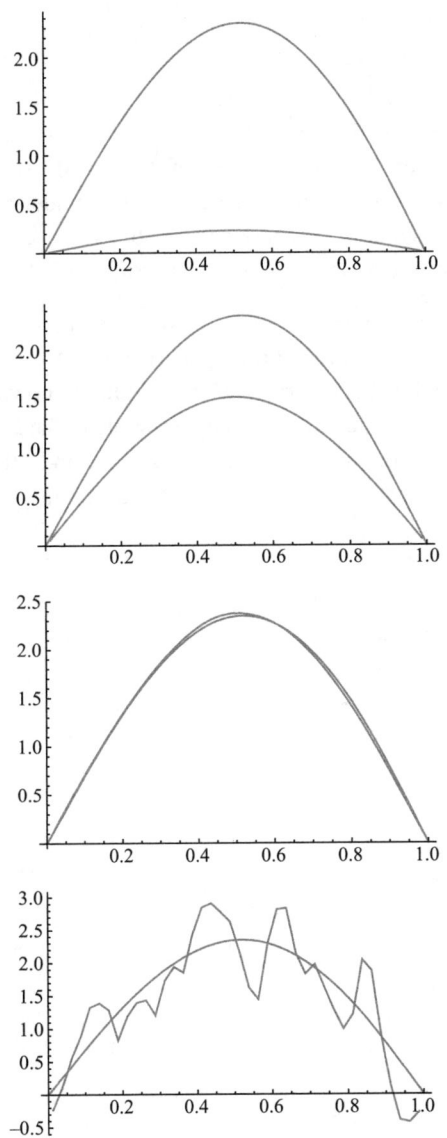

**Figure 4.13.** The regularized solution $S_\lambda y$ and $x^*$ for four values of $t = 1/\lambda$ (from top to bottom): $t = 10, t = 100, t = 3000, t = 1,000,000$.

of $\lambda$ (recall that $\lambda = 1/t$, where $t$ is the independent variable in (4.16)). Figure 4.12 shows the graph of $\|S_\lambda y - x^*\|_{L^2}$ versus $t = 1/\lambda$. The optimal value of $t$ is seen to be approximately $t = 3000$.

Figure 4.13 shows the computed solution $S_\lambda y$, together with $x^*$, for four values of $t$. For very small values of $t$, $S_\lambda y$ is over-regularized and has a very small norm. For large values of $t$, it is under-regularized and contains a large oscillatory error. However, the solution is relatively insensitive to the exact value of $t$ selected (as Figure 4.12 shows).

In presenting Example 4.45, we used our knowledge of the exact solution to choose a good value of the regularization parameter. Of course, in a real problem, it is necessary to select $\lambda$ without knowing the true solution. Methods such as the discrepancy principle or the L-curve strategy can be formulated for asymptotic regularization (and other continuous regularization methods), though we will not discuss them here.

# Chapter 5

# Tikhonov regularization with seminorms

Tikhonov regularization is a reasonable approach to solving an inverse problem $Tx = y$ because of the fundamental property that, for $y \notin D(T^\dagger)$,

$$\|Tx - y\|_Y \to \inf\{\|Tu - y\|_Y : u \in X\} \implies \|x\|_X \to \infty.$$

The idea of Tikhonov regularization is to make $\|Tx - y\|_Y$ relatively small without allowing $x$ to become too bad, that is, too large. However, in many problem $\|x\|_X$ is not the best way to measure the badness of an approximate solution $x$. In general, we might wish to choose a functional $R : X \to \mathbb{R}$ and make sure that $R(x)$ is not too large by solving

$$\min_{x \in X} \|Tx - y\|_Y^2 + \lambda R(x)$$

for some $\lambda > 0$. In order to analyze this approach, we have to restrict the functionals $R$ that will be allowed. We will study the choice $R(x) = \|Lx\|_Z^2$, where $L$ is a linear operator mapping $X$ into another Hilbert space $Z$. In many cases, $X$ is a space of functions and $L$ is chosen to be a derivative operator, in which case $\|Lx\|_Z$ is large when $x$ exhibits rapid changes (large gradients). In such a case, the goal is that $x$ be smooth rather than rough or oscillatory.

Typically, in cases where we might wish to choose $L$ as a derivative operator, the underlying space $X$ includes functions that are not differentiable. As a result, $Lx$ cannot be defined for all $x \in X$ but only for $x$ lying in a dense subspace of $X$. This complicates the analysis, as we shall see.

It is frequently the case that $L$ has a nontrivial null space (such as when $L$ is a derivative operator); in such a case, $x \mapsto \|Lx\|_Z$ defines a seminorm on $X$ (or on a subspace of $X$). Therefore, we will call this approach *seminorm regularization*.

Before presenting the theory, we give an example that shows the advantage of going beyond the basic Tikhonov regularization framework.

**Example 5.1.** Consider the problem of Example 3.27 of Section 3.4. The operator $T : L^2(0, 1) \to L^2(0, 1)$ is defined by

$$(Tx)(s) = \int_0^1 k(s, t)x(t)\, dt, \ 0 < s < 1,$$

where

$$k(s, t) = \begin{cases} s(1 - t), & s < t, \\ t(1 - s), & s \geq t, \end{cases}$$

and $Tx^* = y^*$, where $x^*(t) = t$, $y^*(t) = (t - t^3)/6$. Because $x^*$ does not belong to $\mathcal{R}(T^*)$, $x_{\lambda,y^*}$ does not converge quickly to $x^*$. Here we repeat the calculations in the earlier example, except that we define $\hat{x}_{\lambda,y}$ to be the solution of

$$\min_{x \in L^2(0,1)} \|Tx - y\|_{L^2}^2 + \lambda \|Lx\|_{L^2},$$

where $L$ is the first-derivative operator: $Lx = x'$. We define $y$ to be the noisy data shown in Figure 3.7 (the relative error in $y$ as an estimate of $y^*$ is $\delta = 10^{-3}$). Figure 5.1 shows $\hat{x}_{\lambda,y}$ (for $\lambda = 8 \cdot 10^{-8}$) and $x_{\lambda,y}$, the Tikhonov solution (for $\lambda = 6 \cdot 10^{-4}$—there no reason that the same value of $\lambda$ would be a good choice for both methods).

As the results show, seminorm regularization is more effective for this problem.

# 5.1 Densely defined operators

We have already encountered a densely defined operator, namely the generalized inverse $T^\dagger$ of a bounded linear operator $T$ whose range fails to be

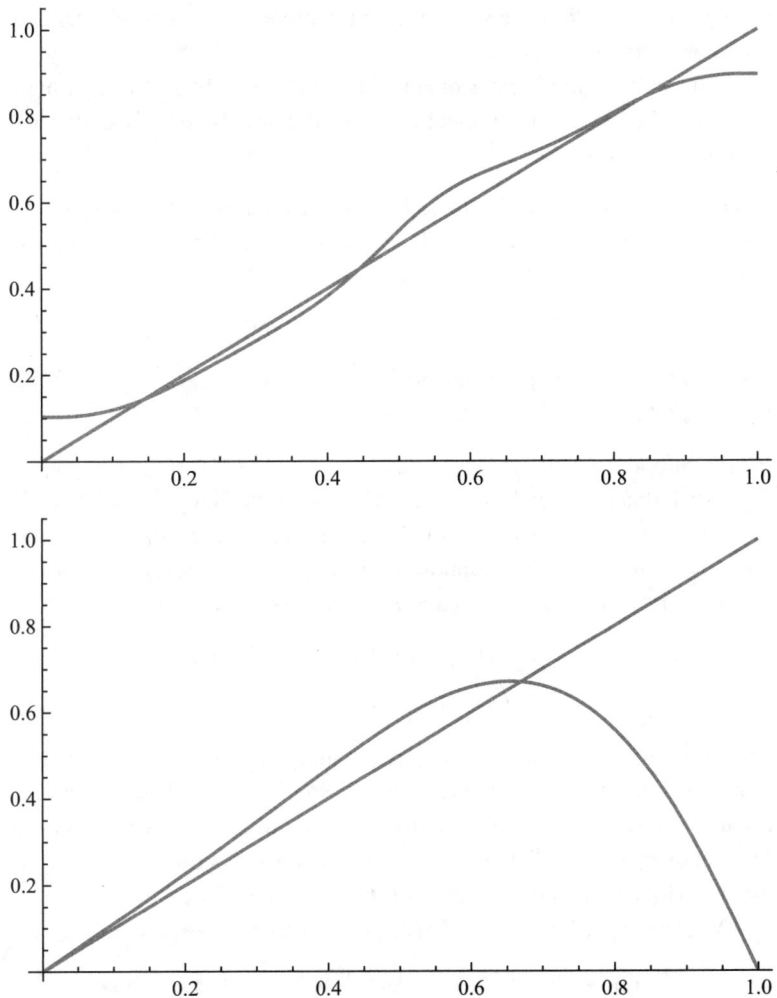

**Figure 5.1.** Example 5.1: (top) $x^*$ and $\hat{x}_{\lambda,y}$; (bottom) $x^*$ and $x_{\lambda,y}$. Both graphs correspond to the noise level $\delta = 10^{-3}$.

closed. In this section, we will develop the properties of densely defined operators that will be needed.

A densely defined linear operator is typically unbounded; if it happens to be bounded, then it can be extended uniquely to an operator defined on the entire space.

**Theorem 5.2.** *Let U and V be Hilbert spaces, let D(L) be a dense subspace of U, and let $L : D(L) \to V$ be a linear operator. If L is bounded, that is, if there exists $M > 0$ such that*

$$\|Lu\|_V \leq M\|u\|_U \text{ for all } u \in D(L),$$

*then there exists a unique bounded linear operator $L_1 : U \to V$ with the properties $L_1|_{D(L)} = L$ and $\|L_1\| = \|L\|$.*

*Proof.* Suppose $L$ is bounded. For any $u \in U \setminus D(L)$, there exists $\{u_n\} \subset D(L)$ such that $u_n \to u$. Since $\{u_n\}$ is Cauchy and $L$ is bounded, it is easy to show that $\{Lu_n\}$ is Cauchy in $V$. Therefore, there exists $v \in V$ such that $Lu_n \to v$. Moreover, supppose $\{u_n\}, \{\hat{u}_n\} \subset D(L)$ satisfy $u_n \to u$ and $\hat{u}_n \to u$, and assume $Lu_n \to v$ and $L\hat{u}_n \to \hat{v}$. We then have

$$\|\hat{v} - v\|_V = \lim_{n \to \infty} \|L\hat{u}_n - Lu_n\|_V \leq \|L\| \lim_{n \to \infty} \|\hat{u}_n - u_n\|_U$$
$$= \|L\|\|u - u\|_U = 0.$$

Therefore, $\hat{v} = v$ and $\lim_{n \to \infty} Lu_n$ is uniquely determined by the condition that $\lim_{n \to \infty} u_n = u$. We can then define $L_1 : U \to V$ by $L_1 u = Lu$ for all $u \in D(L)$ and, for $u \in U \setminus D(L)$, $L_1 u = \lim_{n \to \infty} Lu_n$, where $\{u_n\} \subset D(L)$ converges to $u$. The linearity of $L_1$ then follows immediately from the linearity of $L$ and the linearity of the limit operation.

We then have, for $u \in U \setminus D(L)$, $\{u_n\} \subset D(L)$, $u_n \to u$, that

$$\|L_1 u\|_V = \lim_{n \to \infty} \|Lu_n\|_V \leq \|L\| \lim_{n \to \infty} \|u_n\|_U = \|L\|\|u\|_U.$$

This shows that $L_1$ bounded and that $\|L_1\| \leq \|L\|$. Since $\|L_1\| \geq \|L\|$ is obvious, we see that $\|L_1\| = \|L\|$.  $\square$

We are primarily interested in unbounded densely defined operators, although we will see examples of densely defined operators that turn out to be bounded.

## 5.1.1   The adjoint of a closed densely defined operator

We now define the adjoint of a densely defined operator.

**Definition 5.3.** Let $U$ and $V$ be Hilbert spaces, let $D(L)$ be a dense subspace of $U$, and let $L : D(L) \to V$ be a linear operator. We define $D(L^*)$ by $v \in D(L^*)$ if and only if there exists $w \in U$ such that

$$\langle Lu, v \rangle_V = \langle u, w \rangle_U \text{ for all } u \in D(L).$$

The density of $D(L)$ in $U$ guarantees that there is at most one $w$ satisfying this condition, and we define the *adjoint* $L^* : D(L^*) \to U$ by $L^*v = w$.

It is easy to verify that the adjoint $L^*$ is linear.

We will now show that the adjoint of a densely defined operator is also densely defined, and it is always closed, even if the original operator is not closed. We need the following description of $\mathrm{gr}(L^*)$.

**Theorem 5.4.** *Let $U$ and $V$ be Hilbert spaces, let $D(L)$ be a dense subspace of $U$, and let $L : D(L) \to V$ be a densely defined linear operator. Define a subset $S$ of $V \times U$ by*

$$S = \{(Lu, -u) \in V \times U \; : \; u \in D(L)\} .$$

*Then $\mathrm{gr}(L^*) = S^\perp$.*

*Proof.* We have

$$(v, w) \in \mathrm{gr}(L^*) \Leftrightarrow \langle Lu, v \rangle_V = \langle u, w \rangle_U \text{ for all } u \in D(L)$$
$$\Leftrightarrow \langle Lu, v \rangle_V - \langle u, w \rangle_U = 0 \text{ for all } u \in D(L)$$
$$\Leftrightarrow \langle (Lu, -u), (v, w) \rangle_{V \times U} = 0 \text{ for all } u \in D(L).$$

This last condition is equivalent to $(v, w) \in S^\perp$. $\qquad\square$

**Corollary 5.5.** *Let $U$ and $V$ be Hilbert spaces, let $D(L)$ be a dense subspace of $U$, and let $L : D(L) \to V$ be a linear operator. Then $L^*$ is closed.*

*Proof.* Let $S$ be the set in the previous theorem. Since $\mathrm{gr}(L^*) = S^\perp$, it follows immediately that $\mathrm{gr}(L^*)$ is a closed set and hence $L^*$ is a closed operator. $\qquad\square$

If $L$ is both densely defined and closed, then $L^*$ must also be densely defined (and closed, as already proved by the preceding corollary).

**Corollary 5.6.** *Let $U$ and $V$ be Hilbert spaces, let $D(L)$ be a dense subspace of $U$, and let $L : D(L) \to V$ be a closed linear operator. Then $L^*$ is also closed and densely defined.*

*Proof.* Let $S$ be the set described in the previous theorem. Since $\mathrm{gr}(L)$ is closed, so is the set $S$, and hence $\mathrm{gr}(L^*)^\perp = S^{\perp\perp} = S$. To show that $D(L^*)$ is dense in $V$, it suffices to show that if $v \in D(L^*)^\perp$, then $v = 0$. If $v \in D(L^*)^\perp$, then $(v, 0) \in \mathrm{gr}(L^*)^\perp = S$. But then $(v, 0) = (Lu, -u)$ for some $u \in D(L)$, which means that $u = 0$, which then implies that $v = Lu = 0$, as desired. Thus $D(L^*)^\perp = \{0\}$ and hence $D(L^*)$ is dense in $V$.  $\square$

We can also use the geometric description of $\mathrm{gr}(L)$ to prove the following result.

**Corollary 5.7.** *Let $U$ and $V$ be Hilbert spaces, let $D(L)$ be a dense subspace of $U$, and let $L : D(L) \to V$ be a closed linear operator. Then $L^{**} = L$.*

*Proof.* To prove this result, we express the representation $\mathrm{gr}(L^*) = S^\perp$ of Theorem 5.4 in a way that extends easily to represent $\mathrm{gr}(L^{**})$. To this end, let us define $A : U \times V \to V \times U$ by $A(u, v) = (v, -u)$. Then the set $S$ described in Theorem 5.4 is $A(\mathrm{gr}(L)) = \{A(u, v) : (u, v) \in \mathrm{gr}(L)\}$ and we have $\mathrm{gr}(L^*) = (A(\mathrm{gr}(L)))^\perp$.

The operator $A$ is *unitary*, as is easily checked:

$$\langle A(u_1, v_1), A(u_2, v_2) \rangle_{V \times U}$$
$$= \langle (u_1, v_1), (u_2, v_2) \rangle_{U \times V} \text{ for all } (u_1, v_1), (u_2, v_2) \in U \times V.$$

A unitary operator has the that for any subspace $M$ of $U \times V$, $(A(M))^\perp = A(M^\perp)$. We leave it to the reader to verify this property. It follows that $\mathrm{gr}(L^*) = A(\mathrm{gr}(L)^\perp)$.

To represent $\mathrm{gr}(L^{**})$, we define the analogous operator $A_1 : V \times U \to U \times V$ by $A_1(v, u) = (u, -v)$. Then it is easily checked that $A_1$

is also unitary and moreover that $A_1 = -A^{-1}$. Now we can compute $\mathrm{gr}(L^{**})$:

$$\begin{aligned}
\mathrm{gr}(L^{**}) = (A_1(\mathrm{gr}(L^*)))^\perp &= (A_1((A(\mathrm{gr}(L)))^\perp))^\perp \\
&= A_1((A(\mathrm{gr}(L)))^{\perp\perp}) \\
&= A_1(A(\mathrm{gr}(L))) \\
&= -\mathrm{gr}(L) \\
&= \mathrm{gr}(L).
\end{aligned}$$

We used the fact that $\mathrm{gr}(L)$ is closed, which implies that $A(\mathrm{gr}(L))$ is closed and hence $(A(\mathrm{gr}(L)))^{\perp\perp} = A(\mathrm{gr}(L))$. Also, we used the fact that $-\mathrm{gr}(L) = \mathrm{gr}(L)$ because $\mathrm{gr}(L)$ is a subspace.

We have shown that $\mathrm{gr}(L^{**}) = \mathrm{gr}(L)$, which implies that $L^{**} = L$. $\qquad\square$

## 5.1.2 The graph norm and $L^*L$

Given a closed densely defined linear operator $L : D(L) \to V, D(L) \subset U$, we will need to understand the operator $L^*L$, which has domain

$$D(L^*L) = \{u \in D(L) : Lu \in D(L^*)\}.$$

Clearly $D(L^*L) \subset D(L)$ and it turns out that $D(L^*L)$ is dense in $U$ (and hence in $D(L)$). However, this last fact is not easy to prove directly. We will need a stronger norm on $D(L)$.

**Definition 5.8.** Let $U$ and $V$ be Hilbert spaces, let $D(L)$ be a dense subspace of $U$, and let $L : D(L) \to V$ be a closed linear operator. The *graph norm* on $D(L)$ is defined by

$$\|u\|^2_{\mathrm{gr}(L)} = \|u\|^2_U + \|Lu\|^2_V \text{ for all } u \in D(L).$$

The associated inner product is

$$\langle u, w \rangle_{\mathrm{gr}(L)} = \langle u, w \rangle_U + \langle Lu, Lw \rangle_V \text{ for all } u, w \in D(L).$$

The proof that $\langle \cdot, \cdot \rangle_{\mathrm{gr}(L)}$ actually defines a valid inner product is straightforward. The more significant fact is that $D(L)$ is complete under the graph norm. The assumption that $L$ is closed is crucial.

**Theorem 5.9.** *Let $U$ and $V$ be Hilbert spaces, let $D(L)$ be a dense subspace of $U$, and let $L : D(L) \to V$ be a closed linear operator. Then $D(L)$ is a Hilbert space under the graph norm.*

*Proof.* We show that $D(L)$ is complete. Suppose $\{u_n\}$ is a sequence in $D(L)$ that is Cauchy with respect to the graph norm. Then $\{u_n\}$ and $\{Lu_n\}$ are Cauchy in $U$ and $V$, respectively. Since $U$ and $V$ are Hilbert spaces, there exist $u \in U$ and $v \in V$ such that $u_n \to u$ (with respect to the $U$-norm) and $Lu_n \to v$. But then, since $L$ is closed, it follows that $u \in D(L)$ and $v = Lu$, which yields

$$\|u_n - u\|_{\mathrm{gr}(L)}^2 = \|u_n - u\|_U^2 + \|Lu_n - Lu\|_V^2 \to 0 \text{ as } n \to \infty.$$

This shows that $u_n \to u$ with respect to the graph norm, and hence that $D(L)$ is complete. $\qquad\square$

Now we consider the operator $I + L^*L$, where $I$ is the identity operator on $U$; the operator $I + L^*L$ is defined on $D(L^*L)$, and our goal is to prove that this domain is dense in $U$.

**Theorem 5.10.** *Let $U$ and $V$ be Hilbert spaces, let $D(L)$ be a dense subspace of $U$, and let $L : D(L) \to V$ be a closed linear operator. Then $I + L^*L$ defines a bijection from $D(L^*L)$ (under the graph norm) onto $U$. Moreover, $(I + L^*L)^{-1}$ is bounded, with $\|(I + L^*L)^{-1}\| \leq 1$.*

*Proof.* We have

$$\langle (I + L^*L)u, u \rangle_U = \langle u, u \rangle_U + \langle Lu, Lu \rangle_V = \|u\|_{\mathrm{gr}(L)} \text{ for all } u \in D(L^*L),$$

which implies that $I + L^*L$ is injective. To show that $I + L^*L$ is surjective, let $w \in U$ and consider the equation $(I + L^*L)u = w$. We must show that there is a solution $u \in D(L^*L)$. Under the assumption that $u \in D(L^*L)$, we have

$$(I + L^*L)u = w \Leftrightarrow \langle (I + L^*L)u, x \rangle_U = \langle w, x \rangle_U \text{ for all } x \in D(L)$$

$$\Leftrightarrow \langle u, x \rangle_U + \langle Lu, Lx \rangle_V = \langle w, x \rangle_U \text{ for all } x \in D(L)$$

$$\Leftrightarrow \langle u, x \rangle_{\mathrm{gr}(L)} = \langle w, x \rangle_U \text{ for all } x \in D(L).$$

In the first step, we used the fact that $D(L)$ is dense in $U$ to guarantee that the two equations are equivalent. We have

$$|\langle w, x \rangle_U| \leq \|w\|_U \|x\|_U \leq \|w\|_U \|x\|_{\mathrm{gr}(L)} \text{ for all } x \in D(L)$$

(since obviously $\|x\|_U \leq \|x\|_{\mathrm{gr}(L)}$ for all $x \in D(L)$). Thus $x \mapsto \langle w, x \rangle_U$ defines a bounded linear functional on $D(L)$. By the Riesz representation theorem (Theorem A.21), for each $w \in U$, there exists a unique $u \in D(L)$ such that

$$\langle u, x \rangle_{\mathrm{gr}(L)} = \langle w, x \rangle_U \text{ for all } x \in D(L)$$

and which satisfies $\|u\|_{\mathrm{gr}(L)} = \|w\|_{D(L)^*} \leq \|w\|_U$. We show that, in fact, this $u$ belongs to $D(L^*L)$. We have

$$\langle u, x \rangle_U + \langle Lu, Lx \rangle_V = \langle w, x \rangle_U \text{ for all } x \in D(L)$$

$$\Rightarrow \langle Lu, Lx \rangle_V = \langle w - u, x \rangle_U \text{ for all } x \in D(L),$$

which shows that $v = Lu$ belongs to $D(L^*)$ and $L^*v = w - u$. It follows that $u \in D(L^*L)$ and $L^*Lu = w - u$, that is,

$$(I + L^*L)u = w.$$

We have now shown that $I + L^*L$ defines a bijection from $D(L^*L)$ onto $U$. With $u = (I + L^*L)^{-1}w$, we saw above that $\|u\|_{\mathrm{gr}(L)} \leq \|w\|_U$ for all $w \in U$, which shows that

$$\|(I + L^*L)^{-1}\| \leq 1. \qquad \square$$

We can now prove the result we need.

**Corollary 5.11.** *Let $U$ and $V$ be Hilbert spaces, let $D(L)$ be a dense subspace of $U$, and let $L : D(L) \to V$ be a closed linear operator. Then $D(L^*L)$ is a dense subspace of $U$.*

*Proof.* It suffices to show that the orthogonal complement of $D(L^*L)$ in $U$ is trivial. Suppose $w \in D(L^*L)^\perp$ and let $u = (I + L^*L)^{-1}w$. Then, since $u \in D(L^*L)$ and $w \in D(L^*L)^\perp$, we have

$$\langle u, w \rangle_U = 0 \implies \langle u, (I + L^*L)u \rangle_U = 0 \implies \|u\|^2_{\mathrm{gr}(L)} = 0 \implies u = 0.$$

But then $w = (I + L^*L)u = 0$, and we have shown that $D(L^*L)^\perp = \{0\}$. Thus $D(L^*L)$ is dense in $U$. $\qquad\qquad\qquad\qquad\qquad\qquad\qquad\qquad\square$

For a bounded linear operator $T$ from one Hilbert space to another, we say that $T$ is self-adjoint if $T^* = T$. We make the same definition for a densely defined linear operator; however, there is a subtlety because of the domains of the operators.

**Definition 5.12.** Let $U$ be a Hilbert space and let $L : D(L) \to U$ be a densely defined linear operator. We say that $L$ is *symmetric* if

$$\langle Lu, w \rangle_U = \langle u, Lw \rangle_U \text{ for all } u, w \in D(L).$$

We say that $L$ is *self-adjoint* if $L^* = L$.

If $L$ is symmetric, then, by definition, $D(L) \subset D(L^*)$. The condition $L^* = L$ implies in particular that $D(L^*) = D(L)$; a symmetric operator that is not self-adjoint has the property that $D(L) \subsetneq D(L^*)$.

**Corollary 5.13.** *Let $U$ and $V$ be Hilbert spaces, let $D(L)$ be a dense subspace of $U$, and let $L : D(L) \to V$ be a closed linear operator. Then $L^*L$ is a closed self-adjoint operator.*

*Proof.* Since $L^*L$ is densely defined, it has an adjoint operator. We see that

$$\langle L^*Lu, w \rangle_U = \langle Lu, Lw \rangle_V = \langle u, L^*Lw \rangle_U \text{ for all } u, w \in D(L^*L).$$

It follows that $L^*L$ is symmetric and that $D(L^*L) \subset D((L^*L)^*)$. It remains to show that $D((L^*L)^*) = D(L^*L)$.

Consider the operators

$$K : D(L^*L) \to U, \ K = I + L^*L,$$
$$K_1 : D((L^*L)^*) \to U, \ K_1 = I + (L^*L)^*.$$

We know that $K_1$ and $K$ agree on $D(L^*L)$ and that $K$ is a bijection. It follows immediately that $K_1$ is a surjection. If $u \in \mathcal{N}(K_1)$, then we have

$$\langle (I + (L^*L)^*)u, w \rangle_U = 0 \text{ for all } w \in D(L^*L).$$

But, for all $x \in U$, $(I + L^*L)^{-1}x \in D(L^*L)$. It follows that

$$\langle (I + (L^*L)^*)u, (I + L^*L)^{-1}x \rangle_U = 0 \text{ for all } x \in U$$
$$\Rightarrow \langle u, (I + L^*L)(I + L^*L)^{-1}x \rangle_U = 0 \text{ for all } x \in U$$
$$\Rightarrow \langle u, x \rangle_U = 0 \text{ for all } x \in U$$
$$\Rightarrow u = 0.$$

Therefore, $K_1$ is injective as well. But now, for $w$ in $D((L^*L)^*)$, there exists $u \in D(L^*L)$ such that $K_1 w = Ku$. Since $K_1|_{D(L^*L)} = K$, this yields $K_1 w = K_1 u$ and hence $w = u \in D(L^*L)$ because $K_1$ is injective. Thus $D((L^*L)^*) \subset D(L^*L)$, and we have proved that $L^*L$ is self-adjoint. Since every adjoint operator is closed, this implies that $L^*L$ is closed. $\quad\square$

One of the fundamental properties of a bounded linear operator $T :$ $X \to Y$ is the fact that

$$\mathcal{R}(T)^\perp = \mathcal{N}(T).$$

The following result shows that this property extends to closed densely defined operators.

**Lemma 5.14.** *Let $U$ and $V$ be Hilbert spaces, let $D(L)$ be a dense subspace of $U$, and let $L : D(L) \to V$ be a closed linear operator. Then $\mathcal{R}(L)^\perp = \mathcal{N}(L^*)$.*

*Proof.* Suppose first that $v \in \mathcal{R}(L)^\perp$. Then

$$\langle Lu, v \rangle_V = 0 \text{ for all } u \in D(L),$$

which implies that

$$\langle Lu, v \rangle_V = \langle u, 0 \rangle_U \text{ for all } u \in D(L).$$

This shows that $v \in D(L^*)$ and $L^*v = 0$, so $v \in \mathcal{N}(L^*)$.

Conversely, if $v \in \mathcal{N}(L^*)$, then necessarily $v \in D(L^*)$ and

$$\langle Lu, v \rangle_V = \langle u, L^*v \rangle_U = 0 \text{ for all } u \in D(L).$$

This shows that $v \in \mathcal{R}(L)^\perp$.                                      $\square$

We can now extend Theorem 3.2 to closed densely defined operators.

**Theorem 5.15.** *Let $X$ be a Hilbert space, let $D(A)$ be a dense subspace of $X$, and suppose $A : D(A) \to X$ is a self-adjoint linear operator (which implies that $A$ is closed). Assume that there exists $\gamma > 0$ such that*

$$\langle Ax, x \rangle_X \geq \gamma \|x\|_X^2 \text{ for all } x \in D(A). \tag{5.1}$$

*Then $A$ is a bijection from $D(A)$ to $X$, $A^{-1}$ is bounded, and $\|A^{-1}\| \leq \gamma^{-1}$.*

*Proof.* Once again, $\mathcal{N}(A)$ is obviously trivial. Suppose $u \in \overline{\mathcal{R}(A)}$, say $\{x_n\} \subset D(A)$ and $Ax_n \to u \in X$. As in the proof of Theorem 3.2, $\{x_n\}$ is bounded and hence, without loss of generality, there exists $x \in X$ such that $x_n \to x$ weakly. It follows that $(x_n, Ax_n) \to (x, u)$ weakly. Since $A$ is closed, $\mathrm{gr}(A)$ is closed and hence (being a subspace) closed with respect to weak sequential convergence (Theorem A.35). Therefore, $(x, u) \in \mathrm{gr}(A)$, which implies that $x \in D(A)$ and $u = Ax$. Thus $u \in \mathcal{R}(A)$ and we have shown that $\mathcal{R}(A)$ is closed.

We have $\mathcal{R}(A)^\perp = \mathcal{N}(A^*) = \mathcal{N}(A) = \{0\}$ and this, together with the fact that $\mathcal{R}(A)$ is closed, shows that $\mathcal{R}(A) = X$. It follows that $A$ defines a bijection from $D(A)$ onto $X$, and hence $A^{-1}$ is well-defined. Finally, just as in the proof of Theorem 3.2, (5.1) shows that

$$\|x\|_X \leq \gamma^{-1} \|Ax\|_X \text{ for all } x \in D(A)$$

and this in turn implies that $\|A^{-1}y\|_X \leq \gamma^{-1} \|y\|_X$ for all $y \in X$.       $\square$

## 5.1.3   An example: the derivative operator

We will illustrate the results about closed, densely defined operators by considering $L : D(L) \to L^2(0, 1)$ defined by $Lx = x'$, where $x'$ is the weak derivative of $x$ (see Appendix B). In order that $x'$ belong to $L^2(0, 1)$, it is necessary to take $D(L)$ to be $H^1(0, 1)$ or a subspace of it; we will consider two options. We first define $D(L) = H_0^1(0, 1)$, which is dense in $L^2(0, 1)$. Then, for any $x \in H_0^1(0, 1), y \in H^1(0, 1)$,

$$\langle Lx, y \rangle_{L^2} = \langle x', y \rangle_{L^2} = \int_0^1 x'y = xy\big|_0^1 - \int_0^1 xy'$$

$$= -\int_0^1 xy' \text{ (since } x(0) = x(1) = 0)$$

$$= \langle x, -y' \rangle_{L^2}.$$

This shows that $H^1(0, 1) \subset D(L^*)$ and $L^*y = -y'$ for all $y \in H^1(0, 1)$. In fact, $D(L^*) = H^1(0, 1)$, as we will now show. If $y \in D(L^*)$, then there exists $w \in L^2(0, 1)$ such that

$$\langle Lx, y \rangle_{L^2} = \langle x, w \rangle_L^2 \text{ for all } x \in H_0^1(0, 1)$$

$$\Leftrightarrow \langle x', y \rangle_{L^2} = \langle x, w \rangle_L^2 \text{ for all } x \in H_0^1(0, 1).$$

Let us define $\hat{w} = -w$ and notice that $C_0^\infty(0, 1)$ is a subspace of $H_0^1(0, 1)$. We then have

$$\int_0^1 \hat{w}u = -\int_0^1 yu' \text{ for all } u \in C_0^\infty(0, 1),$$

which shows that $y$ is weakly differentiable and $y' = \hat{w}$. Moreover, we already know that $\hat{w} \in L^2(0, 1)$ and thus we see that $y \in H^1(0, 1)$ and $L^*y = w = -\hat{w} = -y'$. It follows that $D(L^*) = H^1(0, 1)$ and $L^*y = -y'$ for all $y \in D(L^*)$.

We then have

$$D(L^*L) = \left\{ u \in H_0^1(0, 1) : u' \in H^1(0, 1) \right\} = H_0^1(0, 1) \cap H^2(0, 1).$$

Also, as we have seen, $I + L^*L$ defines a bijection from $D(L^*L)$ onto $U = L^2(0, 1)$ with a bounded inverse. Therefore, for each $f \in L^2(0, 1)$, there is a unique solution $u \in D(L^*L) = H_0^1(0, 1) \cap H^2(0, 1)$ of $(I + L^*L)u = f$, and this solution depends continuously on $f$. Combining the restriction that $u \in D(L^*L)$ (which implies boundary condition $u(0) = u(1) = 0$) with the equation $(I + L^*L)u = f$ (which is a second-order differential equation) yields the two-point boundary value problem (BVP)

$$-u'' + u = f \text{ in } (0, 1),$$
$$u(0) = u(1) = 0.$$

This BVP has a unique solution in the classical sense ($u \in C^2[0, 1]$) for each continuous function $f$; we now see that it has a unique solution for each $f \in L^2(0, 1)$ when the differential equation is interpreted in the weak sense.

We could have defined $L$ differently by choosing $D(L) = H^1(0, 1)$. If $y$ belongs to $D(L^*)$, then there exists $w \in L^2(0, 1)$ such that

$$\langle Lx, y \rangle_{L^2} = \langle x, w \rangle_{L^2} \text{ for all } x \in H^1(0, 1)$$
$$\Rightarrow \int_0^1 yx' = -\int_0^1 \hat{w}x \text{ for all } x \in C_0^\infty(0, 1),$$

where $\hat{w} = -w$ (here we use the fact that $C_0^\infty(0, 1)$ is a subspace of $H^1(0, 1)$). It follows as before that $y$ is weakly differentiable and $y' = \hat{w} \in L^2(0, 1)$; thus $y \in H^1(0, 1)$. Now, however, $D(L^*)$ is not all of $H^1(0, 1)$. We know that, for $u, y \in H^1(0, 1)$,

$$\int_0^1 u'y = u(1)y(1) - u(0)y(0) - \int_0^1 uy',$$

while the above calculuation shows that for $u \in H^1(0, 1)$, $y \in D(L^*)$, we have

$$\int_0^1 u'y = -\int_0^1 uy'.$$

It follows that, if $y \in D(L^*)$, then

$$u(1)y(1) - u(0)y(0) = 0 \text{ for all } u \in H^1(0, 1)$$

must hold. If we take $u(t) = t$, then (because $u(0) = 0$, $u(1) = 1$) we see that $y(1) = 0$ must hold. Similarly, taking $u(t) = 1 - t$ shows that $y(0) = 0$. Thus we have shown that

$$y \in D(L^*) \Rightarrow y \in H^1(0, 1), \ y(0) = y(1) = 0;$$

that is,

$$y \in D(L^*) \implies y \in H_0^1(0, 1).$$

The converse (if $y \in H_0^1(0, 1)$, then $y \in D(L^*)$) is easy to verify; thus we see that $D(L^*) = H_0^1(0, 1)$ and $L^*y = -y'$ for all $y \in D(L^*)$. It follows that

$$D(L^*L) = \left\{ u \in H^1(0, 1) : u' \in H_0^1(0, 1) \right\}$$
$$= \left\{ u \in H^2(0, 1) : u'(0) = u'(1) = 0 \right\}.$$

The fact that $I + L^*L$ is invertible now means that the two-point BVP

$$-u'' + u = f \text{ in } (0, 1),$$
$$u'(0) = u'(1) = 0$$

has a unique solution in $H^2(0, 1)$ for each $f \in L^2(0, 1)$. (Once again, it can be shown that the same BVP has a unique solution $u \in C^2[0, 1]$ for each $f \in C[0, 1]$).

## 5.2 Best approximate solutions using seminorms

Let $X$, $Y$, and $Z$ be Hilbert spaces, let $T : X \to Y$ be a bounded linear operator, and let $L : D(L) \to Z$, where $D(L) \subset X$, be a closed densely defined linear operator. We are still interested in the inverse problem $Tx = y$, but instead of seeking the minimum-norm least-squares solution, we now define the best approximate solution as the solution of

$$\min \|Lx\|_Z$$
$$\text{s.t. } x \in D(L), \ \|Tx - y\|_Y = \inf\{\|Tu - y\|_Y : u \in D(L)\}.$$
(5.2)

We need the critical assumption about the relationship between $T$ and $L$ that there exists $\gamma > 0$ such that

$$\|Tx\|_Y^2 + \|Lx\|_Z^2 \geq \gamma \|x\|_X^2 \text{ for all } x \in D(L).$$
(5.3)

This implies, in particular, that $\mathcal{N}(T) \cap \mathcal{N}(L) = \{0\}$. This last condition is certainly necessary; if it did not hold, (5.2) could not have a unique solution. The stronger condition (5.3) will be critical in establishing a satisfactory theory, as we will see.

The following simple lemma implies that a solution of

$$x \in D(L), \quad \|Tx - y\|_Y = \inf\{\|Tu - y\|_Y \, : \, u \in D(L)\}$$

is simply a least-squares solution of $Tx = y$ that happens to lie in $D(L)$.

**Lemma 5.16.** $\overline{\mathcal{R}\left(T|_{D(L)}\right)} = \overline{\mathcal{R}(T)}.$

*Proof.* Obviously $\mathcal{R}\left(T|_{D(L)}\right) \subset \mathcal{R}(T)$ and hence $\overline{\mathcal{R}\left(T|_{D(L)}\right)} \subset \overline{\mathcal{R}(T)}$. Suppose $y \in \overline{\mathcal{R}(T)}$, say $y = \lim_{n\to\infty} Tx_n$, where $\{x_n\} \subset X$. Since $D(L)$ is dense in $X$ by assumption, for each $n$ there exists $u_n \in D(L)$ such that $\|u_n - x_n\|_X < 1/n$. Then

$$\|y - Tu_n\|_Y \le \|y - Tx_n\|_Y + \|Tx_n - Tu_n\|_Y$$
$$\le \|y - Tx_n\|_Y + \|T\|\|x_n - u_n\|_Y$$
$$\le \|y - Tx_n\|_Y + \frac{1}{n}\|T\|.$$

Since $\|y - Tx_n\|_Y \to 0$ and $\|T\|/n \to 0$ as $n \to \infty$, it follows that $\|y - Tu_n\|_Y \to 0$ as $n \to \infty$ and hence $y \in \overline{\mathcal{R}\left(T|_{D(L)}\right)}$. Therefore, $\overline{\mathcal{R}(T)} \subset \overline{\mathcal{R}\left(T|_{D(L)}\right)}$. $\square$

We now draw a few elementary conclusions.

**Corollary 5.17.** *The following conditions are equivalent:*

*1.* $x \in D(L), \; \|Tx - y\|_Y = \inf\{\|Tu - y\|_Y \, : \, u \in D(L)\}.$

*2.* $x \in D(L), \; Tx = \text{proj}_{\overline{\mathcal{R}(T|_{D(L)})}} y.$

*3.* $x \in D(L), \; T^*Tx = T^*y.$

*Proof.* We use the preceding lemma, which implies that

$$\text{proj}_{\overline{\mathcal{R}(T|_{D(L)})}} y = \text{proj}_{\overline{\mathcal{R}(T)}} y.$$

From our earlier results about least-squares problems, we know that $x$ satisfies the first condition if and only if $x \in D(L)$ and $Tx = \text{proj}_{\overline{\mathcal{R}(T|_{D(L)})}} y$. Therefore, the first and second conditions are equivalent. However, $Tx = \text{proj}_{\overline{\mathcal{R}(T|_{D(L)})}} y$ implies that $Tx = \text{proj}_{\overline{\mathcal{R}(T)}} y$ and hence that $x \in D(L)$ is a least-squares solution of $Tx = y$. Therefore, $T^*Tx = T^*y$ must hold, and we see that the second condition implies the third. On the other hand, if the third condition holds, then $x \in D(L)$ is a least-squares solution of $Tx = y$, and hence

$$Tx = \text{proj}_{\overline{\mathcal{R}(T)}} y = \text{proj}_{\overline{\mathcal{R}(T|_{D(L)})}} y.$$

This shows that the second condition must also hold. $\quad\square$

We can now express (5.2) in the simpler form

$$\begin{aligned} &\min \|Lx\|_Z \\ &\text{s.t. } x \in D(L), \ T^*Tx = T^*y. \end{aligned} \tag{5.4}$$

**Corollary 5.18.** $Tx = y$ has a least-squares solution belonging to $D(L)$ if and only if $y \in \mathcal{R}(T|_{D(L)}) \oplus \mathcal{R}(T)^\perp$.

*Proof.* We know that $Tx = y$ has a least-squares solution belonging to $D(L)$ if and only if $y \in \mathcal{R}(T|_{D(L)}) \oplus \mathcal{R}(T|_{D(L)})^\perp$. However, Lemma 5.16 implies that $\mathcal{R}(T|_{D(L)})^\perp = \mathcal{R}(T)^\perp$, and the result follows. $\quad\square$

**Corollary 5.19.** If $y \in \mathcal{R}(T|_{D(L)}) \oplus \mathcal{R}(T)^\perp$ and $\hat{x} \in D(L)$ is one least-squares solution of $Tx = y$, then the set of all least-squares solutions of $Tx = y$ that lie in $D(L)$ is

$$\hat{x} + (\mathcal{N}(T) \cap D(L)) = (\hat{x} + \mathcal{N}(T)) \cap D(L).$$

*Proof.* Since $\mathcal{N}(T|_{D(L)}) = \mathcal{N}(T) \cap D(L)$, the first representation of the solution set is immediate. It is then straightforward to verify that $\hat{x} + (\mathcal{N}(T) \cap D(L))$ and $(\hat{x} + \mathcal{N}(T)) \cap D(L)$ are equal. $\quad\square$

We now wish to study (5.4) under the assumption that $y \in \mathcal{R}(T|_{D(L)}) \oplus \mathcal{R}(T)^\perp$.

**Theorem 5.20.** *If $y \in \mathcal{R}(T|_{D(L)}) \oplus \mathcal{R}(T)^{\perp}$, then (5.4) has a unique solution $\bar{x}$, which is characterized by the conditions*

$$\bar{x} \in D(L),$$

$$T^*T\bar{x} = T^*y,$$

$$\langle L\bar{x}, Lu \rangle_Z = 0 \text{ for all } u \in \mathcal{N}(T) \cap D(L).$$

*Proof.* Assume that $y \in \mathcal{R}(T|_{D(L)}) \oplus \mathcal{R}(T)^{\perp}$; then there exists $\hat{x} \in D(L)$ such that $T^*T\hat{x} = T^*y$ and the set of all least-squares solutions (in $D(L)$) of $Tx = y$ is $\hat{x} + (\mathcal{N}(T) \cap D(L))$. Now let $\{x_n\} \subset \hat{x} + (\mathcal{N}(T) \cap D(L))$ be a minimizing sequence for $\|Lx\|_Z$; that is, assume that

$$\|Lx_n\|_Z \to \inf\{\|Lx\|_Z \,:\, x \in D(L), T^*Tx = T^*y\} \text{ as } n \to \infty.$$

We have

$$\|x_n\|_X^2 \leq \gamma^{-1}\left(\|Tx_n\|_Y^2 + \|Lx_n\|_Z^2\right) \text{ for all } n,$$

which shows that $\{x_n\}$ is bounded in $X$ ($\|Tx_n\|_Y$ is constant with respect to $n$, and $\{\|Lx_n\|\}$ is bounded because it is convergent). Assume, without loss of generality, that $x_n \to \bar{x}$ weakly in $X$. Then, since $x_n \in \hat{x} + \mathcal{N}(T)$ for all $n$, it follows that $\bar{x} \in \hat{x} + \mathcal{N}(T)$ as well. We show that $\bar{x} \in D(L)$. Since $\{Lx_n\}$ is bounded in $Z$, we can assume without loss of generality that $Lx_n \to z$ weakly in $Z$. But then, since the subspace $\mathrm{gr}(L)$ of $X \times Z$ is closed, it is closed with respect to weak sequential convergence (Theorem A.35) and it follows that $(\bar{x}, z) \in \mathrm{gr}(L)$, which means that $\bar{x} \in D(L)$ and $z = L\bar{x}$. Thus we have shown that $\bar{x} \in (\hat{x} + \mathcal{N}(T)) \cap D(L)$. Therefore, $\bar{x}$ is a least-squares solution (in $D(L)$) of $Tx = y$. Moreover, since $Lx \mapsto \|Lx\|_Z$ is lower semicontinuous with respect to weak sequential convergence (Theorem A.38), we have

$$\|L\bar{x}\|_Z \leq \liminf_{n \to \infty} \|Lx_n\|_Z = \inf\{\|Lx\|_Z \,:\, x \in D(L), \ T^*Tx = T^*y\}.$$

Thus we see that $\bar{x}$ is a solution of (5.4).

We can now prove the given characterization of $\bar{x}$ and use it to show that $\bar{x}$ is unique. For a given $\bar{x} \in \hat{x} + (\mathcal{N}(T) \cap D(L))$ and any $u \in \mathcal{N}(T) \cap D(L)$, define $\phi : \mathbb{R} \to \mathbb{R}$ by

$$\phi(t) = \|L(\bar{x} + tu)\|_Z^2 = \|L\bar{x}\|_Z^2 + 2t\langle L\bar{x}, Lu \rangle_Z + t^2 \|Lu\|_Z^2.$$

Then $\bar{x}$ is a solution of (5.4) if and only if $t = 0$ minimizes $\phi(t)$ for each $u$ in $\mathcal{N}(T) \cap D(L)$, that is, if and only if $\phi'(0) = 0$ for all $u \in \mathcal{N}(T) \cap D(L)$. We have

$$\phi'(t) = 2\langle L\bar{x}, Lu \rangle_Z + 2t\|Lu\|_Z^2$$
$$\Rightarrow \phi'(0) = 2\langle L\bar{x}, Lu \rangle_Z.$$

Thus $\bar{x}$ is a solution of (5.4) if and only if $\bar{x} \in \hat{x} + (\mathcal{N}(T) \cap D(L))$ and

$$\langle L\bar{x}, Lu \rangle_Z = 0 \text{ for all } u \in \mathcal{N}(T) \cap D(L),$$

which is equivalent to the stated conditions.

Finally, if $\hat{x}$ is a second vector satisfying $\hat{x} \in D(L)$, $T^*T\hat{x} = T^*y$, and $\langle L\hat{x}, Lu \rangle_Z = 0$ for all $u \in \mathcal{N}(T) \cap D(L)$, then it follows that $\bar{x} - \hat{x} \in \mathcal{N}(T^*T) = \mathcal{N}(T)$ and $\langle L(\bar{x} - \hat{x}), Lu \rangle_Z = 0$ for all $u \in \mathcal{N}(T) \cap D(L)$. But then $\langle L(\bar{x} - \hat{x}), L(\bar{x} - \hat{x}) \rangle_Z = 0$, which implies that $L(\bar{x} - \hat{x}) = 0$ and hence that $\bar{x} - \hat{x} \in \mathcal{N}(L)$. Since $\mathcal{N}(T) \cap \mathcal{N}(L)$ is trivial, this shows that $\bar{x} - \hat{x} = 0$, that is, $\bar{x} = \hat{x}$, as desired. $\qquad\square$

In general, we cannot replace

$$\langle L\bar{x}, Lu \rangle_Z = 0 \text{ for all } u \in \mathcal{N}(T) \cap D(L)$$

by the more concise condition $L^*L\bar{x} \in (\mathcal{N}(T) \cap D(L))^\perp$ because it is not guaranteed that $\bar{x} \in D(L^*L)$.

We now define a variant on the generalized inverse $T^\dagger$.

**Definition 5.21.** We define $T_L^\dagger : D(T_L^\dagger) \to X$ by the condition that $\bar{x} = T_L^\dagger y$ is the unique solution of

$$\min \|Lx\|_Z$$
$$\text{s.t. } x \in D(L), \ T^*Tx = T^*y$$

for each $y \in D(T_L^\dagger) = \mathcal{R}(T|_{D(L)}) \oplus \mathcal{R}(T)^\perp$.

**Theorem 5.22.** $T_L^\dagger$ *is linear and densely defined.*

*Proof.* Since $\mathcal{R}(T|_{D(L)})$ is dense in $\mathcal{R}(T)$, it follows that

$$D(T_L^\dagger) = \mathcal{R}(T|_{D(L)}) \oplus \mathcal{R}(T)^\perp$$

is dense in $\overline{\mathcal{R}(T)} \oplus \mathcal{R}(T)^\perp = Y$. Thus $T_L^\dagger$ is densely defined. Suppose $y_1, y_2 \in D(T_L^\dagger)$ and define $x_1 = T_L^\dagger y_1$, $x_2 = T_L^\dagger y_2$. Then, for each $\alpha_1, \alpha_2 \in \mathbb{R}$,

$$T^*T(\alpha_1 x_1 + \alpha_2 x_2) = \alpha_1 T^*T x_1 + \alpha_2 T^*T x_2 = \alpha_1 T^* y_1 + \alpha_2 T^* y_2$$
$$= T^*(\alpha_1 y_1 + \alpha_2 y_2).$$

Also,

$$x_1, x_2 \in D(L) \implies \alpha_1 x_1 + \alpha_2 x_2 \in D(L)$$

and, for all $u \in \mathcal{N}(T) \cap D(L)$,

$$\langle L(\alpha_1 x_1 + \alpha_2 x_2), Lu \rangle_Z = \alpha_1 \langle L x_1, Lu \rangle_Z + \alpha_2 \langle L x_2, Lu \rangle_Z = 0$$

because $\langle L x_1, Lu \rangle_Z = 0$ for all $u \in \mathcal{N}(T) \cap D(L)$ and similarly for $x_2$. This shows that

$$T_L^\dagger(\alpha_1 y_1 + \alpha_2 y_2) = \alpha_1 x_1 + \alpha_2 x_2 = \alpha_1 T_L^\dagger y_1 + \alpha_2 T_L^\dagger y_2.$$

Thus we have shown that $T_L^\dagger$ is linear. $\qquad\square$

The densely defined operator $T_L^\dagger$ is generally not closed with respect to the default topologies (that is, the norm topologies on $X$ and $Y$). To see this, let us suppose that $\mathcal{N}(T)$ is trivial, so that $x = T_L^\dagger y$ is defined by the conditions $x \in D(L)$, $T^*T x = T^* y$. Furthermore, we suppose that $D(L)$ is a proper subspace of $X$ and $\mathcal{R}(T)$ is not closed. Choose $\{x_n\} \subset D(L)$, $x \in X \setminus D(L)$, $x_n \to x$, and define $y_n = T x_n$, $y = T x$. Notice that $y_n \to y \in \mathcal{R}(T) \setminus \mathcal{R}(T|_{D(L)})$. Then we have $x_n = T_L^\dagger y_n$ for all $n$ and

$$\{y_n\} \subset D(T_L^\dagger), \quad y_n \to y, \quad T_L^\dagger y_n \to x.$$

However, $y \notin D(T_L^\dagger)$, which shows that $T_L^\dagger$ is not closed.

In order to be able to perform useful analysis, we define a stronger norm on $D(L)$; this new norm is equivalent to the graph norm, but it is tailored to the problem at hand. The norm is defined by

$$\|x\|_*^2 = \|T x\|_Y^2 + \|L x\|_Z^2 \text{ for all } x \in D(L),$$

and the corresponding inner product is

$$\langle x, u \rangle_* = \langle T x, T u \rangle_Y + \langle L x, L u \rangle_Z \text{ for all } x, u \in D(L).$$

**Theorem 5.23.** *The norm $\| \cdot \|_*$ is equivalent to $\| \cdot \|_{\mathrm{gr}(L)}$ and therefore $D(L)$ is a Hilbert space under $\langle \cdot, \cdot \rangle_*$.*

*Proof.* We have, for all $x \in D(L)$,

$$\|Tx\|_Y^2 + \|Lx\|_Z^2 \geq \gamma \|x\|_X^2 \Rightarrow \|Tx\|_Y^2 + (1+\gamma)\|Lx\|_Z^2 \geq \gamma \left( \|x\|_X^2 + \|Lx\|_Z^2 \right)$$

$$\Rightarrow \|Tx\|_Y^2 + \|Lx\|_Z^2 \geq \frac{\gamma}{1+\gamma} \left( \|x\|_X^2 + \|Lx\|_Z^2 \right)$$

$$\Rightarrow \|x\|_*^2 \geq \frac{\gamma}{1+\gamma} \|x\|_{\mathrm{gr}(L)}^2.$$

Also,

$$\|Tx\|_Y^2 + \|Lx\|_Z^2 \leq \|T\|^2 \|x\|_X^2 + \|Lx\|_Z^2$$

$$\Rightarrow \|Tx\|_Y^2 + \|Lx\|_Z^2 \leq \max\{1, \|T\|^2\} \left( \|x\|_X^2 + \|Lx\|_Z^2 \right)$$

$$\Rightarrow \|x\|_*^2 \leq \max\{1, \|T\|^2\} \|x\|_{\mathrm{gr}(L)}^2.$$

This proves that $\| \cdot \|_*$ and $\| \cdot \|_{\mathrm{gr}(L)}$ are equivalent. $\qquad\square$

**Theorem 5.24.** *Both $L$ and $T|_{D(L)}$ are bounded when the $*$-norm is used on $D(L)$.*

*Proof.* For all $x \in D(L)$,

$$\|Tx\|_Z \leq \sqrt{\|Tx\|_Y^2 + \|Lx\|_Z^2} = \|x\|_*,$$

$$\|Lx\|_Z \leq \sqrt{\|Tx\|_Y^2 + \|Lx\|_Z^2} = \|x\|_*,$$

and we see that both $T|_{D(L)}$ and $L$ are bounded, with each having an operator norm less than or equal to 1. $\qquad\square$

We write $T_L : D(L) \to Y$ for $T|_{D(L)}$. We know that $\overline{\mathcal{R}(T_L)} = \overline{\mathcal{R}(T)}$. Therefore, since $\mathcal{R}(T_L) \subset \mathcal{R}(T) \subset \overline{\mathcal{R}(T)}$, it follows that if $\mathcal{R}(T)$ fails to be closed, then so does $\mathcal{R}(T_L)$. This means that if $Tx = y$ is a true inverse problem, then so is $T_L x = y$.

All the theory developed for the generalized inverse $T^\dagger$ applies to the operator $T_L$; in particular, $T_L$ has a generalized inverse $(T_L)^\dagger$ defined on $\mathcal{R}(T_L) \oplus \mathcal{R}(T)^\perp$. We wish to show that $(T_L)^\dagger$ is precisely the operator $T_L^\dagger$

defined earlier. We know that $x = (T_L)^\dagger$ is characterized by

$$x \in D(L), \ T_L^* T_L x = T_L^* y, \ x \in \mathcal{N}(T_L)^\perp, \tag{5.5}$$

while $x = T_L^\dagger y$ is characterized by

$$x \in D(L), \ T^* T x = T^* y, \ \langle Lx, Lu \rangle_Z = 0 \text{ for all } u \in \mathcal{N}(T) \cap D(L). \tag{5.6}$$

However, in (5.5), $T_L^*$ and $\mathcal{N}(T_L)^\perp$ must be defined with respect to $\langle \cdot, \cdot \rangle_*$. To reduce the possibility of confusion, we introduce the notation $A^\sharp$ for the adjoint of a linear operator $A$ defined on $D(L)$ when the inner product $\langle \cdot, \cdot \rangle_*$ is applied. We will also use the notation $S^\pm$ for the orthogonal complement with respect to $\langle \cdot, \cdot \rangle_*$ of a subspace $S$ of $D(L)$. (Therefore, the notations $A^*$ and $S^\perp$ will always refer to $\langle \cdot, \cdot \rangle_X$.)

We require one preliminary result before we can compute adjoints with respect to $\langle \cdot, \cdot \rangle_*$. The operator $T^* T + L^* L$ maps $D(L^* L)$ into $X$, and we can regard $T^* T + L^* L$ as a densely defined operator on $X$ (with the $X$-norm) or on $D(L)$ (with the $*$-norm).

**Theorem 5.25.** $T^* T + L^* L$ *defines a bijection from $D(L^* L)$ onto $X$.*

1.  *If $T^* T + L^* L$ is regarded as a densely defined operator on $X$, then the inverse is a bounded linear operator from $X$ into $X$, with $\left\| (T^* T + L^* L)^{-1} \right\| \leq \gamma^{-1}$ (with the $X$-norm on $X$).*

2.  *If $T^* T + L^* L$ is regarded as a densely defined operator on $D(L)$, then the inverse is a bounded linear operator from $X$ into $D(L)$, with*

$$\left\| (T^* T + L^* L)^{-1} \right\| \leq \gamma^{-1/2}$$

*(with the $*$-norm on $D(L)$).*

*Proof.*

1.  We know that $L^* L$ is closed and self-adjoint with respect to the $X$-norm; the same is true of $T^* T + L^* L$ (which is a densely defined operator on $X$; that is, its domain is a dense subspace of $X$) and we have

$$\langle (T^* T + L^* L) x, x \rangle_X \geq \gamma \| x \|_X^2 \text{ for all } x \in D(L^* L).$$

It follows from Theorem 5.15 that $T^*T + L^*L$ defines a bijection from its domain to $X$ and that $\|(T^*T + L^*L)^{-1}\| \leq \gamma^{-1}$.

2. If we regard $T^*T + L^*L$ as densely defined in $D(L)$ rather than $X$, then part 1 implies that it defines a bijection from $D(L^*L)$ onto $X$. (However, we cannot apply Theorem 5.15, because $T^*T + L^*L$ does not map its domain into $D(L)$.) For all $x \in D(L^*L)$, $\|x\|_*^2 = \langle (T^*T + L^*L)x, x \rangle_X$. Also, for all $u \in X$, $(T^*T + L^*L)^{-1}u \in D(L^*L)$ and hence

$$
\begin{aligned}
\|(T^*T &+ L^*L)^{-1}u\|_*^2 \\
&= \langle (T^*T + L^*L)(T^*T + L^*L)^{-1}u, (T^*T + L^*L)^{-1}u \rangle_X \\
&= \langle u, (T^*T + L^*L)^{-1}u \rangle_X \\
&\leq \|u\|_X \|(T^*T + L^*L)^{-1}u\|_X \\
&\leq \gamma^{-1} \|u\|_X^2,
\end{aligned}
$$

which yields the desired bound on $\|(T^*T + L^*L)^{-1}\|$.    □

In the next section, we will need the following result. The proof is a simple modification of the proof of Theorem 5.25, so we present it here.

**Theorem 5.26.** *For each* $\lambda > 0$, $T^*T + \lambda L^*L$ *defines a bijection from* $D(L^*L)$ *onto* $X$.

1. *If* $T^*T + \lambda L^*L$ *is regarded as a densely defined operator on* $X$, *then the inverse is a bounded linear operator from* $X$ *into* $X$, *with*

$$
\|(T^*T + \lambda L^*L)^{-1}\| \leq (\gamma \min\{1, \lambda\})^{-1} = \begin{cases} \gamma^{-1}, & \lambda \geq 1, \\ \gamma^{-1}\lambda^{-1}, & 0 < \lambda < 1 \end{cases}
$$

*(with the $X$-norm on $X$).*

2. *If* $T^*T + \lambda L^*L$ *is regarded as a densely defined operator on* $D(L)$, *then the inverse is a bounded linear operator from* $X$ *into* $D(L)$, *with*

$$
\|(T^*T + \lambda L^*L)^{-1}\| \leq \gamma^{-1/2} \min\{1, \lambda\}^{-1} = \begin{cases} \gamma^{-1/2}, & \lambda \geq 1, \\ \gamma^{-1/2}\lambda^{-1}, & 0 < \lambda < 1 \end{cases}
$$

*(with the $*$-norm on $X$).*

*Proof.* We have

$$\langle (T^*T + \lambda L^*L)x, x\rangle_X \geq \min\{1, \lambda\}\langle (T^*T + L^*L)x, x\rangle_X$$
$$\geq \gamma \min\{1, \lambda\}\|x\|_X^2 \text{ for all } x \in D(L^*L).$$

It follows from Theorem 5.15 that $T^*T + \lambda L^*L$ defines a bijection from its domain to $X$ and that $\|(T^*T + L^*L)^{-1}\| \leq (\gamma \min\{1, \lambda\})^{-1}$.

To prove the second part of the result, we temporarily define the norm

$$\|x\|_\lambda^2 = \|Tx\|_Y^2 + \lambda \|Lx\|_Z^2 \text{ for all } x \in D(L)$$

and note that

$$\|x\|_\lambda^2 = \langle (T^*T + \lambda L^*L)x, x\rangle_X \text{ for all } x \in D(L^*L).$$

Then $\|\cdot\|_\lambda$ is equivalent to $\|\cdot\|_*$:

$$\min\{1, \lambda\}\|x\|_*^2 \leq \|x\|_\lambda^2 \leq \max\{1, \lambda\}\|x\|_*^2 \text{ for all } x \in D(L). \qquad (5.7)$$

We can then repeat the proof of the second part of Theorem 5.25 to prove that

$$\left\|(T^*T + \lambda L^*L)^{-1}x\right\|_\lambda \leq (\gamma \min\{1, \lambda\})^{-1/2}\|x\|_X \text{ for all } x \in X.$$

It follows from (5.7) that

$$\|u\|_* \leq \min\{1, \lambda\}^{-1/2}\|u\|_\lambda \text{ for all } u \in D(L),$$

and hence we have

$$\left\|(T^*T + \lambda L^*L)^{-1}x\right\|_* \leq \min\{1, \lambda\}^{-1/2}\left\|(T^*T + \lambda L^*L)^{-1}x\right\|_\lambda$$
$$\leq \gamma^{-1/2}\min\{1, \lambda\}^{-1}\|x\|_X \text{ for all } x \in X. \qquad \square$$

We can now compute the needed adjoints with respect to $\langle \cdot, \cdot\rangle_*$. Recall that $T_L$ and $L$ are bounded operators from $D(L)$ into $Y$ and $Z$, respectively.

**Theorem 5.27.**

1. $T_L^\sharp = (T^*T + L^*L)^{-1}T^*$. *This implies that* $T_L^\sharp y \in D(L^*L)$ *for all* $y \in Y$.

2. $L^\sharp : Z \to D(L)$ *is a bounded linear operator, with* $L^\sharp z = (T^*T + L^*L)^{-1}L^*z$ *for all* $z \in D(L^*)$. *This implies that* $L^\sharp z \in D(L^*L)$ *for all* $z \in D(L^*)$ *and also that the operator* $(T^*T + L^*L)^{-1}L^*$ *extends from* $D(L^*)$ *to a bounded linear operator defined on all of* $Z$.

*Proof.*

1. For all $x \in D(L^*L)$, $y \in Y$, we have

$$\langle T_L x, y \rangle_Y = \langle Tx, y \rangle_Y = \langle x, T^*v \rangle_X$$
$$= \langle (T^*T + L^*L)x, (T^*T + L^*L)^{-1}T^*y \rangle_X$$
$$= \langle x, (T^*T + L^*L)^{-1}T^*y \rangle_*.$$

Therefore,

$$\langle T_L x, y \rangle_Y = \langle x, (T^*T + L^*L)^{-1}T^*y \rangle_* \text{ for all } x \in D(L^*L), y \in Y.$$

Since both sides of the equation are continuous in $x$ (under the $*$-norm), it can be extended to

$$\langle T_L x, y \rangle_Y = \langle x, (T^*T + L^*L)^{-1}T^*y \rangle_* \text{ for all } x \in D(L), y \in Y,$$

as desired.

2. A similar argument shows that

$$\langle Lx, z \rangle_Z = \langle x, (T^*T + L^*L)^{-1}L^*z \rangle_* \text{ for all } x \in D(L^*L), z \in D(L^*),$$

which can be extended by continuity in $x$ to

$$\langle Lx, z \rangle_Z = \langle x, (T^*T + L^*L)^{-1}L^*z \rangle_* \text{ for all } x \in D(L), z \in D(L^*).$$

This shows that $L^\sharp|_{D(L^*)} = (T^*T + L^*L)^{-1}L^*$. Since $L^\sharp : Z \to D(L)$, as the adjoint of a bounded linear operator, is known to be bounded, the result follows. $\qquad\square$

Now we compute $\mathcal{N}(T)^\perp$.

**Theorem 5.28.** $\mathcal{N}(T_L)^\perp = \{x \in D(L) : \langle Lx, Lu \rangle_Z = 0 \text{ for all } u \in \mathcal{N}(T) \cap D(L)\}$.

*Proof.* We have

$$x \in \mathcal{N}(T_L)^{\pm} \Leftrightarrow x \in D(L), \ \langle x, u \rangle_* = 0 \text{ for all } u \in \mathcal{N}(T) \cap D(L)$$

$$\Leftrightarrow x \in D(L), \ \langle Tx, Tu \rangle_Y + \langle Lx, Lu \rangle_Z = 0 \text{ for all } u \in \mathcal{N}(T)$$
$$\cap \ D(L)$$

$$\Leftrightarrow x \in D(L), \ \langle Lx, Lu \rangle_Z = 0 \text{ for all } u \in \mathcal{N}(T) \cap D(L).$$

The last step follows because $Tu = 0$ for all $u \in \mathcal{N}(T) \cap D(L)$. $\qquad\square$

Since $T_L = T|_{D(L)}$, we have already seen that $\mathcal{R}(T_L)^{\perp} = \mathcal{R}(T)^{\perp}$, and hence

$$D\left((T_L)^{\dagger}\right) = \mathcal{R}(T_L) \oplus \mathcal{R}(T)^{\perp} = D(T_L^{\dagger}).$$

We know from the theory developed in Section 2.3 that, for $y \in \mathcal{R}(T_L) \oplus \mathcal{R}(T)^{\perp}$, $x = (T_L)^{\dagger} y$ is characterized by

$$x \in D(L), \ T_L^{\sharp} T_L x = T_L^{\sharp} y, \ x \in \mathcal{N}(T_L)^{\pm}.$$

We also know that $x = T_L^{\dagger} y$ is characterized by

$$x \in D(L), \ T^* T x = T^* y, \ \langle Lx, Lu \rangle_Z = 0 \text{ for all } u \in \mathcal{N}(T) \cap D(L).$$

The previous theorem shows that, for $x \in D(L)$,

$$x \in \mathcal{N}(T_L)^{\pm} \ \Leftrightarrow \ \langle Lx, Lu \rangle_Z = 0 \text{ for all } u \in \mathcal{N}(T) \cap D(L).$$

Finally, we have $T_L^{\sharp} = (T^* T + L^* L)^{-1} T^*$ and hence

$$T_L^{\sharp} T_L x = T_L^{\sharp} y \Leftrightarrow x \in D(L), (T^* T + L^* L)^{-1} T^* T x = (T^* T + L^* L)^{-1} T^* y$$
$$\Leftrightarrow x \in D(L), T^* T x = T^* y.$$

Thus we have shown that $(T_L)^{\dagger} y = T_L^{\dagger} y$ for all $y \in D\left((T_L)^{\dagger}\right) = D(T_L^{\dagger})$.

**Theorem 5.29.** $(T_L)^{\dagger} = T_L^{\dagger}$.

We now know from the general theory that $T_L^{\dagger}$ is closed (albeit with respect to the $*$-norm on $D(L)$) and that $T_L^{\dagger}$ is bounded if and only if $\mathcal{R}(T_L) = \mathcal{R}(T|_{D(L)})$ is closed. Moreover, as we have seen, if $Tx = y$ is a true inverse problem, then so is $T_L x = y$; that is, if $\mathcal{R}(T)$ fails to be closed, then so does $\mathcal{R}(T_L)$.

The following theorem, a variation on Theorem 5.26, will be used in the next section.

**Theorem 5.30.** *For each $\lambda > 0$, the operator $T^\sharp T + \lambda L^\sharp L$ is a bounded invertible operator from $D(L)$ onto $D(L)$ (under the $*$-norm) and $(T^\sharp T + \lambda L^\sharp L)^{-1}$ is bounded. Moreover, $T^\sharp T + \lambda L^\sharp L$ defines a bijection from $D(L^*L)$ onto itself, and therefore $(T^\sharp T + \lambda L^\sharp L)^{-1}x \in D(L^*L)$ for all $x \in D(L^*L)$.*

*Proof.* We know that $T^\sharp T + \lambda L^\sharp L$ is bounded under the $*$-norm, and it is obviously self-adjoint. We have, for each $x \in D(L)$,

$$\langle (T^\sharp T + \lambda L^\sharp L)x, x \rangle_* = \langle T^\sharp Tx, x \rangle_* + \lambda \langle L^\sharp Lx, x \rangle_*$$
$$= \langle Tx, Tx \rangle_Y + \lambda \langle Lx, Lx \rangle_Z$$
$$\geq \min\{1, \lambda\}\|x\|_*^2.$$

Therefore, by Theorem 3.2, it follows that $T^\sharp T + \lambda L^\sharp L$ is invertible with a bounded inverse.

If $x \in D(L^*L)$, then

$$(T^\sharp T + \lambda L^\sharp L)x = (T^*T + L^*L)^{-1}(T^*Tx + \lambda L^*Lx).$$

Since $(T^*T + L^*L)^{-1}$ maps $X$ into $D(L^*L)$, this shows that $(T^\sharp T + \lambda L^\sharp L)x$ belongs to $D(L^*L)$ for each $x \in D(L^*L)$. Conversely, given any $x \in D(L^*L)$, the vector

$$y = (T^*T + \lambda L^*L)^{-1}(T^*T + L^*L)x$$

is well-defined and belongs to $D(L^*L)$ (since $(T^*T + \lambda L^*L)^{-1}$ also maps $X$ onto $D(L^*L)$). It follows that

$$(T^\sharp T + \lambda L^\sharp L)y = (T^*T + L^*L)^{-1}(T^*T + \lambda L^*L)y$$
$$= (T^*T + L^*L)^{-1}(T^*T + \lambda L^*L)(T^*T + \lambda L^*L)^{-1}$$
$$(T^*T + L^*L)x$$
$$= x.$$

This shows that $T^\sharp T + \lambda L^\sharp L$ defines a bijection from $D(L^*L)$ onto $D(L^*L)$. $\square$

We now illustrate some of the analysis of this section with two simple examples.

**Example 5.31.** Let us define $T : L^2(0, 1) \to L^2(0, 1)$ by

$$(Tx)(s) = \begin{cases} \int_0^s x(t)\,dt, & 0 \le s \le \frac{1}{2}, \\ 0, & \text{otherwise.} \end{cases}$$

The null space of $T$ is the set of all functions that are zero (almost everywhere) on the interval $(0, 1/2)$. We define the regularization operator $L : D(L) \to L^2(0, 1)$ by $D(L) = H^1(0, 1)$, $Lx = x'$. Then

$$\mathcal{N}(T) \cap D(L) = \{u \in H^1(0, 1) \; : \; u = 0 \text{ a.e. in } (0, 1/2)\}.$$

We define $x_0 \in H^1(0, 1)$, $x_0(t) = t$, and $y = Tx_0$; then

$$y(s) = \begin{cases} \frac{1}{2}s^2, & 0 \le s \le \frac{1}{2}, \\ 0, & \frac{1}{2} \le s \le 1. \end{cases}$$

Since $y \in \mathcal{R}(T|_{D(L)})$, it follows that $y \in D(T_L^\dagger)$ and hence $x = T_L^\dagger y$ exists. We can compute $x$ by applying Theorem 5.20; $x$ is characterized by the conditions that $x \in D(L)$, $T^*Tx = T^*y$, and

$$\langle Lx, Lu\rangle_{L^2} = 0 \text{ for all } u \in \mathcal{N}(T) \cap D(L). \tag{5.8}$$

Before continuing, the reader may wish to try to determine $x$ by elementary reasoning (it must satisfy $Tx = y$ and it must have the smallest derivative, in the $L^2$ sense, of all solutions of this equation).

A direct calculation shows that $T^*$ is defined by

$$(T^*v)(r) = \begin{cases} \int_r^{1/2} v(s)\,ds, & 0 \le r \le \frac{1}{2}, \\ 0, & \frac{1}{2} \le r \le 1, \end{cases}$$

which implies that

$$(T^*y)(r) = \begin{cases} \frac{1}{6}\left(\frac{1}{8} - r^3\right), & 0 \le r \le \frac{1}{2}, \\ 0, & \frac{1}{2} \le r \le 1 \end{cases}$$

and

$$(T^*Tx)(r) = \begin{cases} \int_r^{1/2} \int_0^s x(t)\,dt\,ds, & 0 \le r \le \frac{1}{2}, \\ 0, & \frac{1}{2} \le r \le 1. \end{cases}$$

It follows that if $u = T^*Tx$, then

$$u''(r) = \begin{cases} -x(r), & 0 \le r \le \frac{1}{2}, \\ 0, & \frac{1}{2} \le r \le 1. \end{cases}$$

Since $T^*y = T^*Tx$, this gives us a formula for $x$ on $(0, 1/2)$:

$$x(r) = -\frac{d^2}{dr^2}\left((T^*y)(r)\right) = r, \ 0 \le r \le \frac{1}{2}.$$

However, the equation $T^*Tx = T^*y$ does not determine $x$ on $(1/2, 1)$; the value of $x$ on this interval must be determined from condition (5.8). We have

$$\langle Lx, Lu \rangle_{L^2} = 0 \text{ for all } u \in \mathcal{N}(T) \cap D(L)$$

$$\Leftrightarrow \langle x', u' \rangle_{L^2} = 0 \text{ for all } u \in \mathcal{N}(T) \cap D(L)$$

$$\Leftrightarrow \int_{1/2}^{1} x'(t)u'(t)\,dt = 0 \text{ for all } u \in \mathcal{N}(T) \cap D(L).$$

The last step follows from the fact that, for each $u \in \mathcal{N}(T) \cap D(L)$, $u = 0$ a.e. on $(0, 1/2)$. Although we will not prove it formally, it is fairly easy to see that this last condition holds if and only if $x' = 0$ a.e. on $(1/2, 1)$. (If this does not hold, there would exist a small subinterval $(a, b) \subset (1/2, 1)$ such that $x'(t)$ is of one sign—positive or negative—a.e. on it. We could then construct a function $u \in \mathcal{N}(T) \cap D(L)$ with the property that $u'(t) > 0$ for $t \in (a, b)$ and $u'(t) = 0$ for $t \notin (a, b)$; we could, for example, choose $u$ to be continuous and piecewise linear. Then $\langle x', u' \rangle_{L^2} \ne 0$, a contradiction.)

The condition that $x' = 0$ a.e. on $(1/2, 1)$, together with the fact that $x$ must lie in $H^1(0, 1)$, implies that $x$ must equal the constant $1/2$ on $(1/2, 1)$ (because $x$ must be continuous at $r = 1/2$ in order to belong to $H^1(0, 1)$). It follows that $x$ must be the following function:

$$x(r) = \begin{cases} r, & 0 \le r \le \frac{1}{2}, \\ \frac{1}{2}, & \frac{1}{2} \le r \le 1. \end{cases}$$

In the following example, $T_L^\dagger y$ fails to exist.

**Example 5.32.** We define $T : H^1(0, 1) \to L^2(0, 1)$ to be the identity operator: $Tx = x$ for all $x \in H^1(0, 1)$. Then $\mathcal{R}(T) = H^1(0, 1)$, and since $H^1(0, 1)$ is a proper dense subspace of $L^2(0, 1)$, we see that $\mathcal{R}(T)$ fails to be closed. We define $L : D(L) \to L^2(0, 1)$ by $Lx = x''$, where $D(L) = H^2(0, 1)$, a dense subspace of $H^1(0, 1)$ (and also of $L^2(0, 1)$).

We now define $y \in L^2(0, 1)$ by $y(s) = |1 - 2s|$; then $y \in H^1(0, 1)$ (and hence $y \in \mathcal{R}(T)$), but $y \notin H^2(0, 1)$ and therefore $y \notin \mathcal{R}(T_L)$. In fact, $y \in \overline{\mathcal{R}(T_L)} \setminus \mathcal{R}(T_L)$, which implies that $y \notin D(T_L^\dagger)$. It follows that there is no least-squares solution of $T_L x = y$; $T_L x$ can be made arbitrarily close to $y$, but $\|Lx\|_{L^2}$ increases without bound as $T_L x$ approaches $y$.

# 5.3   Tikhonov regularization with seminorms

We could now apply the theory of Tikhonov regularization that we developed in Chapter 3 to the problem

$$\min_{x \in D(L)} \|Tx - y\|_Y^2 + \lambda \|x\|_*^2. \tag{5.9}$$

However, this is unnecessarily complicated. We have

$$\|Tx - y\|_Y^2 + \lambda \|x\|_*^2 = \|Tx - y\|_Y^2 + \lambda \|Tx\|_Y^2 + \lambda \|Lx\|_Z^2.$$

Since $\|Tx\|_Y$ is already controlled by requiring $Tx$ to be close to $y$ (that is, by including the $\|Tx - y\|_Y^2$ term), there is nothing to be gained by including the $\lambda \|Tx\|_Y^2$ term in the functional to be minimized. Instead of (5.9), we will study the problem

$$\min_{x \in D(L)} \|Tx - y\|_Y^2 + \lambda \|Lx\|_Z^2. \tag{5.10}$$

In the remainder of this section, we will always use $\langle \cdot, \cdot \rangle_*$ and $\| \cdot \|_*$ on $D(L)$ unless we explicitly indicate otherwise.

Mimicking our earlier approach, we define $\hat{M}_\lambda : D(L) \to Y \times Z$ by

$$\hat{M}_\lambda x = (Tx, \sqrt{\lambda} Lx).$$

We then have

$$\|\hat{M}_\lambda x - (y, 0)\|_{Y \times Z}^2 = \|(Tx - y, \sqrt{\lambda} Lx)\|_{Y \times Z}^2 = \|Tx - y\|_Y^2 + \lambda \|Lx\|_Z^2$$

and hence solving (5.10) is equivalent to solving $\hat{M}_\lambda x = (y, 0)$ in the least-squares sense.

We begin by proving that (5.10) has a unique solution for each $\lambda > 0$ and $y \in Y$.

**Theorem 5.33.** *For each $\lambda > 0$,*

1. *$\mathcal{N}(\hat{M}_\lambda)$ is trivial;*

2. *$\mathcal{R}(\hat{M}_\lambda)$ is closed;*

3. *for each $(y, z) \in Y \times Z$, $\hat{M}_\lambda^\sharp(y, z) = T^\sharp y + \sqrt{\lambda} L^\sharp z$.*

*Proof.* First, we have

$$\hat{M}_\lambda x = (0, 0) \ \Rightarrow \ Tx = 0, \sqrt{\lambda} Lx = 0 \ \Rightarrow \ \|x\|_*^2 = \|Tx\|_Y^2 + \|Lx\|_Z^2 = 0,$$

which shows that $x = 0$. Thus $\mathcal{N}(\hat{M}_\lambda) = \{0\}$.

Now suppose $\{x_n\} \subset D(L)$ and $\hat{M}_\lambda x_n \to (y, z)$. It follows that $Tx_n \to y$ and $Lx_n \to \lambda^{-1/2} z$. Therefore, $\{Tx_n\}$, $\{Lx_n\}$ are Cauchy sequences (in $Y$ and $Z$, respectively), which implies that $\{x_n\}$ is Cauchy in $\|\cdot\|_*$. Hence there exists $x \in D(L)$ such that $x_n \to x$ in the $*$-norm. Since both $T$ and $L$ are bounded with respect to the $*$-norm, it follows that $Tx_n \to Tx$ and $Lx_n \to Lx$, which in turn implies that $y = Tx$, $z = \lambda^{1/2} Lx$. Thus $(y, z) = \hat{M}_\lambda x$ and hence $(y, z) \in \mathcal{R}(\hat{M}_\lambda)$. We have now shown that $\mathcal{R}(\hat{M}_\lambda)$ is closed.

Finally, we have, for all $x \in D(L)$ and all $(y, z) \in Y \times Z$,

$$\begin{aligned}
\langle \hat{M}_\lambda x, (y, z) \rangle_{Y \times Z} = \langle (Tx, \sqrt{\lambda} Lx), (y, z) \rangle_{Y \times Z} &= \langle Tx, y \rangle_Y + \langle \sqrt{\lambda} Lx, z \rangle_Z \\
&= \langle x, T^\sharp y \rangle_* + \langle x, \sqrt{\lambda} L^\sharp z \rangle_* \\
&= \langle x, T^\sharp y + \sqrt{\lambda} L^\sharp z \rangle_*,
\end{aligned}$$

which shows that $\hat{M}_\lambda^\sharp(y, z) = T^\sharp y + \sqrt{\lambda} L^\sharp z$. $\qquad \square$

**Corollary 5.34.** *For each $y \in Y$ and each $\lambda > 0$, $\hat{M}_\lambda x = (y, 0)$ has a unique least-squares solution $\bar{x}$, which is also the unique solution of (5.10). The solution $\bar{x}$ is characterized as the unique solution of*

$$\hat{M}_\lambda^\sharp \hat{M}_\lambda \bar{x} = \hat{M}_\lambda^\sharp(y, 0). \tag{5.11}$$

*Moreover,* $\bar{x} \in D(L^*L)$, *and (5.11) is equivalent to*

$$\left(T^*T + \lambda L^*L\right)\bar{x} = T^*y. \tag{5.12}$$

*Proof.* The first conclusion follows immediately from the previous theorem, and we have already noted that the least-squares problem for $\hat{M}_\lambda x = (y, 0)$ is equivalent to (5.10). We show that $\bar{x} \in D(L^*L)$ and that (5.11) is equivalent to (5.12). We have

$$\hat{M}_\lambda^\sharp \hat{M}_\lambda \bar{x} = \hat{M}_\lambda^\sharp(T\bar{x}, \sqrt{\lambda}L\bar{x}) = T^\sharp T\bar{x} + \lambda L^\sharp L\bar{x}$$

and

$$\hat{M}_\lambda^\sharp(y, 0) = T^\sharp y.$$

Therefore (5.11) implies $T^\sharp T\bar{x} + \lambda L^\sharp L\bar{x} = T^\sharp y$ and hence that

$$\bar{x} = (T^\sharp T\bar{x} + \lambda L^\sharp L)^{-1} T^\sharp y.$$

By Theorem 5.27, $T^\sharp y \in D(L^*L)$ for all $y \in Y$. By Theorem 5.30, $(T^\sharp T\bar{x} + \lambda L^\sharp L)^{-1}$ maps $D(L^*L)$ into itself, and this shows that $\bar{x} \in D(L^*L)$. Finally, knowing that $\bar{x} \in D(L^*L)$ allows us to apply Theorem 5.27 to rewrite the normal equation:

$$T^\sharp T\bar{x} + \lambda L^\sharp L\bar{x} = T^\sharp y$$

$$\Leftrightarrow (T^*T + L^*L)^{-1} T^*T\bar{x} + \lambda(T^*T + L^*L)^{-1} L^*L\bar{x} = (T^*T + L^*L)^{-1} T^*y$$

$$\Leftrightarrow (T^*T + L^*L)^{-1} \left(T^*T + \lambda L^*L\right)\bar{x} = (T^*T + L^*L)^{-1} T^*y$$

$$\Leftrightarrow \left(T^*T + \lambda L^*L\right)\bar{x} = T^*y.$$

(We needed $\bar{x} \in D(L^*L)$ to use the representation $L^\sharp = (T^*T + L^*L)^{-1}L^*$.) $\qquad\square$

We now define $\hat{N}_\lambda = T^*T + \lambda L^*L$ and $\hat{x}_{\lambda,y} = \hat{N}_\lambda^{-1}T^*y$. We know from Theorem 5.25 that $\hat{x}_{\lambda,y} \in D(L^*L)$ for all $y \in Y$ and all $\lambda > 0$.

**Example 5.35.** This is a continuation of Example 5.1, in which $X = Y = L^2(0, 1)$ and $L : D(L) \to L^2(0, 1)$ is defined by $Lx = x'$. In presenting Example 5.1, we did not define (restrict) $D(L)$, which means that we implicitly took it to be the largest possible subspace: $D(L) = H^1(0, 1)$.

As we saw in Section 5.1.3, this implies that

$$D(L^*) = H_0^1(0, 1) = \left\{ u \in H^1(0, 1) \ : \ u(0) = u(1) = 0 \right\}$$

and

$$D(L^*L) = \left\{ u \in H^2(0, 1) \ : \ u'(0) = u'(1) = 0 \right\}.$$

In Example 5.1, the exact solution was taken to be $x^*(t) = t$, which does not belong to $D(L^*L)$. On the other hand, as noted above, $\hat{x}_{\lambda,y}$ does belong to $D(L^*L)$ for each $\lambda > 0$. This implies that if $x_{\lambda,y}$ succeeds in approximating $x^*$ (which we show to be true in the following two sections), it does so from within $D(L^*L)$. For this particular $L$, this means that the function $\hat{x}_{\lambda,y}$ has derivative zero at the endpoints of the interval $[0, 1]$. This is clearly seen in the left graph in Figure 5.1 on page 227.

## 5.4 Convergence for exact data

We now wish to study the convergence of $\hat{x}_{\lambda,y}$ as $\lambda \to 0^+$ in the case that $y \in D(T_L^\dagger)$. As in Chapter 3, we will need a number of preliminary results. These are similar to what we have seen before; however, some of the proofs are more delicate because of the need to work on $D(L)$ under the $*$-norm.

**Theorem 5.36.** *For all $\lambda > 0$ and $y \in Y$, $\hat{x}_{\lambda,y} \in \mathcal{N}(T_L)^\pm$.*

*Proof.* We prove that $\langle L\hat{x}_{\lambda,y}, Lu \rangle_Z = 0$ for all $u \in \mathcal{N}(T_L)$. Since $\hat{x}_{\lambda,y}$ belongs to $D(L^*L)$, we have $\langle L\hat{x}_{\lambda,y}, Lu \rangle_Z = \langle L^*L\hat{x}_{\lambda,y}, u \rangle_X$. Now,

$$\begin{aligned}
L^*L\hat{N}_\lambda^{-1}T^* &= \lambda^{-1}(\lambda L^*L)\hat{N}_\lambda^{-1}T^* = \lambda^{-1}(\hat{N}_\lambda - T^*T)\hat{N}_\lambda^{-1}T^* \\
&= \lambda^{-1}(T^* - T^*T\hat{N}_\lambda^{-1}T^*) \\
&= \lambda^{-1}T^*(I - T\hat{N}_\lambda^{-1}T^*).
\end{aligned}$$

Therefore, if $u \in \mathcal{N}(T_L)$, we have

$$
\begin{aligned}
\langle L\hat{x}_{\lambda,y}, Lu \rangle_Z &= \langle L^* L\hat{N}_{\lambda}^{-1} T^* y, u \rangle_X \\
&= \lambda^{-1} \langle T^*(I - T\hat{N}_{\lambda}^{-1} T^*) y, u \rangle_X \\
&= \lambda^{-1} \langle (I - T\hat{N}_{\lambda}^{-1} T^*) y, Tu \rangle_Y \\
&= 0
\end{aligned}
$$

since $Tu = 0$. This shows that $\hat{x}_{\lambda,y} \in \mathcal{N}(T_L)^{\perp}$. $\qquad\square$

Now we consider the operators related to $\hat{N}_{\lambda}$.

**Theorem 5.37.** *Suppose $\lambda > 0$.*

1. *$L\hat{N}_{\lambda}^{-1} T^*$ is a bounded operator mapping $Y$ into $Z$, and*

$$
\|L\hat{N}_{\lambda}^{-1} T^*\| \le \frac{1}{2} \lambda^{-1/2}.
$$

2. *$T\hat{N}_{\lambda}^{-1} L^*$ is a bounded linear operator mapping $D(L^*)$ into $Y$ and hence extends to a bounded operator defined on all of $Z$. Moreover,*

$$
\|T\hat{N}_{\lambda}^{-1} L^*\| \le \frac{1}{2} \lambda^{-1/2}.
$$

3. *$\|T\hat{N}_{\lambda}^{-1} T^*\| \le 1$.*

*Proof.* Since $\hat{x}_{\lambda,y}$ is the unique least-squares solution of $\hat{M}_{\lambda} x = (y, 0)$, we have

$$
\begin{aligned}
&\|\hat{M}_{\lambda}\hat{x}_{\lambda,y}\|_{Y \times Z}^2 + \|\hat{M}_{\lambda}\hat{x}_{\lambda,y} - (y, 0)\|_{Y \times Z}^2 = \|(y, 0)\|_{Y \times Z}^2 \\
\Rightarrow\ &\|T\hat{x}_{\lambda,y}\|_Y^2 + \lambda\|L\hat{x}_{\lambda,y}\|_Z^2 + \|T\hat{x}_{\lambda,y} - y\|_Y^2 + \lambda\|L\hat{x}_{\lambda,y}\|_Z^2 = \|y\|_Y^2 \\
\Rightarrow\ &\|T\hat{x}_{\lambda,y}\|_Y^2 + \|T\hat{x}_{\lambda,y} - y\|_Y^2 + 2\lambda\|L\hat{x}_{\lambda,y}\|_Z^2 = \|y\|_Y^2.
\end{aligned}
$$

We immediately see that, for all $y \in Y$,

$$
\|T\hat{x}_{\lambda,y}\|_Y \le \|y\|_Y \ \Rightarrow\ \|T\hat{N}_{\lambda}^{-1} T^* y\|_Y \le \|y\|_Y,
$$

which shows that $\|T\hat{N}_{\lambda}^{-1} T^*\| \le 1$ and proves the third result.

By Lemma 3.6, we have

$$
\frac{1}{2}\|y\|_Y^2 \le \|T\hat{x}_{\lambda,y}\|_Y^2 + \|T\hat{x}_{\lambda,y} - y\|_Y^2 \quad \text{for all } y \in Y
$$

and hence, for all $y \in Y$,

$$\frac{1}{2}\|y\|_Y^2 + 2\lambda\|L\hat{x}_{\lambda,y}\|_Z^2 \le \|y\|_Y^2 \Rightarrow \|L\hat{x}_{\lambda,y}\|_Z^2 \le \frac{1}{4}\lambda^{-1}\|y\|_Y^2$$

$$\Rightarrow \|L\hat{x}_{\lambda,y}\|_Z \le \frac{1}{2}\lambda^{-1/2}\|y\|_Y$$

$$\Rightarrow \|L\hat{N}_\lambda^{-1}T^*y\|_Z \le \frac{1}{2}\lambda^{-1/2}\|y\|_Y.$$

This shows that $\|L\hat{N}_\lambda^{-1}T^*\| \le (1/2)\lambda^{-1/2}$, which proves the first result. It should be noted that $L\hat{N}_\lambda^{-1}T^*$ is bounded even though $L$ is not assumed to be bounded. It follows that $(L\hat{N}_\lambda^{-1}T^*)^*$ is defined on all of $Z$ and $(L\hat{N}_\lambda^{-1}T^*)^* = T\hat{N}_\lambda^{-1}L^*$ on $D(L^*)$. Since

$$\|(L\hat{N}_\lambda^{-1}T^*)^*\| = \|T\hat{N}_\lambda^{-1}L^*\|,$$

the proof is complete. $\qquad\qquad\qquad\square$

In the case of ordinary Tikhonov regularization, the trivial case occurs when $y \in \mathcal{R}(T)^\perp$; then $T^\dagger y = 0$ and $x_{\lambda,y} = 0$ for all $\lambda > 0$. The following theorem identifies the trivial case for seminorm regularization.

**Theorem 5.38.** *If there exist* $\lambda_1, \lambda_2 > 0$, $\lambda_1 \ne \lambda_2$, *such that* $\hat{x}_{\lambda_1,y} = \hat{x}_{\lambda_2,y}$, *then* $\mathrm{proj}_{\overline{\mathcal{R}(T_L)}}y \in \mathcal{R}(T|_{\mathcal{N}(L)})$ *and* $\hat{x}_{\lambda,y} = T_L^\dagger y$ *for all* $\lambda > 0$.

*Proof.* If $\hat{x}_{\lambda_1,y} = \hat{x}_{\lambda_2,y}$, then

$$(T^*T + \lambda_1 L^*L)\hat{x}_{\lambda_1,y} = T^*y = (T^*T + \lambda_2 L^*L)\hat{x}_{\lambda_1,y}$$

$$\Rightarrow (\lambda_1 - \lambda_2)L^*L\hat{x}_{\lambda_1,y} = 0$$

$$\Rightarrow L^*L\hat{x}_{\lambda_1,y} = 0$$

$$\Rightarrow \hat{x}_{\lambda_1,y} \in \mathcal{N}(L^*L) = \mathcal{N}(L).$$

But then it follows that $T^*T\hat{x}_{\lambda_1,y} = T^*y$, which shows that $\bar{x} = \hat{x}_{\lambda_1,y} = \hat{x}_{\lambda_2,y}$ is a least-squares solution of $Tx = y$ satisfying $\|L\bar{x}\|_Z = 0$. Obviously $\bar{x}$ solves

$$\min_{x \in D(L)} \|Tx - y\|_Y^2 + \lambda\|Lx\|_Z^2$$

for each $\lambda > 0$, as well as

$$\min \|Lx\|_Z$$
$$\text{s.t. } x \in D(L), \ T^*Tx = T^*y.$$

Thus $\hat{x}_{\lambda,y} = T_L^\dagger y = \bar{x}$ for all $\lambda > 0$ and

$$\text{proj}_{\overline{\mathcal{R}(T_L)}} y = T\bar{x} \in \mathcal{R}(T|_{\mathcal{N}(L)}). \qquad \square$$

**Corollary 5.39.** *If $y \notin \mathcal{R}(T|_{\mathcal{N}(L)}) \oplus \mathcal{R}(T)^\perp$, then $\hat{x}_{\lambda_1,y} \neq \hat{x}_{\lambda_2,y}$ for all $\lambda_1 \neq \lambda_2$.*

We now present several theorems analogous to results about ordinary Tikhonov regularization in Chapter 3. In each case, the earlier proof requires only minor changes to yield the result.

**Theorem 5.40.** *If $y \notin \mathcal{R}(T|_{\mathcal{N}(L)}) \oplus \mathcal{R}(T)^\perp$, then*

1. *$\|L\hat{x}_{\lambda,y}\|_X$ is a strictly decreasing function of $\lambda$;*

2. *$\|T\hat{x}_{\lambda,y} - y\|_Y$ is a strictly increasing function of $\lambda$.*

*Proof.* Cf. Theorem 3.12. $\qquad \square$

**Theorem 5.41.** *For each $y \in Y$,*

$$T\hat{x}_{\lambda,y} \to \bar{y} = \text{proj}_{\overline{\mathcal{R}(T_L)}} y, \ \|T\hat{x}_{\lambda,y} - y\|_Y \to \|\bar{y} - y\|_Y \ as \ \lambda \to 0^+.$$

*Proof.* Cf. Theorem 3.13. $\qquad \square$

**Theorem 5.42.** *For each $y \in D(T_L^\dagger)$, $\|L\hat{x}_{\lambda,y}\|_Z \leq \|LT_L^\dagger\|_Z$ for all $\lambda > 0$. If $y \notin \mathcal{R}(T|_{\mathcal{N}(L)}) \oplus \mathcal{R}(T)^\perp$, then $\|L\hat{x}_{\lambda,y}\|_Z < \|LT_L^\dagger\|_Z$ for all $\lambda > 0$.*

*Proof.* Cf. Theorem 3.14. $\qquad \square$

**Theorem 5.43.** *For all $y \in Y$, $\lambda^{1/2}\hat{N}_\lambda^{-1}T^*y \to 0$ as $\lambda \to 0^+$.*

*Proof.* Cf. Theorem 3.21. $\qquad \square$

We can now prove the main theorem of this section.

**Theorem 5.44.** *If $y \in D(T_L^\dagger)$, then $\hat{x}_{\lambda,y} \to T_L^\dagger y$ in the $*$-norm (and hence in the weaker X-norm) as $\lambda \to 0^+$.*

*Proof.* By Theorems 5.41 and 5.42, $\{\|T\hat{x}_{\lambda,y}\|_Y : \lambda > 0\}$ and $\{\|L\hat{x}_{\lambda,y}\|_Z : \lambda > 0\}$ are bounded; it follows that $\{\hat{x}_{\lambda,y} : \lambda > 0\}$ is bounded in the $*$-norm and hence every sequence $\{\hat{x}_{\lambda_n,y}\}$, where $\lambda_n \to 0^+$, has a weakly convergent subsequence. By the usual reasoning, it suffices to prove that if $\lambda_n \to 0^+$ and $\hat{x}_{\lambda_n,y} \to \bar{x} \in D(L)$ weakly (with respect to the $*$-topology), then $\bar{x} = T_L^{\dagger}y$ and $\hat{x}_{\lambda_n,y} \to \bar{x}$ strongly in the $*$-norm.

Therefore, we assume that $\lambda_n \to 0^+$ and $\hat{x}_{\lambda_n,y} \to \bar{x} \in D(L)$ weakly. Since $T$ and $L$ are bounded with respect to the $*$-norm, it follows that $T\hat{x}_{\lambda_n,y} \to T\bar{x}$ weakly and $L\hat{x}_{\lambda_n,y} \to L\bar{x}$ weakly. We already know that $T\hat{x}_{\lambda_n,y} \to \bar{y} = \text{proj}_{\overline{\mathcal{R}(T)}}y$ strongly; it follows that $T\bar{x} = \bar{y}$ and hence $\bar{x} \in D(L)$ is a least-squares solution of $Tx = y$. Also, $\hat{x}_{\lambda_n,y} \in \mathcal{N}(T_L)^{\perp}$ for all $n$; since $\mathcal{N}(T_L)^{\perp}$ is closed and hence weakly closed, it follows that $\bar{x} \in \mathcal{N}(T_L)^{\perp}$. This shows that $\bar{x} = T_L^{\dagger}y$.

It remains only to show that $\hat{x}_{\lambda_n,y} \to \bar{x}$ in the $*$-norm; to prove this, it suffices to show that $\|\hat{x}_{\lambda_n,y}\|_* \to \|\bar{x}\|_*$. We already know that $T\hat{x}_{\lambda_n,y} \to T\bar{x} = \bar{y}$ strongly, and hence $\|T\hat{x}_{\lambda_n,y}\|_Y \to \|T\bar{x}\|_Y$. We know that $\{\|L\hat{x}_{\lambda_n,y}\|_Z\}$ is nondecreasing as $n$ increases and $\|L\hat{x}_{\lambda_n,y}\|_Z \leq \|L\bar{x}\|_Z$ for all $n$. It follows that $\{\|L\hat{x}_{\lambda_n,y}\|_Z\}$ converges and

$$\lim_{n\to\infty} \|L\hat{x}_{\lambda_n,y}\|_Z \leq \|L\bar{x}\|_Z.$$

On the other hand, $L\hat{x}_{\lambda_n,y} \to L\bar{x}$ weakly in $Z$ and hence, since the norm is lower semicontinuous with respect to weak sequential convergence, we have

$$\|L\bar{x}\|_Z \leq \liminf_{n\to\infty} \|L\hat{x}_{\lambda_n,y}\|_Z = \lim_{n\to\infty} \|L\hat{x}_{\lambda_n,y}\|_Z \leq \|L\bar{x}\|_Z,$$

which shows that $\|L\hat{x}_{\lambda_n,y}\|_Z \to \|L\bar{x}\|_Z$. We now see that

$$\|\hat{x}_{\lambda_n,y}\|_*^2 = \|T\hat{x}_{\lambda_n,y}\|_Y^2 + \|L\hat{x}_{\lambda_n,y}\|_Z^2 \to \|T\bar{x}\|_Y^2 + \|L\bar{x}\|_Z^2 = \|\bar{x}\|_*^2. \qquad \square$$

Finally, as in the case of ordinary Tikhonov regularization, we see that for $y \notin D(T_L^{\dagger})$, $\hat{x}_{\lambda,y}$ blows up as $\lambda \to 0^+$, but now in the sense that $\|L\hat{x}_{\lambda,y}\|_Z$ is unbounded.

**Theorem 5.45.** *If $y \notin D(T_L^{\dagger})$, then $\|L\hat{x}_{\lambda,y}\|_Z \to \infty$ and hence $\|\hat{x}_{\lambda,y}\|_* \to \infty$ as $\lambda \to 0^+$.*

*Proof.* If the conclusion fails, then there exists $\{\lambda_n\} \subset \mathbb{R}^+$ such that $\lambda_n \to 0^+$ and $\{L\hat{x}_{\lambda_n,y}\}$ is bounded in $Z$. We already know that $T\hat{x}_{\lambda_n,y} \to \bar{y} = \text{proj}_{\overline{\mathcal{R}(T)}} y$ as $n \to \infty$ and hence that $\{T\hat{x}_{\lambda_n,y}\}$ is also bounded, implying that $\{\hat{x}_{\lambda_n,y}\}$ is bounded in $D(L)$ under the $*$-norm. Therefore, without loss of generality, there exists $\bar{x}$ in $D(L)$ such that $\hat{x}_{\lambda_n,y} \to \bar{x}$ weakly. But then $T\hat{x}_{\lambda_n,y} \to T\bar{x}$ weakly, implying that $\bar{y} = T\bar{x}$ and hence that $y \in \mathcal{R}(T_L) \oplus \mathcal{R}(T) = D(T_L^\dagger)$. This contradiction shows that $\|L\hat{x}_{\lambda,y}\|_Z$ must go to infinity as $\lambda \to 0^+$. $\qquad \square$

We will use the notation $\hat{x}_{0,y} = T_L^\dagger y$ for each $y \in D(T_L^\dagger)$.

To illustrate Theorem 5.45, we will construct an example of an operator $T$, a regularization operator $L$, and an exact solution $x^*$ that does not lie in $D(L)$. The corresponding data $y^* = Tx^*$ does not belong to $D(T_L^\dagger)$ and therefore $L\hat{x}_{\lambda,y^*}$ blows up as $\lambda \to 0$.

**Example 5.46.** Consider the operator $T : L^2(0, 1) \to L^2(0, 1)$ defined by

$$(Tx)(s) = \int_0^1 k(s, t)x(t)\, dt, \quad 0 < s < 1,$$

where

$$k(s, t) = \frac{1}{\sqrt{2\pi}\sigma} e^{-(s-t)^2/(2\sigma^2)}, \quad \sigma = 0.05.$$

Assuming that a smooth solution of $Tx = y$ is sought, we define the regularization operator $L : D(L) \to L^2(0, 1)$ by $Lx = x'$. As we discussed in Section 5.1.3, there are different choices for $D(L)$; for this example, we define $D(L) = H_0^1(0, 1)$. We also define $x^*(t) = t$ and $y^* = Tx^*$. Since $x^*$ does not belong to $D(L)$, it follows that $y^* \notin D(T_L^\dagger)$ and that $\|L\hat{x}_{\lambda,y^*}\|_Z \to \infty$ as $\lambda \to 0$. This is illustrated in Figure 5.2, where we compare $\hat{x}_{\lambda,y^*}$, for $\lambda = 10^{-3}, 10^{-6}, 10^{-9}, 10^{-12}$, to the exact solution $x^*$. The values of $\|L\hat{x}_{\lambda,y^*}\|_Z$ in Figure 5.2 are approximately 1.91, 7.76, 11.1, and 41.9.

When choosing a regularization operator $L$ for a problem, it must be the case that $D(L)$ contains $x^*$, and it should also be the case that $\|Lx^*\|_Z$ is not too large. If either of these conditions fails to be satisfied, then

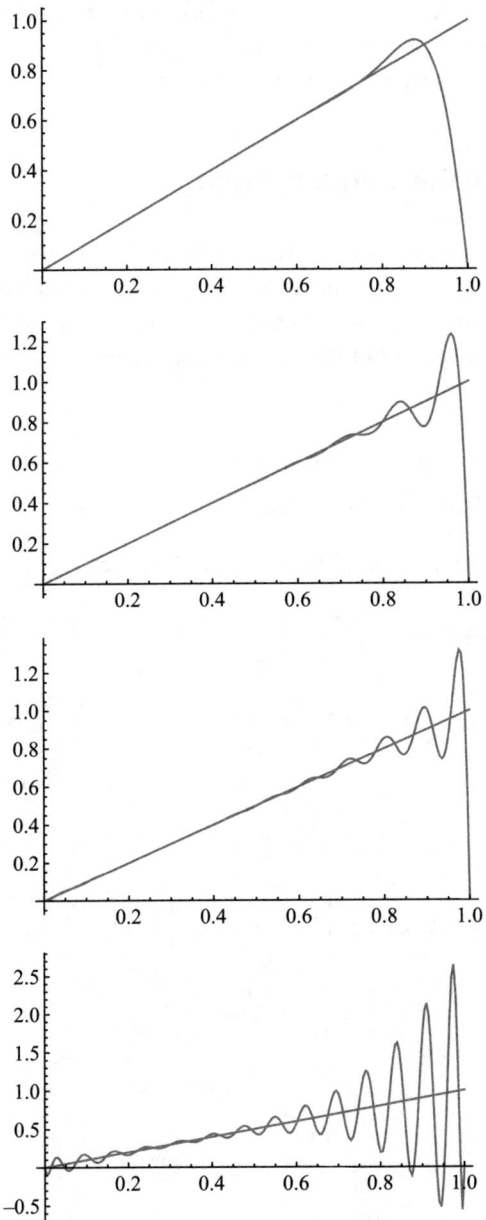

**Figure 5.2.** Example 5.46: $x^*$ and $\hat{x}_{\lambda,y}$ for (from top to bottom) $\lambda = 10^{-3}$, $\lambda = 10^{-6}$, $\lambda = 10^{-9}$, $\lambda = 10^{-12}$.

the underlying idea of seminorm regularization (to choose $x \in D(L)$ to make $T_L x$ close to $y$ without making $\|Lx\|_Z$ too large) is violated, since one could not choose $x = x^*$ even if $x^*$ were known.

## 5.5　Rates of convergence

We now consider an exact data vector $y^* \in D(T_L^\dagger)$ and the corresponding exact solution $\hat{x}^* = T_L^\dagger y^*$ (that is, $\hat{x}^* = \hat{x}_{0,y^*}$). We wish to determine conditions under which $\hat{x}_{\lambda,y} \to \hat{x}_{0,y^*}$ as $y \to y^*$ and $\lambda \to 0^+$. As in Chapter 3, we separate the total error into the perturbation error and the regularization error:

$$\hat{x}_{\lambda,y} - \hat{x}_{0,y^*} = \hat{x}_{\lambda,y} - \hat{x}_{\lambda,y^*} + \hat{x}_{\lambda,y^*} - \hat{x}_{0,y^*}$$
$$\Rightarrow \|\hat{x}_{\lambda,y} - \hat{x}_{0,y^*}\|_* \le \|\hat{x}_{\lambda,y} - \hat{x}_{\lambda,y^*}\|_* + \|\hat{x}_{\lambda,y^*} - \hat{x}_{0,y^*}\|_*.$$

We begin with the perturbation error. We have

$$\|T(\hat{x}_{\lambda,y} - \hat{x}_{\lambda,y^*})\|_Y = \|T\hat{N}_\lambda^{-1}T^*(y - y^*)\|_Y \le \|T\hat{N}_\lambda^{-1}T^*\| \|y - y^*\|_Y$$
$$\le \|y - y^*\|_Y,$$

$$\|L(\hat{x}_{\lambda,y} - \hat{x}_{\lambda,y^*})\|_Z = \|L\hat{N}_\lambda^{-1}T^*(y - y^*)\|_Z \le \|L\hat{N}_\lambda^{-1}T^*\| \|y - y^*\|_Y$$
$$\le \frac{1}{2}\lambda^{-1/2}\|y - y^*\|_Y$$

and therefore

$$\|\hat{x}_{\lambda,y} - \hat{x}_{\lambda,y^*}\|_* \le \left[ \|y - y^*\|_Y^2 + \frac{1}{4}\lambda^{-1}\|y - y^*\|_Y^2 \right]^{1/2}$$
$$\le \sqrt{1 + \frac{1}{4\lambda}} \|y - y^*\|_*$$
$$\le \frac{\sqrt{1 + 4\lambda}}{2\sqrt{\lambda}} \|y - y^*\|_*$$
$$\le \frac{1}{\sqrt{\lambda}} \|y - y^*\|_* \text{ for all } \lambda \in \left(0, \frac{3}{4}\right).$$

Thus we conclude that

$$\|\hat{x}_{\lambda,y} - \hat{x}_{\lambda,y^*}\|_* = O(\lambda^{-1/2}\|y - y^*\|_Y),$$

with $\|\hat{x}_{\lambda,y} - \hat{x}_{\lambda,y^*}\|_* \le \lambda^{-1/2}\|y - y^*\|_Y$ for all $\lambda$ sufficiently small. This shows that the perturbation error converges to zero provided $\|y - y^*\|_Y$ goes to zero faster than $\sqrt{\lambda}$ does; in other words, the perturbation error goes to zero provided

$$\delta, \lambda \to 0 \quad \text{and} \quad \frac{\delta^2}{\lambda} \to 0,$$

where $\|y - y^*\|_Y \le \delta$. Since we already showed (Theorem 5.44 in Section 5.4) that the regularization error converges to zero as $\lambda \to 0$, we obtain the following theorem.

**Theorem 5.47.** *Let $y^* \in D(T_L^\dagger)$ and assume $\{y_n\} \subset Y$ satisfies $y_n \to y^*$. Let $\{\delta_n\} \subset \mathbb{R}^+$ satisfy $\|y_n - y^*\|_Y \le \delta_n$ for all $n$ and $\delta_n \to 0$ as $n \to \infty$, and suppose $\{\lambda_n\} \subset \mathbb{R}^+$ satisfies*

$$\lambda_n \to 0, \quad \frac{\delta_n^2}{\lambda_n} \to 0 \ as \ n \to \infty.$$

*Then $\hat{x}_{\lambda_n,y_n} \to \hat{x}_{0,y^*}$ (with respect to the $*$-norm) as $n \to \infty$.*

Neither Theorem 5.44 nor Theorem 5.47 gives any guarantee on how fast the convergence occurs. To obtain a rate of convergence, we have to bound the regularization error; this analysis follows the same pattern as in the case of ordinary Tikhonov regularization. If $\hat{x}_{0,y^*} \in D(L^*L)$, then we have

$$
\begin{aligned}
\hat{N}_\lambda(\hat{x}_{\lambda,y^*} - \hat{x}_{0,y^*}) &= \hat{N}_\lambda \hat{x}_{\lambda,y^*} - T^*T\hat{x}_{0,y^*} - \lambda L^*L\hat{x}_{0,y^*} \\
&= T^*y^* - T^*y^* - \lambda L^*L\hat{x}_{0,y^*} \\
&= -\lambda L^*L\hat{x}_{0,y^*},
\end{aligned}
$$

and this yields the basic identity that we need:

$$\hat{x}_{0,y^*} \in D(L^*L) \implies \hat{x}_{\lambda,y^*} - \hat{x}_{0,y^*} = -\lambda \hat{N}_\lambda^{-1} L^*L\hat{x}_{0,y^*}.$$

We obtain

$$\|L(\hat{x}_{\lambda,y^*} - \hat{x}_{0,y^*})\|_Z = \lambda\|L\hat{N}_\lambda^{-1}L^*L\hat{x}_{0,y^*}\|_Z \le \lambda\|L\hat{N}_\lambda^{-1}L^*\|\,\|L\hat{x}_{0,y^*}\|_Z,$$

$$\|T(\hat{x}_{\lambda,y^*} - \hat{x}_{0,y^*})\|_Y = \lambda\|T\hat{N}_\lambda^{-1}L^*L\hat{x}_{0,y^*}\|_Y \le \frac{1}{2}\lambda^{1/2}\|L\hat{x}_{0,y^*}\|_Z,$$

where the last inequality follows from $\|T\hat{N}_\lambda^{-1}L^*\| \le (1/2)\lambda^{-1/2}$.

We will need the following result about the operator $L\hat{N}_\lambda^{-1}L^*$.

**Theorem 5.48.** *For each $\lambda > 0$, $L\hat{N}_\lambda^{-1}L^*$ extends to a bounded linear operator defined on all of $Z$, and $\|L\hat{N}_\lambda^{-1}L^*\| \le \lambda^{-1}$.*

*Proof.* For any $z \in Z$, consider the equation $\hat{M}_\lambda x = (0, z)$, which has a unique least-squares solution $u_{\lambda,z} \in D(L)$ characterized by

$$\hat{M}_\lambda^\sharp \hat{M}_\lambda u_{\lambda,z} = \hat{M}_\lambda^\sharp(0, z).$$

For $z \in D(L^*)$, we have $\hat{M}_\lambda^\sharp(0, z) = \sqrt{\lambda}(T^*T + L^*L)^{-1}L^*z$. We have

$$\hat{M}_\lambda^\sharp \hat{M}_\lambda u_{\lambda,z} = \hat{M}_\lambda^\sharp(Tu_{\lambda,z}, \sqrt{\lambda}Lu_{\lambda,z}) = (T^\sharp T + \lambda L^\sharp L)u_{\lambda,z}$$

and thus

$$u_{\lambda,z} = \sqrt{\lambda}(T^\sharp T + \lambda L^\sharp L)^{-1}(T^*T + L^*L)^{-1}L^*z.$$

Since $(T^*T + L^*L)^{-1}$ maps $X$ into $D(L^*L)$ and $(T^\sharp T + \lambda L^\sharp L)^{-1}$ maps $D(L^*L)$ onto itself, this shows that $u_{\lambda,z} \in D(L^*L)$. Therefore

$$\hat{M}_\lambda^\sharp \hat{M}_\lambda u_{\lambda,z} = \hat{M}_\lambda^\sharp(0, z)$$

$$\Rightarrow (T^*T + L^*L)^{-1}(T^*T + \lambda L^*L)u_{\lambda,z} = \sqrt{\lambda}(T^*T + L^*L)^{-1}L^*z$$

$$\Rightarrow (T^*T + \lambda L^*L)u_{\lambda,z} = \sqrt{\lambda}L^*z$$

$$\Rightarrow u_{\lambda,z} = \sqrt{\lambda}(T^*T + \lambda L^*L)^{-1}L^*z,$$

which shows that $u_{\lambda,z} = \sqrt{\lambda}\hat{N}_\lambda^{-1}L^*z$. We then have

$$\|\hat{M}_\lambda u_{\lambda,z}\|_{Y\times Z}^2 + \|\hat{M}_\lambda u_{\lambda,z} - (0,z)\|_{Y\times Z}^2 = \|(0,z)\|_{Y\times Z}^2$$

$$\Rightarrow \|Tu_{\lambda,z}\|_Y^2 + \lambda\|Lu_{\lambda,z}\|_Z^2 + \|Tu_{\lambda,z}\|_Y^2 + \|\sqrt{\lambda}Lu_{\lambda,z} - z\|_Z^2 = \|z\|_Z^2$$

$$\Rightarrow \lambda\|Lu_{\lambda,z}\|_Z^2 \leq \|z\|_Z^2$$

$$\Rightarrow \|Lu_{\lambda,z}\|_Z \leq \lambda^{-1/2}\|z\|_Z$$

$$\Rightarrow \|L\hat{N}_\lambda^{-1}L^*z\|_Z \leq \lambda^{-1}\|z\|_Z.$$

This holds for all $z \in D(L^*)$, which shows that $\|L\hat{N}_\lambda^{-1}L^*\| \leq \lambda^{-1}$. Since $D(L^*)$ is a dense subspace of $Z$, it follows that $L\hat{N}_\lambda^{-1}L^*$ has a unique bounded extension defined on all of $Z$ that satisfies the same bound on the norm.                                                               □

We now have

$$\|T(\hat{x}_{\lambda,y^*} - \hat{x}_{0,y^*})\|_Y \leq \frac{1}{2}\lambda^{1/2}\|L\hat{x}_{0,y^*}\|_Z,$$

$$\|L(\hat{x}_{\lambda,y^*} - \hat{x}_{0,y^*})\|_Z \leq \|L\hat{x}_{0,y^*}\|_Z$$

and therefore

$$\|\hat{x}_{\lambda,y^*} - \hat{x}_{0,y^*}\|_* = \left[\|T(\hat{x}_{\lambda,y^*} - \hat{x}_{0,y^*})\|_Y^2 + \|L(\hat{x}_{\lambda,y^*} - \hat{x}_{0,y^*})\|_Z^2\right]^{1/2}$$

$$\leq \left[\frac{\lambda}{4}\|L\hat{x}_{0,y^*}\|_Z^2 + \|L\hat{x}_{0,y^*}\|_Z^2\right]^{1/2}$$

$$= \sqrt{1 + \frac{\lambda}{4}}\|L\hat{x}_{0,y^*}\|_Z$$

$$\leq 2\|L\hat{x}_{0,y^*}\|_Z \text{ for all } \lambda \text{ sufficiently small.}$$

This shows only that $\|\hat{x}_{\lambda,y^*} - \hat{x}_{0,y^*}\|_*$ is bounded, and we already know (Theorem 5.44) that $\|\hat{x}_{\lambda,y^*} - \hat{x}_{0,y^*}\|_* \to 0$ as $\lambda \to 0^+$. As in the case of ordinary Tikhonov regularization, we are unable to give a useful bound on the rate of convergence unless we make further assumptions about the exact solution $\hat{x}_{0,y^*}$.

It is not difficult to see how to restrict the class of solutions so as to allow us to bound the rate of convergence. Recall the identity

$$\hat{x}_{\lambda,y^*} - \hat{x}_{0,y^*} = -\lambda \hat{N}_\lambda^{-1} L^* L \hat{x}_{0,y^*}.$$

If we can replace $L^* L \hat{x}_{0,y^*}$ with $T^* w_{0,y^*}$ or $T^* T u_{0,y^*}$, then we obtain useful bounds.

**Theorem 5.49.** *If* $y^* \in D(T_L^\dagger)$, $\hat{x}_{0,y^*} \in D(L^*L)$, *and* $L^* L \hat{x}_{0,y^*} \in \mathcal{R}(T^*)$, *then* $\|\hat{x}_{\lambda,y^*} - \hat{x}_{0,y^*}\|_* = o(\lambda^{1/2})$ *as* $\lambda \to 0^+$.

*Proof.* If $L^* L \hat{x}_{0,y^*} = T^* w_{0,y^*}$, then $\hat{x}_{\lambda,y^*} - \hat{x}_{0,y^*} = -\lambda \hat{N}_\lambda^{-1} T^* w_{0,y^*}$ and

$$\begin{aligned}
\|T(\hat{x}_{\lambda,y^*} - \hat{x}_{0,y^*})\|_Y &= \lambda \|T \hat{N}_\lambda^{-1} T^* w_{0,y^*}\|_Y \\
&\leq \lambda \|T \hat{N}_\lambda^{-1} T^*\| \|w_{0,y^*}\|_Y \leq \lambda \|w_{0,y^*}\|_Y, \\
\|L(\hat{x}_{\lambda,y^*} - \hat{x}_{0,y^*})\|_Z &= \lambda \|L \hat{N}_\lambda^{-1} T^* w_{0,y^*}\|_Z = \lambda^{1/2} \|\lambda^{1/2} L \hat{N}_\lambda^{-1} T^* w_{0,y^*}\|_Y.
\end{aligned}$$

By Theorem 5.43, $\|\lambda^{1/2} L \hat{N}_\lambda^{-1} T^* w_{0,y^*}\|_Y \to 0$ as $\lambda \to 0^+$, and thus we see that both $\|T(\hat{x}_{\lambda,y^*} - \hat{x}_{0,y^*})\|_Y$ and $\|L(\hat{x}_{\lambda,y^*} - \hat{x}_{0,y^*})\|_Z$ are $o(\lambda^{1/2})$. It follows that

$$\|\hat{x}_{\lambda,y^*} - \hat{x}_{0,y^*}\|_* = o(\lambda^{1/2}). \qquad \square$$

**Theorem 5.50.** *If* $y^* \in D(T_L^\dagger)$, $\hat{x}_{0,y^*} \in D(L^*L)$, *and* $L^* L \hat{x}_{0,y^*} \in \mathcal{R}(T^* T|_{D(L)})$, *then* $\|\hat{x}_{\lambda,y^*} - \hat{x}_{0,y^*}\|_* = O(\lambda)$.

*Proof.* If we restrict the operator $L \hat{N}_\lambda^{-1} T^* T$ to $D(L^*L)$, then we have

$$L \hat{N}_\lambda^{-1} T^* T = L \hat{N}_\lambda^{-1} (\hat{N}_\lambda - \lambda L^* L) = L - \lambda L \hat{N}_\lambda^{-1} L^* L.$$

Moreover, we know that $L \hat{N}_\lambda^{-1} L^*$ extends to a bounded operator $B_\lambda : Z \to Z$, with $\|B_\lambda\| = \|L \hat{N}_\lambda^{-1} L^*\| \leq \lambda^{-1}$. Therefore, the bounded operator $L \hat{N}_\lambda^{-1} T^* T$ has the representation $L - \lambda B_\lambda L$ that is valid on $D(L)$.

Now suppose that $L^*L\hat{x}_{0,y^*} = T^*Tu_{0,y^*}$, where $u_{0,y^*} \in D(L)$. Then

$$\|T(\hat{x}_{\lambda,y^*} - \hat{x}_{0,y^*})\|_Y = \lambda\|T\hat{N}_\lambda^{-1}L^*L\hat{x}_{0,y^*}\|_Y = \lambda\|T\hat{N}_\lambda^{-1}T^*Tu_{0,y^*}\|_Y$$
$$\leq \lambda\|T\hat{N}_\lambda^{-1}T^*\|\,\|Tu_{0,y^*}\|_Y$$
$$\leq \lambda\|Tu_{0,y^*}\|_Y$$

$$\|L(\hat{x}_{\lambda,y^*} - \hat{x}_{0,y^*})\|_Z = \lambda\|L\hat{N}_\lambda^{-1}L^*L\hat{x}_{0,y^*}\|_Z = \lambda\|L\hat{N}_\lambda^{-1}T^*Tu_{0,y^*}\|_Z$$
$$= \lambda\|Lu_{0,y^*} - \lambda B_\lambda Lu_{0,y^*}\|_Z$$
$$\leq \lambda\left(\|Lu_{0,y^*}\|_Z + \lambda\|B_\lambda\|\,\|Lu_{0,y^*}\|_Z\right)$$
$$\leq 2\lambda\|Lu_{0,y^*}\|_Z.$$

It now follows that

$$\|\hat{x}_{\lambda,y^*} - \hat{x}_{0,y^*}\|_*^2 \leq \lambda^2\|Tu_{0,y^*}\|_Y^2 + 4\lambda^2\|Lu_{0,y^*}\|_Z^2$$
$$\leq 4\lambda^2\left(\|Tu_{0,y^*}\|_Y^2 + \|Lu_{0,y^*}\|_Z^2\right) = 4\lambda^2\|u_{0,y^*}\|_*^2$$

and hence that $\|\hat{x}_{\lambda,y^*} - \hat{x}_{0,y^*}\|_* \leq 2\lambda\|u_{0,y^*}\|_*$. $\qquad\square$

The final results, for noisy data, follow exactly as in Chapter 3.

**Theorem 5.51.** *Suppose $y^* \in D(T_L^\dagger)$.*

1. *Suppose $\hat{x}_{0,y^*} \in D(L^*L)$ and $L^*L\hat{x}_{0,y^*} \in \mathcal{R}(T^*)$. If $\delta = \|y - y^*\|_Y$ and $\lambda = k\delta$, then*

$$\|\hat{x}_{\lambda,y} - \hat{x}_{0,y^*}\|_* = O(\delta^{1/2}).$$

2. *Suppose $\hat{x}_{0,y^*} \in D(L^*L)$ and $L^*L\hat{x}_{0,y^*} \in \mathcal{R}(T^*T|_{D(L)})$. If $\delta = \|y - y^*\|_Y$ and $\lambda = k\delta^{2/3}$, then*

$$\|\hat{x}_{\lambda,y} - \hat{x}_{0,y^*}\|_* = O(\delta^{2/3}).$$

We noted at the beginning of Section 5.3 that we could just apply the theory of Chapter 3 to the problem

$$\min_{x \in D(L)} \|Tx - y\|_Y^2 + \lambda\|x\|_*^2.$$

We pointed out our reason for not doing so, but it is reasonable to ask how the final results would differ had we done so. Theorem 3.23 requires $\hat{x}^* \in \mathcal{R}(T_L^\sharp T_L)$ to obtain the optimal rate of convergence, while Theorem 5.50

requires $\hat{x}^* \in D(L^*L)$ and $L^*L\hat{x}^* \in \mathcal{R}(T^*T|_{D(L)})$ to obtain the same rate of convergence (in the same norm, namely, the $*$-norm). We now show that these two conditions are, in fact, the same.

If $\hat{x}^* \in D(L^*L)$ and $L^*L\hat{x}^* = T^*Tu$ for $u \in D(L)$, then we have

$$L^*L\hat{x}^* = T^*Tu \Rightarrow (T^*T + L^*L)\hat{x}^* = T^*T(u + \hat{x}^*)$$
$$\Rightarrow \hat{x}^* = (T^*T + L^*L)^{-1}T^*T(u + \hat{x}^*)$$
$$\Rightarrow \hat{x}^* = T_L^\sharp T_L \hat{u}, \text{ where } \hat{u} = u + \hat{x}^*.$$

Conversely, suppose $\hat{x}^* = T_L^\sharp T_L \hat{u}$ for some $\hat{u} \in D(L)$. Then, by Theorem 5.25, $\hat{x}^* \in D(L^*L)$ and

$$\hat{x}^* = (T^*T + L^*L)^{-1}T^*T\hat{u} \Rightarrow (T^*T + L^*L)\hat{x}^* = T^*T\hat{u}$$
$$\Rightarrow L^*L\hat{x}^* = T^*T(\hat{u} - \hat{x}^*)$$
$$\Rightarrow L^*L\hat{x}^* = T^*Tu \text{ where } u = \hat{u} - \hat{x}^* \in D(L).$$

Therefore, under the two approaches, the conditions that yield the optimal rate of convergence are the same.

# Epilogue

Over the last half-century, inverse problems have played an increasingly important role in applied mathematics. In most cases, after we learn to model a given experiment mathematically, there arises the need to solve a related inverse problem. For this reason, every applied mathematician should know something about inverse problems.

This book has explored one important class of inverse problems and the most popular approach to resolving problems from it. There are many important topics that we have not covered, and it seems appropriate to close by discussing, albeit briefly, some of them.

**Discretization** Almost every inverse problem that one encounters in practice requires a numerical solution. Since inverse problems, by their nature, are infinite-dimensional, a numerical solution requires discretization, and it is critical that the discretization faithfully represents the important features of the problem. In Section 2.4, we discussed the relationship of a discretized problem to the original infinite-dimensional problem. However, we did not present or analyze any particular methods of discretization. Readers interested in learning about this topic can consult [3, Section 3.3, Section 5.2, and Chapter 9] and [18, Chapter 3].

**Probability-based analysis** In this book, we have implicitly adopted a deterministic point of view: the data is regarded as given and from it we must determine a satisfactory solution. The probabilistic point of view recognizes that the data is subject to random error; that is, the error is a random variable with a certain probability distribution. From this distribution, we wish to determine an estimator of the solution and its probability distribution. This analysis is usually presented in finite dimensions; the problem is to solve $A\mathbf{x} = \mathbf{b}^{\text{obs}}$, where the matrix $A$ is ill-conditioned, $\mathbf{x}^*$ satisfies $A\mathbf{x}^* = \mathbf{b}^*$ (that is, $\mathbf{b}^*$ and $\mathbf{x}^*$ are the exact data and solution, respectively), $\mathbf{b}^{\text{obs}} = \mathbf{b}^* + \epsilon$, and $\epsilon$ is the random noise vector. In this context, we can offer a justification of the use of the least-squares approach; for the purposes of our discussion, we assume that $A$ has full rank. The least-squares solution of $A\mathbf{x} = \mathbf{b}^{\text{obs}}$ is $\mathbf{x}^{\text{est}} = (A^T A)^{-1} A^T \mathbf{b}^{\text{obs}}$. If $\mathbf{b}^{\text{obs}} = \mathbf{b}^* + \epsilon$ and the expected value $E(\epsilon)$ of $\epsilon$ is $\mathbf{0}$, then the least-squares estimator $\mathbf{x}^{\text{est}}$ is *unbiased*, meaning that $E(\mathbf{x}^{\text{obs}}) = \mathbf{x}^*$. Moreover, among all the linear unbiased estimators, the least-squares estimator has the smallest variance (this result assumes that all the components of $\epsilon$ have the same variance). For this reason, the least-squares estimator is called the *best linear unbiased estimator.*

Although there is no other unbiased estimator with a lower variance, nevertheless, the variance of the least-squares estimator is still large if the matrix $A$ is ill-conditioned. For this reason, a biased estimator may be preferrable if it has a lower variance. Tikhonov regularization was invented in this context (independently of Tikhonov) by Hoerl [15], who named the method "ridge regression."[8]

The book [23] by Tarantola is a standard work on the probabilistic approach to inverse problems. For a briefer introduction, the reader can consult [24, Chapter 4].

**BV regularization**   At the beginning of Chapter 5, we described a general approach to regularization: solve

$$\min_{x \in X} \|Tx - y\|_Y^2 + \lambda R(x),$$

---

[8]According to Marquardt and Snee [19],

Hoerl gave the name "ridge regression" [15] to his procedure because of the similarity of its mathematics to methods he used earlier [14], i.e., "ridge analysis," for graphically depicting the characteristics of second order response surface equations in many predictor variables.

(Original citations have been replaced by citations (of the same papers) in our bibliography.)

where $R$ can be any regularization function. One particular choice of $R$, which deserves special mention, results in what is called *total variation* or *bounded variation* regularization (briefly, BV regularization). The most common version of Tikhonov regularization is seminorm regularization, as introduced in Chapter 5, where the regularization function is $R(x) = \|Lx\|_{L^2}$ and the operator $L$ is a derivative operator. This choice is meant to ensure that the computed solution is smooth, which naturally means that this approach would not work well if the true (exact) solution is not smooth. Nonsmooth solutions arise in certain applications, especially in image processing. For instance, if the true solution is an image of a city block consisting of office buildings, then the mathematical description of the (true) image is likely to have discontinuities that correspond to the sharp edges of the buildings.

We will briefly describe BV regularization for a one-dimensional problem in which the unknown is a function on $(0, 1)$. We will motivate the definition of the regularization function (known as the BV seminorm) using an example, namely, the function

$$u(t) = \begin{cases} 0, & 0 < t < \frac{1}{2}, \\ 1, & \frac{1}{2} < t < 1. \end{cases}$$

This function has no weak derivative (see Example B.2 in Appendix B), and therefore neither

$$\int_0^1 (u'(t))^2 \, dt$$

nor

$$\int_0^1 |u'(t)| \, dt$$

exists. However, we will see that a variant of the second measure is meaningful and yields a finite number (1, in fact). To show this, let us consider a sequence $\{u_n\}$ of functions converging (in some sense) to $u$:

$$u_n(t) = \begin{cases} 0, & 0 < t < \frac{1}{2} - \frac{1}{2n}, \\ n\left(t - \frac{1}{2} + \frac{1}{2n}\right), & \frac{1}{2} - \frac{1}{2n} \leq t \leq \frac{1}{2} + \frac{1}{2n}, \\ 1, & \frac{1}{2} + \frac{1}{2n} < t < 1 \end{cases}$$

(see Figure E.1). It is interesting to note that

$$\|u_n'\|_{L^2} = \sqrt{\int_0^1 (u_n'(t))^2 \, dt} = \sqrt{n} \to \infty \text{ as } n \to \infty,$$

while

$$\|u_n'\|_{L^1} = \int_0^1 |u_n'(t)| \, dt = 1 \text{ for all } n \in \mathbb{Z}^+.$$

(The $L^1$ norm of a function $f$ defined on $(0, 1)$ is defined by $\|f\|_{L^1} = \int_0^1 |f(t)| \, dt$, and $f$ is said to belong to $L^1(0, 1)$ if this integral is finite.) This suggests that it might be worthwhile to somehow measure the total change in a function even if it is not weakly differentiable. This is accomplished by the BV seminorm:

$$|u|_{BV} = |u|_{BV(0,1)}$$
$$= \sup\left\{ \int_0^1 u(t)\phi'(t) \, dt \ : \ \phi \in C_0^\infty(0, 1), |\phi(t)| \le 1, 0 < t < 1 \right\}.$$

The space $BV(0, 1)$ is defined by

$$BV(0, 1) = \left\{ u \in L^1(0, 1) \ : \ |u|_{BV} < \infty \right\},$$

and the norm on $BV(0, 1)$ is

$$\|u\|_{BV} = \|u\|_{BV(0,1)} = \|u\|_{L^1} + |u|_{BV}.$$

Elements of $BV(0, 1)$ are called functions of *bounded variation*. If $u \in BV(0, 1)$ is weakly differentiable, then integration by parts can be used to show that

$$|u|_{BV} = \int_0^1 |u'(t)| \, dt.$$

Thus the BV seminorm provides a measure of the size of $u'$ that extends to functions that are too irregular to have a weak derivative. It is important to recognize that $|u|_{BV}$ penalizes large derivatives (or jump discontinuities) much less than does $\|u'\|_{L^2}$ (because the square of a large number is much larger than the number itself). For this reason, the regularization function $R(x) = |x|_{BV}$ has proven to be suitable for problems

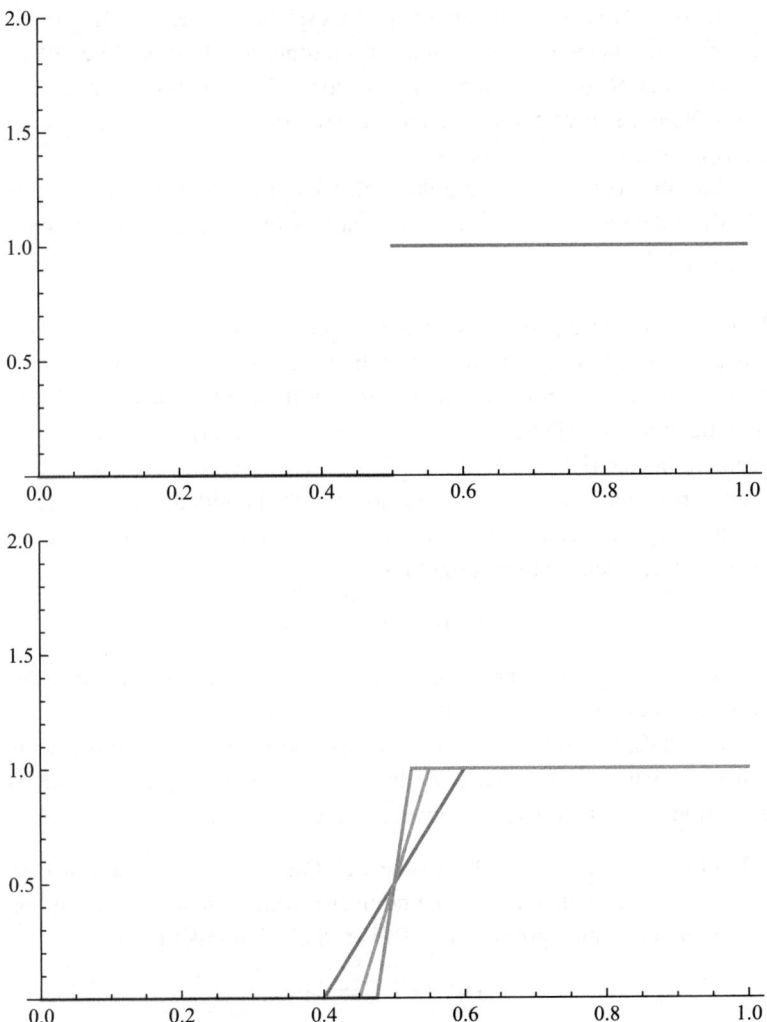

**Figure E.1.** The function $u$ (top) and $u_5$, $u_{10}$, $u_{20}$ (bottom) from the example introduced on page 273.

in which the true solution is not smooth, especially when the important features are characterized by jump discontinuities. It should be noted, though, that $BV(0, 1)$ is not a Hilbert space (it is a Banach space—a complete normed vector space, but one without an inner product), so the analysis of this book does not apply.

An introduction to BV regularization can be found in [24, Chapter 8]. More detailed information about analysis in BV spaces is presented in [1] and [6].

**Nonlinear inverse problems**   We have focused on linear inverse problems, but many important inverse problems are nonlinear. This is commonly the case when the problem arises in the context of a partial differential equation (PDE). One such example is the inverse conductivity problem in electricity.

Suppose we have an object, occupying the (bounded, open) set $\Omega$ in $\mathbb{R}^3$, that conducts electricity. If $u = u(x, y, z)$ is the potential (voltage) at $(x, y, z) \in \Omega$, then $u$ satisfies the PDE

$$-\nabla \cdot (\kappa \nabla u) = 0 \text{ in } \Omega,$$

where $\kappa = \kappa(x, y, z)$ is the conductivity. Here we have assumed that the potential, and hence the current $-\kappa \nabla u$, are in steady state.

The PDE, by itself, does not determine a unique solution $u$ from $\kappa$; it is necessary to add a boundary condition. Here are two possible boundary conditions, and the forward problems that they define:

1. Given the potential (voltage) $u$ on $\partial\Omega$, find $u$ in $\Omega$. The condition $u = g$ on $\partial\Omega$ is called a Dirichlet boundary condition, and the resulting boundary value problem (BVP) is called a Dirichlet problem:

$$-\nabla \cdot (\kappa \nabla u) = 0 \text{ in } \Omega,$$

$$u = g \text{ on } \partial\Omega.$$

2. Given the current $-\kappa \nabla u \cdot n$ across the boundary $\partial\Omega$, find $u \in \Omega$. (Here $\partial u/\partial n = \nabla u \cdot n$ is the derivative in the direction of the unit vector $n$ orthogonal at each point to $\partial\Omega$.) The corresponding

boundary condition $-\kappa\,\partial u/\partial n = h$ on $\partial\Omega$ is called a Neumann condition, and it defines the following Neumann BVP:

$$-\nabla\cdot(\kappa\nabla u) = 0 \text{ in } \Omega,$$

$$-\kappa\nabla u = h \text{ on } \partial\Omega.$$

The Dirichlet problem has a unique solution for each (sufficiently smooth) $g$, while the Neumann problem has infinitely many solutions, any two of which differ by a constant, provided $h$ satisfies the compatibility condition

$$\int_{\partial\omega} h = \int_{\partial\Omega} \kappa\nabla u \cdot n = -\int_{\Omega} \nabla\cdot(\kappa\nabla u) = 0.$$

We can impose an extra condition, for example, that the average potential in $\Omega$ be zero, to select a unique solution. With this understanding, both forward problems are well-posed.

The most common version of the inverse problem, which is called the inverse conductivity problem, assumes that only boundary measurements are available. Specifically, we assume that we can specify (apply) a given voltage on the boundary and measure the resulting current (again, on the boundary). In other words, the Dirichlet-to-Neumann map (the map from the Dirichlet boundary data to the induced Neumann boundary data) is available. From this map, we wish to find the conductivity $\kappa$.

A variant of the inverse conductivity problem, which presents very different (and easier) mathematical issues, assumes that the potential $u$ can be measured in $\Omega$ (that is, in the interior of the object), and again asks for $\kappa$.

The inverse conductivity problem (with boundary measurements) is the basis for electrical impedance tomography, which aims to map the conductivity of a region of the interior of the human body and thus create an image of it (relying on the fact that different types of tissue have different conductivities). Unique dependence of the conductivity on the Dirichlet-to-Neumann map has been proved and much effort has been expended on developing practical algorithms, but this remains a very difficult inverse problem.

The important point for our discussion is that either version of the inverse problem is nonlinear. This is in spite of the fact that $-\nabla \cdot (\kappa \nabla u) = 0$ is normally classified as a linear PDE—in this classification, one is assuming that $\kappa$ is known and fixed. Actually, though, the mapping from $\kappa$ to $u$ is nonlinear, as should be clear upon a little thought, and so is the mapping from $u$ back to $\kappa$.

Most inverse problems involving PDEs are nonlinear, even when the governing PDE is linear, because the typical inverse problem seeks one or more coefficients in the differential equation. (Since the coefficients multiply the solution of the PDE, or one of its derivatives, the relationships between coefficients and solution are nonlinear.) This kind of inverse problem is often called a parameter identification problem, and frequently a distributed parameter identification problem (to emphasize that the parameters sought are functions, not scalars).

Readers who would like to learn more about inverse problems in PDEs can consult [17].

# Appendix A

# Basic Hilbert space theory

## A.1 Hilbert spaces and linear operators

**Definition A.1.** Let $U$ be an inner product space. If $U$ is complete under the norm induced by the inner product, then $U$ is called a *Hilbert space*.

Recall that a normed vector space is *complete* if every Cauchy sequence in the space converges to a vector in the space.

**Example A.2.** Euclidean $n$-space, denoted $\mathbb{R}^n$, is a Hilbert space under the dot product. In fact, every finite-dimensional inner product space is isomorphic to $\mathbb{R}^n$ (for some $n$), so every finite-dimensional inner product space is complete. These are standard results of elementary real analysis.

**Example A.3.** Let

$$C[0, 1] = \{f : [0, 1] \to \mathbb{R} : f \text{ is continuous}\}$$

and define an inner product on $C[0, 1]$ by

$$\langle f, g \rangle_{L^2} = \int_0^1 f(x)g(x) \, dx.$$

This is called the $L^2$ inner product. Then $C[0, 1]$ is an inner product space, but it can be shown that it is not complete. Indeed, if $f : [0, 1] \to \mathbb{R}$ is defined by

$$f(x) = \begin{cases} 1, & 0 \le x \le \frac{1}{2}, \\ 0, & \frac{1}{2} < x \le 1, \end{cases}$$

and $f_N : [0, 1] \to \mathbb{R}$ is the sum of the first $N$ terms of the Fourier sine series of $f$, then $\{f_N\}$ is a sequence in $C[0, 1]$ that converges to $f$ in the $L^2$ norm (so that $\{f_N\}$ is Cauchy in the $L^2$ norm). However, $f \notin C[0, 1]$, and so $C[0, 1]$ is not complete.

**Example A.4.** Define $L^2(0, 1)$ to be the space of Lebesgue measurable functions $f$ that are defined on $[0, 1]$ and are square-integrable:

$$\int_0^1 |f(x)|^2 \, dx < \infty.$$

(To be precise, the elements of $L^2(0, 1)$ are not measureable functions, but equivalence classes of measureable functions, with two functions identified if they differ on a set of measure zero. The same remark applies to the next example.) It can be shown that $L^2(0, 1)$ is complete under the $L^2$ inner product, and hence that $L^2(0, 1)$ is a Hilbert space.

**Example A.5.** More generally, let $\Omega \subseteq \mathbb{R}^n$ (usually $n = 2$ or $n = 3$) be a connected open set, and define $L^2(\Omega)$ to be the space of all square-integrable real-valued measurable functions defined on $\Omega$. Then $L^2(\Omega)$ is a Hilbert space under the inner product

$$\langle f, g \rangle_{L^2} = \int_\Omega fg.$$

Recall that if $V$ is an inner product space, then the Cauchy-Schwarz inequality holds on $V$:

$$|\langle u, v \rangle| \leq \|u\|_V \|v\|_V \quad \text{for all } u, v \in V.$$

**Definition A.6.** Let $U$ and $V$ be Hilbert spaces, and let $L : U \to V$. We say that $L$ is a *linear* operator if

$$L(\alpha u + \beta v) = \alpha L(u) + \beta L(v) \quad \text{for all } \alpha, \beta \in \mathbb{R}, \text{ for all } u, v \in U.$$

A linear operator mapping one finite-dimensional space to another can always be represented in terms of matrix-vector multiplication once bases have been chosen for the domain and codomain spaces. The continuity of such an operator is then apparent. However, linear operators defined on infinite-dimensional spaces need not be continuous. The continuity of such operators is most conveniently analyzed using the operator norm.

**Definition A.7.** Let $U$ and $V$ be Hilbert spaces, and let $L : U \to V$ be a linear operator. We define the *operator norm* $\|L\|$ of $L$ by

$$\|L\| = \sup \left\{ \frac{\|Lu\|_V}{\|u\|_U} : u \in U, u \neq 0 \right\}.$$

We say that $L$ is *bounded* if $\|L\| < \infty$.

**Lemma A.8.** *Let $U$ and $V$ be Hilbert spaces, and let $L : U \to V$ be a linear operator.*

1. *If $L$ is bounded, then*

$$\|Lu\|_V \leq \|L\| \|u\|_U \quad \text{for all } u \in U.$$

2. *The operator $L$ is bounded if and only if there exists a constant $C > 0$ with $\|Lu\|_V \leq C\|u\|_U$ for all $u \in U$. In this case, $\|L\| \leq C$.*

3. *The norm of $L$ can be equivalently defined as*

$$\|L\| = \sup \{\|Lu\|_V : u \in U, \|u\| = 1\}$$

   *or*

$$\|L\| = \inf \{C > 0 : \|Lu\|_V \leq C\|U\|_U \quad \text{for all } u \in U\}.$$

*Proof.* Exercise. $\square$

**Theorem A.9.** *Let $U$ and $V$ be Hilbert spaces, and let $L : U \to V$ be a linear operator. Then $L$ is continuous if and only if $L$ is bounded.*

*Proof.* Suppose first that $L$ is bounded, in which case

$$\|Lu\|_V \leq \|L\| \|u\|_U \quad \text{for all } u \in U.$$

Now, to show that $L$ is continuous, we must show that if $\{u_n\}$ is a sequence in $U$ and $u_n \to u$ (that is, $\|u - u_n\|_U \to 0$), then $Lu_n \to Lu$ (that is, $\|Lu_n - Lu\|_V \to 0$). But this follows immediately from the linearity and boundedness of $L$:

$$\|Lu - Lu_n\|_V = \|L(u - u_n)\|_V \leq \|L\|\|u - u_n\|_U \to 0.$$

Thus $L$ is continuous.

Now suppose that $L$ is not bounded; that is, suppose there exists a sequence $\{u_n\}$ in $U$, with $\|u_n\|_U = 1$ for all $n$, such that $\|Lu_n\|_V \to \infty$. Define $w_n = \alpha_n u_n$, where $\alpha_n = 1/\|Lu_n\|_V$. Then $w_n \to 0$, since

$$\|w_n - 0\| = \|w_n\| = \alpha_n\|u_n\| = \frac{\|u_n\|}{\|Lu_n\|} = \frac{1}{\|Lu_n\|} \to 0.$$

However,

$$\|Lw_n\|_V = \alpha_n\|Lu_n\|_V = \frac{\|Lu_n\|_V}{\|Lu_n\|_V} = 1 \nrightarrow 0.$$

This shows that $L$ is not continuous.                              $\square$

The following two examples of linear operators are of fundamental importance. The first, as we shall see, is a source of ill-posed inverse problems, while the second is the prototypical example of an unbounded (discontinuous) linear operator.

**Example A.10.** Let $\Omega = (0, 1) \times (0, 1)$, and suppose that $k \in L^2(\Omega)$. Define $K : L^2(0, 1) \to L^2(0, 1)$ by

$$(Kx)(s) = \int_0^1 k(s, t)x(t)\, dt.$$

We now show that $K$ is a bounded linear operator. The linearity of $K$ is obvious; it follows from the linearity of integration. We must show that $Kx \in L^2(0, 1)$ for all $x \in L^2(0, 1)$, and that there is a constant $C$ with $\|Kx\| \leq C\|x\|$ for all $x \in L^2(0, 1)$.

Let $x$ belong to $L^2(0, 1)$ and define $y = Kx$. By standard Lebesgue theory, $k(s, \cdot) \in L^2(0, 1)$ for almost every $s \in (0, 1)$. Therefore, we have

$$|y(s)|^2 = \left(\int_0^1 k(s, t)x(t)\, dt\right)^2$$

$$\leq \|k(s, \cdot)\|_{L^2(0,1)}^2 \|x\|_{L^2(0,1)}^2 \text{ (by the Cauchy-Schwarz inequality)},$$

where

$$\|k(s, \cdot)\|_{L^2(0,1)}^2 = \int_0^1 k(s, t)^2 \, dt.$$

Thus,

$$\int_0^1 |y(s)|^2 \, ds \leq \int_0^1 \|k(s, \cdot)\|_{L^2(0,1)}^2 \|x\|_{L^2(0,1)}^2 \, ds$$

$$= \|x\|_{L^2(0,1)}^2 \int_0^1 \|k(s, \cdot)\|_{L^2(0,1)}^2 \, ds$$

$$= \|x\|_{L^2(0,1)}^2 \int_0^1 \int_0^1 k(s, t)^2 \, dt \, ds$$

$$= \|k\|_{L^2(\Omega)}^2 \|x\|^2.$$

This shows both that $y \in L^2(0, 1)$ and that

$$\|y\|_{L^2(0,1)} \leq \|k\|_{L^2(\Omega)} \|x\|_{L^2(0,1)}.$$

Thus $K$ is bounded with $\|K\| \leq \|k\|_{L^2(\Omega)}$.

**Example A.11.** We could try to define $D : L^2(0, 1) \rightarrow L^2(0, 1)$ by

$$Dx = x'.$$

Actually, $D$ can only be defined on a dense subspace of $L^2(0, 1)$, since $x \in L^2(0, 1)$ does not imply that $x' \in L^2(0, 1)$. Ignoring this for now (but see Appendix B for more details), we show that $D$ is unbounded. Define $x_n \in L^2(0, 1)$ by

$$x_n(t) = \sin(n\pi t);$$

a direct calculation shows that $\|x_n\|_{L^2(0,1)} = 1/\sqrt{2}$ for all $n$. However,

$$(Dx_n)(t) = n\pi \cos(n\pi t),$$

and, for all $n$, $\|Dx_n\|_{L^2(0,1)} = n\pi/\sqrt{2}$. Therefore,

$$\frac{\|Dx_n\|_{L^2(0,1)}}{\|x_n\|_{L^2(0,1)}} = n\pi,$$

which is unbounded as $n \rightarrow \infty$. Thus $D$ is not bounded.

## A.2   Subspaces of Hilbert space

**Definition A.12.** A *subspace* $W$ of a Hilbert space $U$ is a nonempty subset of $U$ that is closed under addition and scalar multiplication; that is,

$$x, y \in W \Rightarrow x + y \in W,$$
$$x \in W, \alpha \in \mathbb{R} \Rightarrow \alpha x \in W.$$

A subspace of a Hilbert space is an inner product space (although it might not be complete).

Recall that a subset $S$ of a normed linear space $U$ is *closed* if, whenever $\{w_n\}$ is a sequence in $S$ and $w_n \to v \in U$, it follows that $v \in S$. (Some authors require that a subspace of a Hilbert space be closed by definition; however, in this book, "subspace" and "closed subspace" are not synonymous.) It is straightforward to prove that a closed subspace of a Hilbert space is complete and hence is a Hilbert space in its own right.

If $S$ is a subset of a normed linear space $U$, we denote by $\overline{S}$ the *closure* of $S$, that is, $S$ together with all of its limit points.

**Definition A.13.** Let $U$ be a Hilbert space and $S$ a subset of $U$. The *orthogonal complement* of $S$ is the set

$$S^\perp = \{w \in H : \langle w, u \rangle_U = 0 \text{ for all } u \in S\}.$$

**Lemma A.14.** *Let $S$ be a subset of a Hilbert space $U$. Then $S^\perp$ is a closed subspace of $U$.*

*Proof.* First note that if $u, v \in S^\perp$, then

$$\langle u + v, z \rangle_U = \langle u, z \rangle_U + \langle v, z \rangle_U = 0 + 0 = 0 \quad \text{for all } z \in S,$$

which implies that $u + v \in S^\perp$. Also,

$$\langle \alpha u, z \rangle_U = \alpha \langle u, z \rangle_U = 0 \quad \text{for all } \alpha \in \mathbb{R}, z \in S$$

which implies that $\alpha u \in S^\perp$. The set $S^\perp$ is not empty because $0 \in S^\perp$, and hence we have shown that $S^\perp$ is a subspace.

Now suppose $\{w_n\}$ is a sequence in $S^\perp$ and $w_n \to w \in H$. Then, for $z \in S$, we have

$$|\langle w, z \rangle_U| = |\langle w, z \rangle_U - \langle w_n, z \rangle_U|$$
$$= |\langle w - w_n, z \rangle_U| \le \|w - w_n\|_U \|z\|_U \to 0.$$

Thus $\langle w, z \rangle_U = 0$, that is, $w \in S^\perp$. This shows that $S^\perp$ is closed. $\qquad\square$

Given a closed subspace of a Hilbert space and any vector in the space, there always exists a unique vector in the subspace that is closest to the given vector. To prove this, we need the following lemma.

**Lemma A.15 (Parallelogram law).** *Let $U$ be a Hilbert space. Then*

$$\|v + w\|_U^2 + \|v - w\|_U^2 = 2(\|v\|_U^2 + \|w\|_U^2) \quad \text{for all } v, w \in U.$$

*Proof.* Expand both sides using the inner product. $\qquad\square$

The converse of the parallelogram law holds in the following sense: if $U$ is a normed vector space in which the parallelogram law holds, then the norm is induced by an inner product. However, we will not have need of this fact.

**Theorem A.16 (The projection theorem).** *Let $U$ be a Hilbert space, $M$ a closed subspace of $U$, and $v \in U$. Then there exists a unique $w_0 \in M$ such that*

$$\|v - w_0\|_U = \inf \{\|v - w\|_U : w \in M\}$$

*(that is, $w_0$ is the element of $M$ closest to $v$). Moreover, $w_0$ is characterized by*

$$v - w_0 \in M^\perp.$$

*Proof.* Let

$$\delta = \inf \{\|v - w\|_U : w \in M\}.$$

Then, by the definition of infimum, there exists a sequence $\{w_n\}$ in $M$ such that

$$\lim_{n \to \infty} \|v - w_n\|_U = \delta.$$

We now use the parallelogram law to show that $\{w_n\}$ is Cauchy. We have

$$\|(w_n - v) + (w_m - v)\|_U^2 + \|(w_n - v) - (w_m - v)\|_U^2$$
$$= 2\left(\|w_n - v\|_U^2 + \|w_m - v\|_U^2\right)$$

or

$$\|w_n - w_m\|_U^2 = 2\left(\|w_n - v\|_U^2 + \|w_m - v\|_U^2\right) - 4\left\|v - \frac{w_n + w_m}{2}\right\|_U^2.$$

Note that

$$\|w_n - v\|_U^2 \to \delta^2, \quad \|w_m - v\|_U^2 \to \delta^2,$$

and

$$\left\|v - \frac{w_n + w_m}{2}\right\|_U^2 \geq \delta^2 \text{ for all } m, n$$

(since $(w_n + w_m)/2 \in M$). Thus $\|w_n - w_m\|_U \to 0$ as $m, n \to \infty$; that is, $\{w_n\}$ is Cauchy.

Since $U$ is complete, there exists a vector $w_0 \in U$ such that $w_n \to w_0$, and it follows that $w_0 \in M$ because $M$ is closed. The norm is a continuous function and hence

$$\|v - w_0\|_U = \lim_{n \to \infty} \|v - w_n\|_U = \delta.$$

Thus $w_0$ is an element in $M$ closest to $v$.

Next, we show that $w_0 \in M$ is an element closest to $v$ if and only if $v - w_0$ belongs to $M^\perp$. Every vector in $M$ can be written in the form $w_0 + \alpha z$, where $z \in M$, and, on the other hand, every such vector is in $M$. Now,

$$\|v - (w_0 + \alpha z)\|_U^2 = \|v - w_0\|_U^2 - 2\alpha\langle z, v - w_0\rangle + \alpha^2\|z\|_U^2.$$

This is a quadratic in $\alpha$ that has its minimum at $\alpha = 0$ if and only if $\langle z, v - w_0\rangle = 0$. Since $\alpha = 0$ corresponds to the vector $w_0$, this shows that $w_0$ is the vector closest to $v$ if and only if $\langle z, v - w_0\rangle = 0$ for all $z \in M$.

Finally, we show that $w_0$ is unique. This follows from the Pythagorean theorem: let $w_1 \in M$, and note that $w_0 - w_1 \in M$, so that $v - w_0$ is orthogonal to $w_0 - w_1$. Therefore

$$\|v - w_1\|_U^2 = \|v - w_0 + w_0 - w_1\|_U^2 = \|v - w_0\|_U^2 + \|w_0 - w_1\|_U^2.$$

We see that $\|v - w_1\|_U > \|v - w_0\|_U$ unless $\|w_1 - w_0\|_U = 0$, that is, unless $w_1 = w_0$. This completes the proof. $\qquad\square$

The unique vector $w_0 \in M$ closest to $v$ is called the *projection* of $v$ onto $M$ and is denoted $w_0 = \text{proj}_M v$. It is important to recognize that $M$ must be closed to ensure that $\text{proj}_M v$ exists. For a general subspace $M$ (not assumed to be closed), we can only assume that $w_0 = \text{proj}_{\overline{M}} v$ exists.

The projection theorem leads to an important representation theorem for Hilbert space.

**Definition A.17.** Let $H$ be a Hilbert space, and let $U$, $V$ be subspaces of $H$. We say that $H$ is the *direct sum* of $U$ and $V$ if, given $x \in H$, there exist unique $u \in U$ and $v \in V$ with $x = u + v$. In this case, we write $H = U \oplus V$.

**Theorem A.18.** *Let $H$ be a Hilbert space and $U$ a closed subspace of $H$. Then $H = U \oplus U^\perp$.*

*Proof.* Let $v \in H$, and define

$$u = \text{proj}_U v, \quad w = v - u.$$

Then, by the projection theorem, $w \in U^\perp$, and obviously $v = u + w$. It remains only to show that this representation of $v$ is unique. But this also follows from the projection theorem, since $u \in U$ and $v - u \in U^\perp$ imply that $u = \text{proj}_U v$. The uniqueness of the projection completes the proof. $\qquad\square$

Here is another important consquence of the projection theorem.

**Lemma A.19.** *Let $U$ be a Hilbert space and let $S$ be a subspace of $U$. Then $S^{\perp\perp} = \left(S^\perp\right)^\perp = \overline{S}$.*

*Proof.* Suppose first that $x \in S$. Then $\langle x, u \rangle_U = 0$ for all $u \in S^\perp$, which implies that $x \in (S^\perp)^\perp$. Thus $S \subset (S^\perp)^\perp$; moreover, since $(S^\perp)^\perp$ is closed, it follows that $\overline{S} \subset (S^\perp)^\perp$.

Now suppose that $x \in (S^\perp)^\perp$ and define $\overline{x} = \mathrm{proj}_{\overline{S}} x$. We already know that $\overline{S} \subset (S^\perp)^\perp$ and hence, since $(S^\perp)^\perp$ is a subspace, it follows that $x - \overline{x} \in (S^\perp)^\perp$. On the other hand, by the projection theorem, $x - \overline{x} \in \overline{S}^\perp = S^\perp$. It follows that

$$x - \overline{x} \in S^\perp \cap (S^\perp)^\perp = \{0\}$$

and hence that $x = \overline{x}$. It follows that $x \in \overline{S}$ and we see that $(S^\perp)^\perp \subset \overline{S}$.  $\square$

# A.3    Linear functionals

Many properties of an infinite-dimensional vector space are best studied using the real-valued linear functions defined on the space. In the case of a Hilbert space, there is a simple representation for every such function.

**Definition A.20.** A linear operator $f : U \to \mathbb{R}$, where $U$ is a Hilbert space, is called a *linear functional*. The *dual space* $U^*$ of $U$ is the space of all bounded linear functionals defined on $U$.

The second part of the following result is called the *Riesz representation theorem*.

**Theorem A.21.** *Let $U$ be a Hilbert space.*

1. *If $v \in U$ is fixed and $f_v : U \to \mathbb{R}$ is defined by $f_v(u) = \langle u, v \rangle_U$, then $f \in U^*$.*

2. *For each $f \in U^*$, there exists a unique $v \in U$ such that $f = \langle \cdot, v \rangle_U$, and $\|f\|_{U^*} = \|u\|_U$.*

*Thus $U^*$ may be identified with $U$, and we say that a Hilbert space is its own dual.*

*Proof.*

1. Given $v \in U$, it is clear that $f_v$ is linear, and

$$|\langle u, v \rangle_U| \leq \|v\|_U \|u\|_U \quad \text{for all } u \in U$$

by the Cauchy-Schwarz inequality. Thus $f_v$ is bounded with $\|f_v\|_{U^*} \leq \|v\|_U$, and hence $f_v \in U^*$. Moreover,

$$f_v(v) = \langle v, v \rangle_U = \|v\|_U \|v\|_U.$$

This shows that $\|f_v\|_{U^*} = \|v\|_U$.

2. Now suppose $f \in U^*$. If $f = 0$ (that is, $f(u) = 0$ for all $u$), then we can take $v = 0$ and the result follows. So suppose $f \neq 0$ and let $M \subseteq U$ be defined by

$$M = \{u \in U : f(u) = 0\}.$$

It is straightforward to show that $M$ is closed and hence $U = M \oplus M^\perp$. Since $f \neq 0$, there exists $w \in M^\perp$, $w \neq 0$. We show first that $M^\perp$ is one-dimensional; in fact,

$$M^\perp = \text{sp}\{w\} = \{\beta w : \beta \in \mathbb{R}\}.$$

To see this, let $z \in M^\perp$, and define $\overline{\beta} = f(z)/f(w)$. Then

$$f(z - \overline{\beta}w) = f(z) - \overline{\beta}f(w) = f(z) - \frac{f(z)}{f(w)}f(w) = 0.$$

This shows that $z - \overline{\beta}w \in M$. But we also have $z - \overline{\beta}w \in M^\perp$ since $M^\perp$ is a subspace. The only vector in both $M$ and $M^\perp$ is 0; thus $z - \overline{\beta}w = 0$ and hence $z = \overline{\beta}w \in \text{sp}\{w\}$. It now follows immediately that

$$\|f\|_{U^*} = \frac{|f(w)|}{\|w\|_U}.$$

Now define $v = \alpha w$, where $\alpha = f(w)/\|w\|_U^2$. Note that

$$f(v) = \alpha f(w) = \frac{f(w)}{\|w\|_U^2}f(w) = \frac{f(w)^2}{\|w\|_U^2} = \|v\|_U^2 = \langle v, v \rangle_U.$$

Given any $u \in U$, we can write $u = \beta v + x$, $x \in M$. Because $x \in M$, $f(x) = 0$, and $\langle v, x \rangle_U = 0$ because $x \in M$ and $v \in M^\perp$. It follows that

$$f(u) = \beta f(v) + f(x) = \beta f(v) = \beta \langle v, v \rangle_U = \beta \langle v, v \rangle_U + \langle x, v \rangle_U$$
$$= \langle \beta v + x, v \rangle_U$$
$$= \langle u, v \rangle_U.$$

Thus $f(u) = \langle u, v \rangle_U$ for all $u \in U$. Moreover,

$$\|v\|_U = \frac{|f(w)|}{\|w\|_U^2} \|w\|_U = \frac{|f(w)|}{\|w\|_U} = \|f\|_{U^*}.$$

It remains only to show that $v$ is unique. If $f(u) = \langle u, v \rangle_U = \langle u, z \rangle_U$ for all $u \in U$, then $\langle u, v - z \rangle_U = 0$ for all $u \in U$. In particular,

$$\|v - z\|_U^2 = \langle v - z, v - z \rangle_U = 0,$$

which shows that $v - z = 0$, that is, $v = z$.                                       $\square$

## A.4  The adjoint of a linear operator

The transpose of a matrix plays a fundamental role in finite-dimensional linear algebra. The analogue for a linear operator on Hilbert space is provided by the following theorem.

**Theorem A.22.** *Let $U$ and $V$ be Hilbert spaces, and let $L : U \to V$ be a bounded linear operator. Then there exists a unique bounded linear operator $L^* : V \to V$, called the* adjoint *of $L$, such that*

$$\langle Lu, v \rangle_V = \langle u, L^*v \rangle_U \quad \text{for all } u \in U, v \in V.$$

*Moreover, $\|L^*\| = \|L\|$.*

*Proof.* Let $v \in U$ be given and consider the linear functional $f : U \to \mathbb{R}$ defined by

$$f(u) = \langle Lu, v \rangle_V \quad \text{for all } u \in U.$$

Note that $f$ is bounded since

$$|f(u)| \le \|Lu\|_V \|v\|_V \le \|L\| \|u\|_U \|v\|_V = (\|L\| \|v\|_V) \|u\|_U \quad \text{for all } u \in U$$

(which shows that $\|f\| \le \|L\| \|v\|_V$). Thus $f \in U^*$ and so, by the Riesz representation theorem, there exists a unique $w \in U$ such that

$$f(u) = \langle u, w \rangle_U \quad \text{for all } u \in U,$$

which means that

$$\langle Lu, v \rangle_V = \langle u, w \rangle_U \quad \text{for all } u \in U.$$

Now, for each $v \in V$ we can find a unique such $w$; therefore, we can define an operator $L^* : V \to U$ satisfying

$$\langle Lu, v \rangle_V = \langle u, L^* v \rangle_V \quad \text{for all } u \in U, v \in V.$$

It remains to prove that $L^*$ is linear and $\|L^*\| = \|L\|$.

We have, on the one hand,

$$\langle Lu, \alpha_1 v_1 + \alpha_2 v_2 \rangle_V = \langle u, L^*(\alpha_1 v_1 + \alpha_2 v_2) \rangle_U,$$

and, on the other hand,

$$\begin{aligned}
\langle Lu, \alpha_1 v_1 + \alpha_2 v_2 \rangle_V &= \alpha_1 \langle Lu, v_1 \rangle_V + \alpha_2 \langle Lu, v_2 \rangle_V \\
&= \alpha_1 \langle u, L^* v \rangle_U + \alpha_2 \langle u, L^* v_2 \rangle_U \\
&= \langle u, \alpha_1 L^* v_1 + \alpha_2 L^* v_2 \rangle_U.
\end{aligned}$$

Thus

$$\langle u, L^*(\alpha_1 v_1 + \alpha_2 v_2) \rangle_U = \langle u, \alpha_1 L^* v_1 + \alpha_2 L^* v_2 \rangle_U.$$

Since this holds for all $u \in U$, we see that

$$L^*(\alpha_1 v_1 + \alpha_2 v_2) = \alpha_1 L^* v_1 + \alpha_2 L^* v_2,$$

and therefore $L^*$ is linear.

Finally, to show that $\|L^*\| = \|L\|$, we first note that, by the Cauchy-Schwarz inequality,

$$\|w\|_V = \sup_{\|v\|_V = 1} \langle w, v \rangle \quad \text{for all } w \in V,$$

and similarly for vectors in $U$. Therefore,

$$\|L\| = \sup_{\|u\|_U=1} \|Lu\| = \sup_{\|u\|_U=1} \sup_{\|v\|_V=1} \langle Lu, v \rangle_V$$

and

$$\|L^*\| = \sup_{\|v\|_V=1} \|Lv\| = \sup_{\|v\|_V=1} \sup_{\|u\|_U=1} \langle u, L^*v \rangle_U.$$

By elementary set theory, the order of the supremum operations can be interchanged, and hence

$$\|L\| = \sup_{\|u\|_U=1} \sup_{\|v\|_V=1} \langle Lu, v \rangle_V = \sup_{\|v\|_V=1} \sup_{\|u\|_U=1} \langle Lu, v \rangle_V$$

$$= \sup_{\|v\|_V=1} \sup_{\|u\|_U=1} \langle u, L^*v \rangle_U = \|L^*\|. \qquad \square$$

It is easy to verify from the definition of adjoint that $L^{**} = (L^*)^* = L$.

**Example A.23.**  Suppose $k \in L^2(\Omega)$, $\Omega = (0, 1) \times (0, 1)$, and define

$$K : L^2(0, 1) \to L^2(0, 1),$$

$$(Kx)(s) = \int_0^1 k(s, t)x(t)\,dt.$$

We now compute $K^*$. We have

$$\langle Kx, y \rangle_{L^2(0,1)} = \int_0^1 (Kx)(s)y(s)\,ds = \int_0^1 \left( \int_0^1 k(s, t)x(t)\,dt \right) y(s)\,ds$$

$$= \int_0^1 \int_0^1 k(s, t)x(t)y(s)\,dt\,ds$$

$$= \int_0^1 \left( \int_0^1 k(s, t)y(s)\,ds \right) x(t)\,dt$$

$$= \langle x, K^*y \rangle_{L^2(0,1)},$$

where

$$(K^*y)(s) = \int_0^1 k(t, s)y(t)\,dt.$$

Thus the adjoint of an integral operator with kernel $k(s, t)$ is the integral operator with kernel $k^*(s, t) = k(t, s)$.

Fundamental to understanding inverse problems are the following subspaces associated with a linear operator.

**Definition A.24.** Suppose $U$ and $V$ are Hilbert spaces and $L : U \to V$ is a bounded linear operator. The *range* of $L$ is

$$\mathcal{R}(L) = \{Lu : u \in U\}$$

and the *null space (kernel)* of $L$ is

$$\mathcal{N}(L) = \{u \in U : Lu = 0\}.$$

**Lemma A.25.** *Suppose $U$ and $V$ are Hilbert spaces and $L$ is a bounded linear operator. Then*

1. *$\mathcal{R}(L)$ is a subspace of $V$;*

2. *$\mathcal{N}(L)$ is a closed subspace of $U$.*

*Proof.*

1. If $v_1, v_2 \in \mathcal{R}(L)$, then, by definition, there exist $u_1, u_2 \in U$ such that $Lu_1 = v_1$, $Lu_2 = v_2$. It follows that, for any $\alpha_1, \alpha_2 \in \mathbb{R}$,

$$\alpha_1 v_1 + \alpha_2 v_2 = \alpha_1 Lu_1 + \alpha_2 Lu_2 = L(\alpha_1 u_1 + \alpha_2 u_2)$$

   by the linearity of $L$, and hence $\alpha_1 v_1 + \alpha_2 v_2 \in \mathcal{R}(L)$. This shows that $\mathcal{R}(L)$ is a subspace of $V$ ($\mathcal{R}(L)$ is not empty because $Lu \in \mathcal{R}(L)$ for all $u \in U$ and $U$ is not empty).

2. The set $\mathcal{N}(L)$ is not empty because $L0 = 0$ always holds for a linear operator and hence $0 \in \mathcal{N}(L)$. If $u_1, u_2 \in \mathcal{N}(L)$ and $\alpha_1, \alpha_2 \in \mathbb{R}$, then

$$L(\alpha u_1 + \alpha_2 u_2) = \alpha Lu_1 + \alpha_2 Lu_2 = 0 + 0 = 0,$$

   which shows that $\mathcal{N}(L)$ is a subspace of $V$. If $\{u_n\}$ is a sequence in $\mathcal{N}(L)$ and $u_n \to u \in U$, then

$$Lu = \lim_{n \to \infty} Lu_n = \lim_{n \to \infty} 0 = 0$$

   by the continuity of $L$. Thus $u \in \mathcal{N}(L)$, and $\mathcal{N}(L)$ is closed.  $\square$

It should be noted that there is no reason to expect $\mathcal{R}(L)$ to be closed. Linear inverse problems, the topic of this book, arise precisely when the range of the operator in question fails to be closed.

**Theorem A.26.** *Suppose $U$ and $V$ are Hilbert spaces and $L$ is a bounded linear operator. Then*

1. $\mathcal{R}(L)^{\perp} = \mathcal{N}(L^*)$;

2. $\mathcal{N}(L)^{\perp} = \overline{\mathcal{R}(L^*)}$.

*Proof.*

1. We have

$$v \in \mathcal{N}(L^*) \Leftrightarrow L^*v = 0 \Leftrightarrow \langle u, L^*v \rangle_U = 0 \text{ for all } u \in U$$
$$\Leftrightarrow \langle Lu, v \rangle_V = 0 \text{ for all } u \in U$$
$$\Leftrightarrow v \in \mathcal{R}(L)^{\perp}.$$

   Thus $\mathcal{R}(L)^{\perp} = \mathcal{N}(L^*)$.

2. By the first result, $\mathcal{R}(L^*)^{\perp} = \mathcal{N}(L)$ (since $L^{**} = L$). Thus

$$\mathcal{N}(L)^{\perp} = \mathcal{R}(L^*)^{\perp\perp} = \overline{\mathcal{R}(L^*)}. \qquad \square$$

# A.5 The open mapping theorem and the uniform boundedness principle

We will have need of two fundamental theorems about linear operators on Hilbert space.

**Theorem A.27 (The open mapping theorem).** *Suppose $U$ and $V$ are Hilbert spaces and $L$ is a bounded, surjective linear operator. Then, if $G$ is an open subset of $U$,*

$$L(G) = \{Lu : u \in G\}$$

*is an open subset of $V$.*

In order to prove this theorem, we need the so-called Baire category theorem. We state the following definitions and the theorem in some generality, but for the purpose of this book, it suffices to replace "topological space" and "metric space" by "normed linear space" (a metric space is a set on which has been defined a distance function (*metric*) $d(\cdot, \cdot)$; in a normed linear space the metric is $d(x, y) = \|x - y\|$).

**Definition A.28.** Let $X$ be a topological space and $M$ a subspace of $X$.

1. We say that $M$ is *nowhere dense* in $X$ if $\overline{M}$ does not contain a nonempty open set.

2. We say that $M$ is of the *first category* if $M$ is the union of a countable number of nowhere dense sets. If $M$ is not of the first category, it is said to be of the *second category*.

**Theorem A.29. (Baire category theorem)** *A nonempty complete metric space is of the second category. That is, a nonempty complete metric space cannot be written as the union of a countable collection of nowhere dense sets.*

*Proof.* Suppose $X$ is a nonempty complete metric space and $\{M_n\}$ is a sequence of nowhere dense sets in $X$ with

$$X = \bigcup_{n=1}^{\infty} M_n.$$

We may as well assume that each $M_n$ is closed, since otherwise we could replace $M_n$ by $\overline{M_n}$ (clearly the closure of a nowhere dense set is also nowhere dense). We now derive a contradiction.

Since $X \setminus M_1$ is open, it contains a closed ball $S_1 = \{x \in X : d(x_1, x) \leq r_1\}$. We assume that $0 < r_1 < 1/2$. Similarly, $X \setminus M_2$ is open and its closure is all of $X$ (since $M_2$ is nowhere dense), so we can find another closed ball

$$S_2 = \{x \in X : d(x_2, x) \leq r_2\}$$

contained in $(X \setminus M_2) \cap S_1$, with $0 < r_2 < 1/4$. Arguing by induction, we find a sequence of closed balls $\{S_n\}$, with

$$S_n = \{x \in X : d(x_n, x) \leq r_n\}, \ S_{n+1} \subseteq S_n, \ S_n \cap M_n = \emptyset, \ 0 < r_n < \frac{1}{2^n}.$$

By construction, the sequence $\{x_n\}$ is Cauchy and hence has a limit $x^* \in X$. Moreover, for any $n$, we have

$$d(x_n, x^*) \leq d(x_n, x_m) + d(x_m, x^*) \text{ for all } m.$$

Now, since the balls $\{S_n\}$ are nested, we have $d(x_n, x_m) \leq r_n$ for all $m \geq n$, and $d(x_m, x^*) \to 0$ as $m \to \infty$. Therefore, taking $m \to \infty$, we see that

$$d(x_n, x^*) \leq r_n,$$

which implies that $x^* \in S_n$ for all $n$. It follows that $x^* \notin M_n$ for all $n$, contradicting our assumption that

$$X = \bigcup_{n=1}^{\infty} M_n. \qquad \qquad \square$$

*Proof.* We can now prove Theorem A.27. Suppose $L : U \to V$ is bounded and surjective. We wish to show that if $G \subseteq U$ is open, then so is

$$L(G) = \{Lu : u \in U\}.$$

We use the notation

$$B_\epsilon(u; U) = \{x \in U : \|x - u\| < \epsilon\}$$

and similarly for $B_\delta(v; V)$. Also, for any sets $X, Y$ in a vector space $Z$, we write

$$\alpha X = \{\alpha x : x \in X\}, \ X \pm Y = \{x \pm y : x \in X, y \in Y\}.$$

Note that, for $\alpha > 0$,

$$\alpha B_\epsilon(0; Z) = B_{\alpha \epsilon}(0; Z).$$

We prove the theorem in the following steps.

1. If $\epsilon > 0$ is given, then

$$\overline{L\left(B_\epsilon(0; U)\right)}$$

contains an open set. To see this, note that the fact that $L$ is surjective implies that

$$V = \bigcup_{n=1}^{\infty} nL(B_\epsilon(0; U)) = \bigcup_{n=1}^{\infty} n\overline{L(B_\epsilon(0; U))}.$$

If $\overline{L\left(B_\epsilon(0; U)\right)}$ does not contain an open set, then neither does any $n\overline{L\left(B_\epsilon(0; U)\right)}$, and we have written $V$ as a countable union of nowhere dense sets, which is impossible since $V$ is a nonempty complete metric space.

2. If $\epsilon > 0$ is given, then there exists $\delta > 0$ such that

$$B_\delta(0; V) \subseteq \overline{L\left(B_\epsilon(0; U)\right)}.$$

To see this, note that

$$B_{\epsilon/2}(0; U) - B_{\epsilon/2}(0; U) \subseteq B_\epsilon(0; U),$$

so, if $S$ is the open subset of $\overline{L\left(B_{\epsilon/2}(0; U)\right)}$ whose existence is guaranteed by step 1, we have

$$S - S \subseteq \overline{L\left(B_{\epsilon/2}(0; U)\right)} - \overline{L\left(B_{\epsilon/2}(0; U)\right)} \subseteq \overline{L(B_\epsilon(0; U))}.$$

Now, $0 \in S - S$ and

$$S - S = \bigcup\{S - x : x \in S\},$$

which shows that $S - S$ is open (the union of open sets—each $S - x$ is open, being the translate of an open set). Therefore,

$$0 \in S - S \subseteq \overline{L\left(B_\epsilon(0; U)\right)},$$

and the assertion follows.

3. Let $\epsilon_0 > 0$ be given, and let $\delta_0 > 0$ correspond to $\epsilon_0$ as in step 2. Then

$$B_{\delta_0}(0; V) \subseteq L\left(B_{2\epsilon_0}(0; U)\right).$$

To prove this, let $y$ be any element of $B_{\delta_0}(0; V)$. Choose a sequence $\{\epsilon_i\}$ of positive numbers with

$$\sum_{i=1}^{\infty} \epsilon_i < \epsilon_0.$$

Choose a sequence $\{\delta_i\}$ with $\delta_i \to 0$ and

$$B_{\delta_i}(0; V) \subseteq \overline{L(B_{\epsilon_i}(0; U))}.$$

Since $L(B_\epsilon(0; U))$ is dense in $\overline{L(B_\epsilon(0; U))}$, there exists $x_0 \in B_{\epsilon_0}(0; U)$ such that

$$\|y - Lx_0\| < \delta_1.$$

It follows that $y - Lx_0 \in B_{\delta_1}(0; V)$, so there exists $x_1 \in B_{\epsilon_1}(0; U)$ such that

$$\|y - Lx_0 - Lx_1\| < \delta_2.$$

By induction, we can choose a sequence $\{x_i\}$ such that

$$x_i \in B_{\epsilon_i}(0; U) \text{ and } \left\| y - \sum_{j=0}^{i} Lx_j \right\| < \delta_{i+1}.$$

Now, the sequence $\{x_i\}$ is absolutely summable, so

$$x = \sum_{i=0}^{\infty} x_i$$

exists and satisfies $\|x\| < 2\epsilon_0$. Moreover,

$$\|y - Lx\| = \left\| y - \sum_{j=0}^{\infty} Lx_j \right\| = \lim_{i \to \infty} \left\| y - \sum_{j=0}^{i} Lx_j \right\| = 0,$$

that is, $y = Lx$. This proves that $y \in L\left(B_{2\epsilon_0}(0; U)\right)$.

4. Now we prove that $L(G)$ is open whenever $G \subseteq U$ is open. For any $x \in G$, there exists $\epsilon > 0$ such that $x + B_\epsilon(0; U) \subseteq G$. By step 3, there exists $\delta > 0$ such that

$$B_\delta(0; V) \subseteq L(B_\epsilon(0; U)).$$

It follows that

$$L(x + B_\epsilon(0; U)) = Lx + L(B_\epsilon(0; U)) \subseteq L(G).$$

But then

$$Lx + B_\delta(0; V) \subseteq L(G).$$

Since this holds for every $x \in G$ (possibly with a different $\delta$ for each $x$), it follows that $L(G)$ is open. $\qquad\square$

The following corollary of the open mapping theorem is also very important.

**Corollary A.30 (Bounded inverse theorem).** *Let $U$ and $V$ be Hilbert spaces and let $L : U \to V$ be a bijective bounded linear operator. Then $L^{-1}$ is bounded.*

*Proof.* Since $L$ is a bijection, $L^{-1} : V \to U$ is well-defined. To prove that $L^{-1}$ is bounded, it suffices to prove that it is continuous, which is equivalent to the condition that the inverse image under $L^{-1}$ of every open set in $U$ is open in $V$. But this is equivalent to the condition that the image under $L$ of every open set in $U$ is open in $V$, which is true by the open mapping theorem. $\qquad\square$

Another important consequence of the open mapping theorem is the closed graph theorem, which is proved in Chapter 2 along with other properties of closed operators.

The following theorem states that for any collection of bounded linear operators defined on a Hilbert space, pointwise boundedness is equivalent to uniform boundedness.

**Theorem A.31 (Uniform boundedness principle).** *Let $U$ and $V$ be Hilbert spaces, let $A$ be any set, and, for each $a \in A$, let $T_a : U \to V$ be* *a bounded linear operator. If, for each $x \in U$,*

$$\sup\{\|T_a x\|_V \ : \ a \in A\} < \infty,$$

*then*

$$\sup\{\|T_a\| \ : \ a \in A\} < \infty.$$

*Proof.* The proof is based on the Baire category theorem. Let us define, for each $n \in \mathbb{Z}^+$,

$$U_n = \{x \in U \ : \ \|T_a x\|_V \le n \text{ for all } a \in A\}.$$

Suppose $\{x_k\} \subset U_n$ and $x_k \to x \in U$. Then, for each $a \in A$, we have

$$T_a x = \lim_{k \to \infty} T_a x_k \ \Rightarrow \ \|T_a x\|_V = \lim_{k \to \infty} \|T_a x_k\|_V \le n.$$

This shows that $x \in U_n$ and hence that $U_n$ is closed. By assumption, every $x \in U$ belongs to some $U_n$, and therefore

$$U = \cup_{n=1}^{\infty} U_n.$$

Since $U$ is complete, the Baire category theorem implies that there exists $m \in \mathbb{Z}^+$ and a closed ball $\overline{B_\epsilon(x_0)}$ that is contained in $U_m$.

Now let $a$ be any element of $A$. For any $x \in U$,

$$x_0 + \frac{\epsilon}{\|x\|_U} x \in \overline{B_\epsilon(x_0)} \subset U_m,$$

which implies that

$$\left\| T_a \left( x_0 + \frac{\epsilon}{\|x\|_U} x \right) \right\|_V \le m$$

$$\Rightarrow \frac{\epsilon}{\|x\|_U} \|T_a x\|_V - \|T_a x_0\|_V \le m \quad \text{(by the reverse triangle inequality)}$$

$$\Rightarrow \frac{\epsilon}{\|x\|_U} \|T_a x\|_V \le 2m \quad \text{(since } \|T_a x_0\|_V \le m)$$

$$\Rightarrow \|T_a x\|_V \le \frac{2m}{\epsilon} \|x\|_U.$$

This holds for all $x \in U$, and therefore $\|T_a\| \le 2m/\epsilon$. Moreover, this conclusion holds for all $a \in A$, and hence

$$\sup\{\|T_a\| \ : \ a \in A\} \le \frac{2m}{\epsilon}. \qquad \square$$

# A.6 Weak sequential convergence in Hilbert space

One of the major ways in which infinite-dimensional spaces differ from their finite-dimensional counterparts is that an infinite-dimensional space admits more than one topology. Every Hilbert space is equipped with the norm topology; however, this is a strong topology in the sense that compactness is rare in infinite-dimensional normed spaces. Indeed, one of the standard theorems of linear analysis is that a normed vector space is finite-dimensional if and only if its closed unit ball is compact. This is particularly easy to prove if the space is a Hilbert space.

**Theorem A.32.** *Let $U$ be a Hilbert space. Then $U$ is finite-dimensional if and only if $\overline{B_1(0)} = \{u \in U \ : \ \|u\|_U \leq 1\}$ is compact.*

*Proof.* If $U$ is finite-dimensional, then $\overline{B_1(0)}$ is a closed and bounded subset of a finite-dimensional space and hence is compact by the Heine-Borel theorem of analysis. Conversely, suppose $U$ is not finite-dimensional. We will show that $\overline{B_1(0)}$ cannot be compact. Construct an orthonormal sequence $\{u_n\}$ in $\overline{B_1(0)}$ as follows. Choose any $u_1 \in \overline{B_1(0)}$ such that $\|u_1\|_U = 1$. Given an orthonormal sequence $\{u_1, u_2, \ldots, u_{n-1}\}$ in $\overline{B_1(0)}$, $\mathrm{sp}\{u_1, u_2, \ldots, u_{n-1}\}^\perp$ is a proper closed subspace of $U$ (if it were not a proper subspace of $U$, then $U$ would be finite-dimensional). Therefore, we can choose $u_n \in \mathrm{sp}\{u_1, u_2, \ldots, u_{n-1}\}^\perp$ such that $\|u_n\|_U = 1$. In this way, we can define inductively an infinite orthonormal sequence in $\mathrm{sp}\{u_1, u_2, \ldots, u_{n-1}\}^\perp$. But such a sequence cannot have a convergent subsequence. A convergent sequence would have to be Cauchy, but we have

$$
\begin{aligned}
\|u_m - u_n\|_U^2 &= \langle u_m - u_n, u_m - u_n \rangle_U \\
&= \langle u_m, u_m \rangle_U - 2\langle u_m, u_n \rangle_U + \langle u_n, u_n \rangle_U \\
&= \|u_m\|_U^2 + \|u_n\|_U^2 \text{ (since } u_m, u_n \text{ are orthogonal)} \\
&= 2 \text{ (since } \|u_m\|_U = \|u_n\|_U = 1),
\end{aligned}
$$

which implies that $\|u_m - u_n\|_U = \sqrt{2}$ for all $m, n$. Therefore, no subsequence of $\{u_n\}$ can be Cauchy. $\qquad\square$

Since compactness is extremely useful in analysis, we often make use of another canonical topology in which compactness is more common.

**Definition A.33.** Let $U$ be a Hilbert space. The weak topology on $U$ is the topology generated by the collection of subsets

$$\left\{ f^{-1}(G) : G \subset \mathbb{R} \text{ is open, } f \in U^* \right\}.$$

Recall that the topology on $U$ generated by a collection of subsets (whose union is $U$) is the collection of all unions of finite intersections of the subsets.

Another characterization of the weak topology is that it is the weakest topology (that is, the topology with the fewest open sets) under which each $f \in U^*$ is continuous. This follows immediately from the definition. From this, we conclude that the norm topology is stronger than the weak topology. Thus, if $u_n \to u$ in norm, then $u_n \to u$ weakly (that is, in the weak topology). This result also follows from the next theorem.

**Theorem A.34.** *Let $U$ be a Hilbert space and let $\{u_n\}$ be a sequence in $U$. Then $\{u_n\}$ converges weakly to $u \in U$ if and only if, for each $f \in U^*$, $f(u_n) \to f(u)$ (that is, if and only if, for each $v \in U$, $\langle u_n, v \rangle_U \to \langle u, v \rangle_U$).*

*Proof.* If $u_n \to u$ weakly, then, since each $f \in U^*$ is continuous with respect to the weak topology, we have $f(u_n) \to f(u)$ for all $f \in U^*$.

On the other hand, suppose $f(u_n) \to f(u)$ for all $f \in U^*$. If $\{u_n\}$ does not converge to $u$ weakly, then there exists a weakly open neighborhood $N$ of $u$ and a subsequence $\{u_{n_k}\}$ with $u_{n_k} \notin N$ for all $k$. Without loss of generality, we may assume that

$$N = \bigcap_{i=1}^{n} f_i^{-1}(W_i),$$

where, for each $i = 1, 2, \ldots, n$, $f_i \in U^*$ and $W_i \subset \mathbb{R}$ is open. From $u_{n_k} \notin N$ for all $k$, it follows that, for each $k$, there exists $i_k$ such that $u_{n_k} \notin f_{i_k}^{-1}(W_{i_k})$. Since there are only finitely many $f_i(W_i)$, there must exist $j$ such that $u_{n_k} \notin f_j^{-1}(W_j)$ for infinitely many values of $k$. But

$f(u_{n_k}) \to f(u)$, which implies that, for every $i$, $f(u_{n_k}) \in W_i$ (a neighborhood of $f(u)$ in $\mathbb{R}$) for all $k$ sufficiently large. This contradiction proves the result. $\square$

The weak topology on an infinite-dimensional Hilbert space $U$ is not metrizable, that is, we cannot define a metric (distance function) $d$ on $U$ so that the weakly open sets are generated by the open balls $B_\epsilon(u) = \{v \in U : d(u, v) < \epsilon\}$ defined by the metric. This implies that the weak topology is not completely characterized by the behavior of sequences. For instance, if $f$ is a function mapping a metric space $X$ into another metric space, then we have the theorem that $f$ is continuous if and only if $x \in X$, $\{x_n\} \subset X$, $x_n \to x$ imply that $f(x_n) \to f(x)$. This theorem does not hold if $f$ maps a general topological space (for example, a Hilbert space under the weak topology) into another.

To deal with general topological spaces, such as a Hilbert space under the weak topology, the concept of sequence can be replaced by the more general concept of *net*.[9] However, we will not need this generality, since the results about weak convergence that we use are restricted to results about weak sequential convergence.

For example, a subspace of a Hilbert space that is closed under the norm topology is also closed under the weak topology. However, we actually use only the following weaker result.

**Theorem A.35.** *Let $U$ be a Hilbert space and let $S$ be a subspace of $U$ that is closed with respect to the norm topology on $U$. Then $S$ is also closed with respect to weak sequential convergence; that is, if $\{u_n\} \subset S$ and $u_n \to u$ weakly, then $u \in S$.*

*Proof.* Suppose $\{x_k\} \subset S$ and $x_k \to x \in X$ weakly. Define $\bar{x} = \text{proj}_S x$. We will show that $x_k \to \bar{x}$ weakly; since the weak limit of a sequence is unique, this will show that $x = \bar{x}$ and hence that $x \in S$. Since $\bar{x} = \text{proj}_S x$, we have $\langle \bar{x} - x, v \rangle_X = 0$ for all $v \in S$ and hence $\langle \bar{x}, v \rangle_X = \langle x, v \rangle_X$ for all

---

[9] A sequence in a set $X$ is a function mapping $\mathbb{Z}^+$ into $X$ (for which we use the special notation $x_k$ instead of the standard functional notation $x(k)$ to denote the images). A net in $X$ is a function mapping a directed set $A$ into $X$; we normally denote a net by $\{x_\alpha : \alpha \in A\}$. We will not define directed set here since we do not use the concept of a net.

$v \in S$. Given any $u \in X$, we can write $u = \bar{u} + z$, where $\bar{u} \in S$ and $z \in S^{\perp}$. Therefore,

$$
\begin{aligned}
\langle x_k - \bar{x}, u \rangle_X &= \langle x_k - \bar{x}, \bar{u} \rangle_X + \langle x_k - \bar{x}, z \rangle_X \\
&= \langle x_k - \bar{x}, \bar{u} \rangle_X \text{ (since } x_k - \bar{x} \in S, z \in S^{\perp}) \\
&= \langle x_k - x, \bar{u} \rangle_X \text{ (since } \langle \bar{x}, \bar{u} \rangle_X = \langle x, \bar{u} \rangle_X) \\
&\to 0 \text{ as } k \to \infty. \qquad \square
\end{aligned}
$$

We also have the result that if $U$ and $V$ are Hilbert spaces and a linear operator $L : U \to V$ is continuous with respect to the norm topologies on $U$ and $V$, then it is also continuous with respect to the weak topologies. However, we will use only the following weaker result.

**Theorem A.36.** *Suppose $U$ and $V$ are Hilbert spaces and $L : U \to V$ is a bounded linear operator. If $\{u_n\}$ is a sequence in $U$ with $u_n \to u \in U$ weakly, then $Lu_n \to Lu$ weakly in $V$.*

*Proof.* We have

$$
\begin{aligned}
u_n \to u \text{ weakly} &\Rightarrow \langle u_n, x \rangle_U \to \langle u, x \rangle_U \text{ for all } x \in U \\
&\Rightarrow \langle u_n, L^*y \rangle_U \to \langle u, L^*y \rangle_U \text{ for all } y \in V \\
&\Rightarrow \langle Lu_n, y \rangle_V \to \langle Lu, y \rangle_V \text{ for all } y \in V \\
&\Rightarrow Lu_n \to Lu \text{ weakly.} \qquad \square
\end{aligned}
$$

The following theorem is perhaps the most important fact about the weak topology: a bounded sequence has a weakly convergent subsequence. This is true in a general Hilbert space, but we shall prove it only for separable Hilbert spaces. Recall that a metric space is *separable* if it contains a countable dense subset.

**Theorem A.37.** *Let $U$ be a separable Hilbert space and suppose $\{u_n\}$ is a bounded sequence in $U$. Then $\{u_n\}$ has a weakly convergent subsequence.*

*Proof.* Suppose $\{u_n\}$ satisfies $\|u_n\| \leq M$ for all $n$. Since $U$ is separable, there exists a dense sequence $\{x_n\}$ in $U$. Now, $\{\langle u_n, x_1 \rangle_U\}$ is a bounded sequence of real numbers, so it has a convergence subsequence. Hence there exists a subsequence $\{u_{n_k^{(1)}}\}$ of $\{u_n\}$ such that

$$\{\langle u_{n_k^{(1)}}, x_1\rangle_U\}$$

is convergent. Since $\{u_{n_k^{(1)}}\}$ is bounded, there exists a subsequence $\{u_{n_k^{(2)}}\}$ of $\{u_{n_k^{(1)}}\}$ such that

$$\{\langle u_{n_k^{(2)}}, x_2\rangle_U\}$$

is convergent. We can continue by induction to obtain subsequences

$$\{u_{n_k^{(\ell)}}\}, \ell = 1, 2, 3, \ldots$$

such that

1. $\{u_{n_k^{(\ell+1)}}\}$ is a subsequence of $\{u_{n_k^{(\ell)}}\}$;

2. $\{\langle u_{n_k^{(\ell)}}, x_\ell\rangle_U\}$ is convergent.

Consider the diagonal subsequence $\{u_{n_k^{(k)}}\}$. We see that

$$\{\langle u_{n_k^{(k)}}, x_j\rangle_U\}$$

is convergent for each $j$, since $\{u_{n_k^{(k)}}\}$ is a subsequence of $\{u_{n_k^{(j)}}\}$ for every $j$ (perhaps after discarding a finite number of terms at the beginning).

Now take any $x \in U$. We wish to show that

$$\{\langle u_{n_k^{(k)}}, x\rangle_U\}$$

is convergent, which we can do by showing that it is Cauchy. Let $\epsilon > 0$ be given. Since $\{x_n\}$ is dense in $U$, there exists $j$ with $\|x - x_j\| < \epsilon/(3M)$. Now,

$$|\langle u_{n_k^{(k)}}, x\rangle_U - \langle u_{n_\ell^{(\ell)}}, x\rangle_U|$$

$$\leq |(\langle u_{n_k^{(k)}}, x\rangle_U - \langle u_{n_k^{(k)}}, x_j\rangle_U| + |(\langle u_{n_k^{(k)}}, x_j\rangle_U - \langle u_{n_\ell^{(\ell)}}, x_j\rangle_U|$$

$$+ |(\langle u_{n_\ell^{(\ell)}}, x_j\rangle_U - \langle u_{n_\ell^{(\ell)}}, x\rangle_U|$$

$$\leq \|u_{n_k^{(k)}}\|\|x - x_j\| + \|u_{n_\ell^{(\ell)}}\|\|x - x_j\| + |\langle u_{n_k^{(k)}}, x_j\rangle_U - \langle u_{n_\ell^{(\ell)}}, x_j\rangle_U|$$

$$< \frac{\epsilon}{3} + \frac{\epsilon}{3} + |\langle u_{n_k^{(k)}}, x_j\rangle_U - \langle u_{n_\ell^{(\ell)}}, x_j\rangle_U|.$$

Since $\{\langle u_{n_k^{(k)}}, x_j \rangle_U\}$ is convergent, it is Cauchy, and so there exists $K$ sufficiently large that, for all $k, \ell \geq K$,

$$|\langle u_{n_k^{(k)}}, x_j \rangle_U - \langle u_{n_\ell^{(\ell)}}, x_j \rangle_U| < \frac{\epsilon}{3}.$$

We then see that, for all $k, \ell \geq K$,

$$|\langle u_{n_k^{(k)}}, x \rangle_U - \langle u_{n_\ell^{(\ell)}}, x \rangle_U| < \epsilon,$$

and so $\{\langle u_{n_k^{(k)}}, x \rangle_U\}$ is Cauchy.

Thus we can define a functional $f : H \to \mathbb{R}$ by

$$f(x) = \lim_{k \to \infty} \langle u_{n_k^{(k)}}, x \rangle_U.$$

Linearity of $f$ is obvious, and we also have $|f(x)| \leq M\|x\|$ for all $x \in U$. Therefore, $f \in U^*$, and by the Riesz representation theorem, there exists a unique $u \in U$ with $f(x) = \langle u, x \rangle_U$ for all $x \in U$. It follows that

$$u_{n_k^{(k)}} \to u \text{ weakly.} \qquad \qquad \square$$

One more fact we need about weak convergence is the lower semicontinuity of the norm with respect to the weak topology. We will describe this property with reference to weak sequential convergence only.

**Theorem A.38.** *Let $U$ be a Hilbert space, let $\{u_n\}$ be a sequence in $U$, and suppose that $\{u_n\}$ converges weakly to $u \in U$. Then*

$$\|u\|_U \leq \liminf_{n \to \infty} \|u_n\|_U.$$

*Proof.* There is nothing to prove if $u = 0$; therefore, we assume that $u \neq 0$. Since $u_n \to u$ weakly, it follows that

$$\lim_{n \to \infty} \langle u_n, u \rangle_U = \langle u, u \rangle_U = \|u\|_U^2.$$

Also, the Cauchy-Schwarz inequality implies that

$$\langle u_n, u \rangle_U \leq \|u_n\|_U \|u\|_U \text{ for all } n.$$

We thus have

$$\liminf_{n \to \infty} \|u_n\|_U \geq \liminf_{n \to \infty} \frac{\langle u_n, u \rangle_U}{\|u\|_U} = \lim_{n \to \infty} \frac{\langle u_n, u \rangle_U}{\|u\|_U} = \frac{\|u\|_U^2}{\|u\|_U} = \|u\|_U. \qquad \square$$

# Appendix B

# Sobolev spaces

In this appendix, we describe some specific Hilbert spaces, namely, the Sobolev spaces defined on an interval, that are used in some of the examples presented in the book. We will not present proofs of the results (such as completeness of the spaces). The reader can consult [4] for basic facts about Sobolev spaces, and [1] for more detailed information.

## B.1 Sobolev spaces on an interval

For simplicity, we use the interval $(0, 1)$ in these examples, although nothing significant would change if we were to allow a general interval $(a, b)$. Recall that $L^2(0, 1)$ is the space of all measurable functions $x$ defined on $(0, 1)$ that are square-integrable:

$$\int_0^1 x^2 < \infty.$$

Here we have the understanding that two functions that differ only on a set of Lebesgue measure zero represent the same element of $L^2(0, 1)$. (So, to be more precise, the elements of $L^2(0, 1)$ are not measureable functions, but rather equivalence classes of measureable functions.) The inner product on $L^2(0, 1)$ is defined by

$$\langle x, y \rangle_{L^2} = \int_0^1 xy \text{ for all } x, y \in L^2(0, 1).$$

It can be shown that $L^2(0, 1)$ is a Hilbert space under the given inner product.

We wish to use derivatives of functions in $L^2(0, 1)$. However, a measurable function need not be continuous, much less differentiable, in order to be integrable (or square-integrable). In fact, given $x \in L^2(0, 1)$, the values $x(t)$ for $t \in (0, 1)$ are not well-defined, because $x = y$ allows $x(t)$ and $y(t)$ to differ on a set of measure zero. In this sense, we cannot regard $x$ as having a definite value at any particular point in $(0, 1)$.

In spite of the lack of regularity of some measurable functions, we can define a notion of derivative that applies to many of them. The weak derivative of a measurable function is defined with reference to integration by parts. For smooth functions, we have

$$\int_0^1 x'y = xy\big|_0^1 - \int_0^1 xy' = x(1)y(1) - x(0)y(0) - \int_0^1 xy'.$$

The left side of this equation involves $x'$, while the right side involves only $x$. The right side does involve $x(0)$ and $x(1)$, which (as explained above) is problematic if $x$ is only assumed to lie in $L^2(0, 1)$; however, we can eliminate this problem by only considering functions $y$ that satisfy $y(0) = y(1) = 0$. For such functions $y$, we have

$$\int_0^1 x'y = -\int_0^1 xy'.$$

We want to require as little traditional smoothness on $x$ as possible; to compensate, we require the test functions $y$ to be as smooth as possible. Thus we define the space $C_0^\infty(0, 1)$ of test functions to consist of all $y : [0, 1] \to \mathbb{R}$ such that all derivatives of $y$ exist (that is, $y$ is infinitely differentiable) and $\operatorname{supp}(y)$ is a compact subset of $(0, 1)$. Here $\operatorname{supp}(y)$ is the closure of the set on which $y$ is nonzero:

$$\operatorname{supp}(y) = \overline{\{t \in [0, 1] \,:\, y(t) \neq 0\}}.$$

The condition that $\operatorname{supp}(y) \subset (0, 1)$ implies that $y(0) = y(1) = 0$ (in fact, that $y(t) = 0$ for all $t$ sufficiently close to either 0 or 1).

If $x$ is integrable on $(0, 1)$ and there exists a measurable function $u$ on $(0, 1)$ such that

$$\int_0^1 uy = -\int_0^1 xy' \text{ for all } y \in C_0^\infty(0, 1),$$

then we say that $x$ is weakly differentiable and call $u$ the *weak derivative* of $x$. (If $u$ satisfies this condition, then it is not difficult to show that it must be integrable—that is, have a finite integral—on every compact subset of $(0, 1)$. Such a function is called *locally integrable*.) If $x$ is continuously differentiable (in the classical sense), then $x$ is obviously weakly differentiable and it can be shown that $x' = u$, where $u$ is the weak derivative of $x$. Henceforth we will use the same notation $x'$ for the weak and classical derivative, which should cause no confusion because the two concepts agree when $x$ has a derivative in the classical sense.

Here are two examples that illustrate the concept of weak derivative.

**Example B.1.** Let $x : [0, 1] \to \mathbb{R}$ be defined by

$$x(t) = \begin{cases} t, & 0 \le t \le \frac{1}{2}, \\ 1 - t, & \frac{1}{2} \le t \le 1. \end{cases}$$

We will show that $x$ is weakly differentiable and that the weak derivative of $x$ is $x' = u$, where

$$u(t) = \begin{cases} 1, & 0 \le t < \frac{1}{2}, \\ -1, & \frac{1}{2} < t \le 1. \end{cases}$$

We have, for any $y \in C_0^\infty(0, 1)$,

$$\int_0^1 x(t)y'(t)\,dt = \int_0^{1/2} ty'(t)\,dt + \int_{1/2}^1 (1 - t)y'(t)\,dt$$

$$= ty(t)\big|_0^{1/2} - \int_0^{1/2} y(t)\,dt + (1 - t)y(t)\big|_{1/2}^1 + \int_{1/2}^1 y(t)\,dt$$

$$= \frac{y(1/2)}{2} - \int_0^{1/2} y(t)\,dt - \frac{y(1/2)}{2} + \int_{1/2}^1 y(t)\,dt$$

$$= -\left( \int_0^{1/2} 1 \cdot y(t)\,dt + \int_{1/2}^1 (-1)y(t)\,dt \right)$$

$$= -\int_0^1 u(t)y(t)\,dt.$$

This gives the desired result.

In the previous example, $x$ is differentiable in the classical sense except at $t = 1/2$, and the weak derivative equals the classical derivative at every other point. In the next example, we have another function that is differentiable (in the classical sense) except at the point $t = 1/2$, and yet the function is not weakly differentiable.

**Example B.2.** Let $x : [0, 1] \to \mathbb{R}$ be defined by

$$x(t) = \begin{cases} 0, & 0 \le t < \frac{1}{2}, \\ 1, & \frac{1}{2} < t \le 1, \end{cases}$$

which is differentiable in the classical sense except at $t = 1/2$. The derivative is zero at every other point. We have, for any $y \in C_0^\infty(0, 1)$,

$$\int_0^1 x(t) y'(t)\, dt = \int_{1/2}^1 y'(t)\, dt$$
$$= y(t)|_{1/2}^1$$
$$= y(1) - y(1/2) = -y(1/2).$$

If $x$ were weakly differentiable with $x' = u$, then $u$ would have to satisfy

$$\int_0^1 u(t) y(t)\, dt = y(1/2) \text{ for all } y \in C_0^\infty(0, 1). \tag{B.1}$$

It is easy to see that no such function $u$ can exist. For example, given any interval $[a, b] \subset (0, 1/2)$ or $[a, b] \subset (1/2, 1)$, we can find an element $y$ of $C_0^\infty(0, 1)$ such that $\operatorname{supp}(y) = [a, b]$ and $y > 0$ on $(a, b)$. It follows that $u$, if it existed, would have to satisfy $u(t) = 0$ for almost every $u \ne 1/2$, which implies that $u = 0$ a.e. But then

$$\int_0^1 u(t) y(t)\, dt = 0 \text{ for all } C_0^\infty(0, 1),$$

contradicting (B.1). Thus $x$ is not weakly differentiable.

In general, if $x$ is continuous and piecewise smooth (that is, $x'$ is piecewise continuous in the classical sense), then $x$ is weakly differentiable and $x'$ is the classical derivative.

We now define subspaces of $L^2(0, 1)$ that are useful in many examples:

$H^1(0, 1) = \{x \in L^2(0, 1) : x$ is weakly differentiable and $x' \in L^2(0, 1)\}$,

$H^2(0, 1) = \{x \in L^2(0, 1) : x, x'$ are weakly differentiable and

$$x', x'' \in L^2(0, 1)\}.$$

They are dense subspaces of $L^2(0, 1)$ and are Hilbert spaces themselves under the stronger norms

$$\langle x, y \rangle_{H^1} = \int_0^1 (xy + x'y'),$$

$$\langle x, y \rangle_{H^2} = \int_0^1 (xy + x'y' + x''y'').$$

It should be obvious how to extend these definitions to $H^k(0, 1)$ for integers $k \geq 3$; however, we will not use larger values of $k$. The spaces $H^k(0, 1)$ are examples of *Sobolev spaces*. It is natural to regard $L^2(0, 1)$ as equal to $H^k(0, 1)$ for $k = 0$.

It can be shown that $C_0^\infty(0, 1)$ is dense in $L^2(0, 1)$, and hence so are larger subspaces such as $C^2[0, 1]$, $C^1[0, 1]$, and $C[0, 1]$. On the other hand, $C_0^\infty(0, 1)$ is not dense in $H^1(0, 1)$ or $H^2(0, 1)$, but $C^\infty[0, 1]$ is dense in both. Also, $C^1[0, 1]$ is dense in $H^1(0, 1)$ and $C^2[0, 1]$ is dense in $H^2(0, 1)$.

We explained that $x \in L^2(0, 1)$ cannot be regarded as having well-defined values $x(t)$. However, the same is not true of $x \in H^1(0, 1)$. The *trace theorem* implies that, for each $t_0 \in [0, 1]$, the evaluation map

$$e_{t_0} : H^1(0, 1) \to \mathbb{R},$$

$$e_{t_0}(x) = x(t_0) \text{ for all } x \in H^1(0, 1)$$

is a bounded linear functional on $H^1(0, 1)$. The linearity of these maps is easy to verify. Here we will verify that $e_0$ is bounded.

**Lemma B.3.** *The evaluation map* $e_0 : C^1[0, 1] \to \mathbb{R}$ *defined by* $e_0(x) = x(0)$ *for all* $x \in C^1[0, 1]$ *extends uniquely to a bounded linear functional defined on all of* $H^1(0, 1)$.

*Proof.* Since $C^1[0, 1]$ is dense in $H^1(0, 1)$, it suffices to prove that $e_0$ is bounded on $C^1[0, 1]$. We argue by contradiction and assume that there exists $\{x_n\} \subset C^1[0, 1]$ such that $\|x_n\|_{H^1(0,1)} = 1$ for all $n$ and $|x_n(0)| \to \infty$ as $n \to \infty$. Let us define $\{u_n\} \subset C^1[0, 1]$ by $u_n = x_n^2$; then $u_n' = 2x_n x_n'$ and we have

$$|u_n(t) - u_n(0)| \le \left| \int_0^t u_n' \right| = 2 \left| \int_0^t x_n x_n' \right| \le 2 \left[ \int_0^t x_n^2 \right]^{1/2} \left[ \int_0^t (x_n')^2 \right]^{1/2}$$

$$\le 2 \left[ \int_0^1 x_n^2 \right]^{1/2} \left[ \int_0^1 (x_n')^2 \right]^{1/2}$$

$$= 2\|x_n\|_{L^2} \|x_n'\|_{L^2} \le 2$$

(since obviously $\|x_n\|_{L^2}, \|x_n'\|_{L^2} \le \|x_n\|_{H^1} = 1$). By assumption $u_n(0) \to \infty$ and hence $u_n(0) \ge 4$ for all $n$ sufficiently large. But then, for all $t \in [0, 1]$ and all $n$ sufficiently large,

$$|u_n(t) - u_n(0)| \le 2 \implies u_n(t) \ge 2,$$

and this implies that

$$\int_0^1 u_n \ge 2.$$

This contradicts $\int_0^1 u_n = \|x_n\|_{L^2}^2 \le 1$ for all $n$, which proves that $e_0$ is bounded. $\qquad\square$

We can now prove that $e_t$ is bounded for all $t \in [0, 1]$.

**Lemma B.4.** *For each $t \in [0, 1]$, the evaluation map $e_t : C^1[0, 1] \to \mathbb{R}$ defined by $e_t(x) = x(t)$ for all $x \in C^1[0, 1]$ extends to a bounded linear functional defined on all of $H^1(0, 1)$.*

*Proof.* Let $C > 0$ satisfy $|x(0)| \le C\|x\|_{H^1}$ for all $x \in C^1[0, 1]$. For all $x \in C^1[0, 1]$ and all $t \in [0, 1]$, we have

$$x(t) = x(0) + \int_0^t x'.$$

Therefore,

$$|x(t)| \leq |x(0)| + \left| \int_0^t x' \right| \leq C\|x\|_{H^1} + \left[ \int_0^t (x')^2 \right]^{1/2} \left[ \int_0^t 1^2 \right]^{1/2}$$

$$\leq C\|x\|_{H^1} + \left[ \int_0^1 (x')^2 \right]^{1/2} \left[ \int_0^1 1^2 \right]^{1/2}$$

$$\leq C\|x\|_{H^1} + \|x'\|_{L^2}$$

$$\leq (C+1)\|x\|_{H^1},$$

since $\|x'\|_{L^2} \leq \|x\|_{H^1}$. $\qquad\qquad\square$

We can now refer to the values $x(t)$ for $x \in H^1(0, 1)$. In particular, we define

$$H_0^1(0, 1) = \left\{ u \in H^1(0, 1) : u(0) = u(1) = 0 \right\}.$$

It can be shown that $H_0^1(0, 1)$ is a closed subspace of $H^1(0, 1)$. It is useful for expressing the solutions of certain differential equations, namely, differential equations for which the boundary conditions $u(0) = u(1) = 0$ are imposed.

## B.2  Sobolev spaces in higher dimensions

Sobolev spaces defined on subsets of $\mathbb{R}^n$ are analogous to the spaces described above. If $\Omega \subset \mathbb{R}^n$ (usually $n = 2$ or $n = 3$ in practical applications) is an open set, then we define $L^2(\Omega)$ to be the space of all measurable functions $u$ mapping $\Omega$ into $\mathbb{R}$ that are square-integrable:

$$\int_\Omega u^2 < \infty.$$

As before, we have to identify functions that are equal except on a set of measure zero (therefore, as in the case of $L^2(0, 1)$, the elements of $L^2(\Omega)$ are equivalence classes of measureable functions). This space is a Hilbert space under the $L^2$ inner product, which is defined by

$$\langle u, v \rangle_{L^2} = \int_\Omega uv \text{ for all } u, v \in L^2(\Omega).$$

It can be shown that $C_0^\infty(\Omega)$, the space of infinitely differentiable functions with compact support inside of the open set $\Omega$, is dense in $L^2(\Omega)$, as are the larger spaces $C^\infty(\overline{\Omega})$ and $C^k(\overline{\Omega})$, $k \geq 0$. (We cannot say that $C^\infty(\Omega)$ is dense in $L^2(\Omega)$ because $C^\infty(\Omega)$ is not a subspace of $L^2(\Omega)$; it includes functions that are not square-integrable because they blow up at the boundary).

Analogous to the case of an interval, we use $C_0^\infty(\Omega)$ as the space of test functions to define weak derivatives. If $u$ is integrable on $\Omega$ and there exists a locally integrable function $v$ such that

$$\int_\Omega v\phi = -\int_\Omega u\frac{\partial \phi}{\partial x_i} \text{ for all } \phi \in C_0^\infty(\Omega),$$

then we say that $v$ is the weak partial derivative of $u$ with respect to $x_i$ (here we are writing $x \in \Omega$ as $x = (x_1, x_2, \ldots, x_n)$) and write

$$\frac{\partial u}{\partial x_i} = v \text{ or } \partial_{x_i} u = v.$$

We will also write $\nabla u$ for the vector whose components are the weak partial derivatives of $u$: $\nabla u = (\partial_{x_1} u, \partial_{x_2} u, \ldots, \partial_{x_n} u)$.

Here are two examples of functions of two variables, one of which is weakly differentiable and the other of which is not. These examples are analogous to Examples B.1 and B.2, and the details are left to the reader.

**Example B.5.** Let $\Omega$ be the unit square, $\Omega = (0, 1) \times (0, 1)$, and define

$$u(x, y) = \begin{cases} x, & 0 < x \leq \frac{1}{2}, \\ 1 - x, & \frac{1}{2} < x < 1. \end{cases}$$

Then $u$ is weakly differentiable with respect to $x$ and

$$\frac{\partial u}{\partial x}(x, y) = g(x, y) = \begin{cases} 1, & 0 < x < \frac{1}{2}, \\ -1, & \frac{1}{2} < x < 1. \end{cases}$$

**Example B.6.** Let $\Omega$ be the unit square, $\Omega = (0, 1) \times (0, 1)$, and define

$$u(x, y) = \begin{cases} 0, & 0 < x \leq \frac{1}{2}, \\ 1, & \frac{1}{2} < x < 1. \end{cases}$$

If $v \in C_0^\infty(\Omega)$, then

$$-\int_\Omega u \frac{\partial v}{\partial x} = \int_0^1 v(1/2, y)\, dy.$$

It is not difficult to show that there is no locally integrable function $g$ defined on $\Omega$ such that

$$\int_0^1 \int_0^1 g(x, y)v(x, y)\, dxdy = \int_0^1 v(1/2, y)\, dy \text{ for all } v \in C_0^\infty(\Omega).$$

Such a function $g$ would have to satisfy $g(x, y) = 0$ for $x \neq 1/2$, and if $g$ is nonzero only on the line segment corresponding to $x = 1/2$, then

$$\int_\Omega gv = 0 \text{ for all } v \in C_0^\infty(\Omega)$$

must hold (since a line segment has zero area).

We now define

$$H^1(\Omega) = \left\{ u \in L^2(\Omega) : \partial_{x_i} u \in L^2(\Omega), i = 1, 2, \ldots, n \right\}.$$

It is a Hilbert space under the inner product

$$\langle u, v \rangle_{H^1} = \int_\Omega (uv + \nabla u \cdot \nabla v).$$

The spaces $C^\infty(\overline{\Omega})$ and $C^1(\overline{\Omega})$ (as well as $C^k(\overline{\Omega})$) for $k > 1$ are dense in $H^1(\Omega)$, but the space $C_0^\infty(\Omega)$ is not. Although elements of $H^1(0, 1)$ need not be continuous, there is a trace theorem that allows us to refer to boundary values. If the boundary $\partial\Omega$ of $\Omega$ is sufficiently smooth, then the mapping

$$u \mapsto u|_{\partial\Omega} \in L^2(\partial\Omega) \text{ for all } u \in C^1(\overline{\Omega})$$

is bounded and hence extends uniquely to a bounded linear operator from $H^1(\Omega)$ to $L^2(\partial\Omega)$. This allows us to refer to the boundary values of $u \in H^1(\Omega)$ (noting, however, that $u|_{\partial\Omega}$ need not be continuous for $u \in H^1(\Omega)$).

Using the trace theorem, we can define

$$H_0^1(\Omega) = \left\{ u \in H^1(\Omega) : u|_{\partial\Omega} = 0 \right\}.$$

This space is a closed subspace of $H^1(\Omega)$, and $C_0^\infty(\Omega)$ is dense in $H_0^1(\Omega)$.

We can also define the Sobolev spaces $H^k(\Omega)$ for $k > 1$. For example, $H^2(\Omega)$ is the space consisting of all $u \in L^2(\Omega)$ such that $\partial_{x_i} u \in L^2(\Omega)$ for all $i = 1, 2, \ldots, n$ and $\partial^2_{x_i x_j} u \in L^2(\Omega)$ for all $i = 1, 2, \ldots, n$, $j = 1, 2, \ldots, n$. The space $H^2(\Omega)$ is a Hilbert space under the inner product

$$\langle u, v \rangle_{H^2} = \int_\Omega uv + \int_\Omega \nabla u \cdot \nabla v + \sum_{i=1}^n \sum_{j=1}^n \int_\Omega (\partial^2_{x_i x_j} u)(\partial^2_{x_i x_j} v).$$

The space $H^2(\Omega)$ naturally arises if we need to refer to the boundary values of the partial derivatives of $u$. In particular, the *normal derivative*

$$\frac{\partial u}{\partial \mathbf{n}} = \nabla u \cdot \mathbf{n},$$

where $\mathbf{n}$ is the outward pointing unit normal vector to $\partial\Omega$, often appears in applications. Since the first partial derivatives of $u \in H^2(\Omega)$ belong to $H^1(\Omega)$, the trace theorem implies that the normal derivative of $u$ is well-defined as an element of $L^2(\partial\Omega)$. On the other hand, if $u$ is merely in $H^1(\Omega)$, the normal derivative of $u$ is not well-defined.

# Bibliography

[1] Hedy Attouch, Giuseppe Buttazzo, and Gérard Michaille. *Variational Analysis in Sobolev and BV Spaces: Applications to PDEs and Optimization.* SIAM, Philadelphia, second edition, 2014.

[2] Richard Courant and David Hilbert. *Methods of Mathematical Physics*, volume 2. Wiley Interscience, New York, 1989.

[3] Heinz W. Engl, Martin Hanke, and Andreas Neubauer. *Regularization of Inverse Problems.* Kluwer Academic Publishers, Dordrecht/Boston/London, 2000.

[4] Gerald B. Folland. *Real Analysis: Modern Techniques and their Applications.* John Wiley & Sons, New York, 1984.

[5] Gerald B. Folland. *Fourier Analysis and Its Applications.* Brooks/Cole, Pacific Grove, 1992.

[6] Enrico Giusti. *Minimal Surfaces and Functions of Bounded Variation.* Springer, New York, 1984.

[7] Mark S. Gockenbach. *Partial Differential Equations: Analytical and Numerical Methods.* SIAM, Philadelphia, second edition, 2010.

[8] Charles W. Groetsch. *The theory of Tikhonov regularization for Fredholm equations of the first kind.* Pitman Advanced Publishing Program, Boston, 1984.

[9] Jacques Hadamard. Sur les problèmes aux derivées partielles et leur signification physique. *Princeton University Bulletin*, 13(4):49–52, 1902.

[10] Jacques Hadamard. *Lectures on Cauchy's Problem in Linear Partial Differential Equations.* Dover, New York, 1952. Translation of *Le problème de Cauchy et les équations aux dérivées partielles linéaires hyperboliques*, Hermann 1932.

[11] Jacques Hadamard. *La théorie des équations aux dérivées partielles.* Editions Scientifiques, 1964.

[12] P. C. Hansen. Analysis of discrete ill-posed problems by means of the L-curve. *SIAM Review*, 34:561–580, 1992.

[13]  P. C. Hansen and D. P. O'Leary. The use of the L-curve in the regularization of discrete ill-posed problems. *SIAM J. Sci. Comput.*, 14:1487–1503, 1993.

[14]  Arthur E. Hoerl. Optimum solution of many variables equations. *Chemical Engineering Progress*, 55(11):69–78, 1959.

[15]  Arthur E. Hoerl. Application of ridge analysis to regression problems. *Chemical Engineering Progress*, 58(3):54–59, 1962.

[16]  Wolfram Research Inc. *Mathematica, Version 10.0.* Wolfram Research, Inc., Champaign, IL, 2015.

[17]  Victor Isakov. *Inverse Problems for Partial Differential Equations.* Springer, New York, 1998.

[18]  Andreas Kirsch. *An Introduction to the Mathematical Theory of Inverse Problems.* Springer-Verlag, New York, 1996.

[19]  Donald W. Marquardt and Ronald D. Snee. Ridge regression in practice. *The American Statistician*, 29(1):3–20, 1975.

[20]  Vladimir Maz'ya and Tatyana Shaposhnikova. *Jacques Hadamard: A Universal Mathematician.* American Mathematical Society, New York, 1999.

[21]  Andreas Neubauer. On converse and saturation results for Tikhonov regularization of linear ill-posed problems. *SIAM J. Numerical Analysis*, 34(2):517–527, 1997.

[22]  H. L. Royden. *Real Analysis.* Macmillan, New York, second edition, 1968.

[23]  Albert Tarantola. *Inverse Problem Theory and Methods for Model Parameter Estimation.* SIAM, Philadelphia, 2005.

[24]  Curtis R. Vogel. *Computational Methods for Inverse Problems.* SIAM, Philadelphia, 2002.

# Index

# About the Author

**Mark S. Gockenbach** received his PhD in Computational and Applied Mathematics from Rice University in 1994. He has held faculty positions at Indiana University (teaching in the ITM/MUCIA-Indiana University cooperative program in Malaysia for two years), the University of Michigan, and Rice University. He is now Professor and Chair of the Department of Mathematical Sciences at Michigan Technological University. Professor Gockenbach has won several awards for teaching, and he currently serves as a volunteer lecturer in the International Mathematical Union's Volunteer Lecturer Program (VLP). As a VLP lecturer, he has taught master's degree courses in Phnom Penh, Cambodia.

Professor Gockenbach's research interests are primarily in inverse problems in partial differential equations. His previous books are *Partial Differential Equations: Analytical and Numerical Methods* (first edition 2002, second edition 2010) and *Understanding and Implementing the Finite Element Method* (2006), both published by the Society for Industrial and Applied Mathematics, and *Finite-Dimensional Linear Algebra* (2010), published by CRCPress.